STUDENT SOLUTIONS MANUAL

TO ACCOMPANY

CHARLES P. McKEAGUE

INTERMEDIATE ALGEBRA

FOURTH EDITION

PREPARED BY

PATRICIA K. BEZONA

VALDOSTA STATE COLLEGE

HBJ

HARCOURT BRACE JOVANOVICH, PUBLISHERS
and its subsidiary, Academic Press
San Diego   New York   Chicago   Austin   Washington, D.C.
London   Sydney   Tokyo   Toronto

Requests for permission to make copies of any part of the work should be mailed to:
Permissions Department, Harcourt Brace Jovanovich, Inc., 8th Floor, Orlando,
Florida 32887.

ISBN: 0-15-541394-5

Printed in the United States of America

PREFACE

This Student Solutions Manual accompanies *Intermediate Algebra,* Fourth Edition, by Charles P. McKeague.

I believe the manual will help students who need more guidance with the math problems than is usually available through class time or individual assistance.

The manual contains solutions to every other odd-numbered problem. The solution should be followed step by step to solve any problem that is confusing to you.

I would like to thank Charles P. McKeague, both for giving me this opportunity and for all his guidance, and Amy Barnett, my editor at Harcourt Brace Jovanovich, for her patience and understanding. Without Hope Ince, my typist, this would never have been; thanks so much. A special thanks to my husband, Ron, my sister, Paula Kurjack, my mother, Theo Reed, and my aunt, Caroline "Kina" DePriest, for their support during this project.

Patricia K. Bezona

CONTENTS

SOLUTIONS TO SELECTED PROBLEMS

CHAPTER 1

Section 1.1

1. $x + 5$

5. $2t < y$

9. $x + y < x - y$

13. $s - t \neq s + t$

17. $6^2 = 6 \cdot 6 = 36$ Base 6, exponent 2

21. $2^3 = 2 \cdot 2 \cdot 2 = 8$ Base 2, exponent 3

25. $10^4 = 10 \cdot 10 \cdot 10 \cdot 10 = 10,000$ Base 10, exponent 4

29. $3 \cdot 5 + 4 = 15 + 4 = 19$ First multiply

33. $\begin{aligned} 2 + 8 \cdot 5 &= 2 + 40 && \text{Multiply} \\ &= 42 && \text{Add} \end{aligned}$

37. $\begin{aligned} 6 + 3 \cdot 4 - 2 &= 6 + 12 - 2 && \text{Multiply} \\ &= 18 - 2 && \text{Add} \\ &= 16 && \text{Subtract} \end{aligned}$

41. $\begin{aligned} (6 + 3)(4 - 2) &= 9 \cdot 2 && \text{Simplify inside parentheses} \\ &= 18 && \text{Multiply} \end{aligned}$

45. $\begin{aligned} (4 \cdot 2)^2 + (5 \cdot 2)^3 &= (8)^2 + (10)^3 && \text{Simplify inside parentheses} \\ &= 64 + 1,000 && \text{Exponents} \\ &= 1,064 \end{aligned}$

49. $\begin{aligned} 5^2 + 3^2 &= 25 + 9 && \text{Exponents} \\ &= 34 && \text{Add} \end{aligned}$

53. $\begin{aligned} (7 - 4)(7 + 4) &= (3)(11) && \text{Simplify inside parentheses} \\ &= 33 && \text{Multiply} \end{aligned}$

57. $\begin{aligned} 2 + 3 \cdot 2^2 + 3^2 &= 2 + 3 \cdot 4 + 9 && \text{Exponents} \\ &= 2 + 12 + 9 && \text{Multiply} \\ &= 23 && \text{Add} \end{aligned}$

61. $\begin{aligned} (2 + 3)(2^2 + 3^2) &= (2 + 3)(4 + 9) && \text{Exponents} \\ &= (5)(13) && \text{Simplify inside parentheses} \\ &= 65 && \text{Multiply} \end{aligned}$

65. $\begin{aligned} (40 - 10) \div 5 + 1 &= (30) \div 5 + 1 && \text{Simplify inside parentheses} \\ &= 6 + 1 && \text{Divide} \\ &= 7 && \text{Add} \end{aligned}$

69. $24 \div 4 + 8 \div 2 = 6 + 4$ Divide
                   $= 10$ Add

73. $5 \cdot 10^3 + 4 \cdot 10^2 + 3 \cdot 10 + 1$
     $= 5 \cdot 1{,}000 + 4 \cdot 100 + 3 \cdot 10 + 1$ Exponents
     $= 5{,}000 + 400 + 30 + 1$ Multiply
     $= 5{,}431$ Add

77. $16 - (4 \cdot 5^2 - 9 \cdot 10) = 16 - (4 \cdot 25 - 9 \cdot 10)$ Exponents
                           $= 16 - (100 - 90)$ Multiply
                           $= 16 - 10$ Subtract
                           $= 6$ Subtract

81. $40 - 10 - 4 - 2 = 30 - 4 - 2$ Subtract
                     $= 26 - 2$ Subtract
                     $= 24$

85. $(3 + 2)(2 \cdot 3^2 + 1) = (3 + 2)(2 \cdot 9 + 1)$ Exponents
                        $= (3 + 2)(18 + 1)$ Multiply
                        $= (5)(19)$ Parentheses
                        $= 95$ Multiply

89. $6[3 + 2(5 \cdot 3 - 10)]$
     $= 6[3 + 2(15 - 10)]$ Multiply
     $= 6[3 + 2(5)]$ Subtract
     $= 6[3 + 10]$ Multiply
     $= 6(13)$ Add
     $= 78$ Multiply

93. $5^3 + 4^3 \div 2^4 - 3^4$
     $= 125 + 64 \div 16 - 81$ Exponents
     $= 125 + 4 - 81$ Divide
     $= 129 - 81$ Add
     $= 48$ Subtract

97. $36 \div 2 - 3 = 18 - 3$ Divide
                 $= 15$ Subtract

101. Let $A = \{0,2,4,6\}$, $B = \{1,2,3,4,5\}$, find $A \cup B$.

     $A \cup B = \{x \,|\, x \in A \ \text{ or } \ x \in B\}$

Solution: We are looking for all elements that are in A **or** in B.

     $A \cup B = \{0,1,2,3,4,5,6\}$

105. Let $B = \{1,2,3,4,5\}$ and $C = \{1,3,5,7\}$, find $B \cap C$.

     $B \cap C = \{x \,|\, x \in B \text{ and } x \in C\}$

Solution: We are looking for all the elements that are in **both** B and C.

     $B \cap C = \{1,3,5\}$

Section 1.1 continued

109. Let $A = \{0,2,4,6\}$, find $\{x \mid x \in A \text{ and } x < 4\}$

Solution: We are looking for all the elements that are in A but less than 4.

$$\{x \mid x \in A \text{ and } x < 4\} = \{0,2\}$$

113. Let $A = \{0,2,4,6\}$ and $C = \{1,3,5,7\}$, find $\{x \mid x \in A \text{ or } x \in C\}$.

Solution: We are looking for all elements that are in A **or** in C.

$$\{x \mid x \in A \text{ or } x \in C\} = \{0,1,2,3,4,5,6,7\}$$

117. If $n(A) = 4$, $n(B) = 5$ and $A \cap B = \emptyset$, find $n(A \cup B)$.

Solution: This statement means there are 4 elements in A, 5 elements in B and the intersection of the two sets is no elements. This means the sets A and B have no elements in common, so if we take the union of these two sets, we would only write these elements once, thus:

$$n(A \cup B) = 4 + 5 - 0$$
$$n(A \cup B) = 9$$

Section 1.2

1. Graph $\{x \mid x < 1\}$

Solution: We want to graph all real numbers less than 1; that is all real numbers below 1 except at the end point 1. In this case, we will use an open circle since 1 is not included in the graph. See the graph on page A14 in your textbook.

5. Graph $\{x \mid x \geq 4\}$

Solution: We want to graph all real numbers greater than or equal to 4; that is all real numbers above 4 and including 4. We use a solid circle at 4 since 4 is included in the graph. See the graph on page A14 in your textbook.

9. Graph $\{x \mid x > 0\}$

Solution: We want to graph all real numbers greater than 0; that is all real numbers above 0 but not including 4. We will use an open circle at 0 since 0 is not included in the graph. See the graph on page A14 in the textbook.

13. Graph $\{x \mid 4 \leq x\}$

Solution: We may write $4 \leq x$ as $x \geq 4$ which will be easier to graph. This problem is graphed exactly the same as problem 5 in this section.

17. Graph $\{x \mid x \leq -3 \text{ or } x \geq 1\}$

Solution: Since these compound inequalities are connected by the word "or", we graph their union. That is, we graph all points on either graph. See the graph on page A15 in your textbook.

3

21. Graph $\{x|-3 < x \text{ and } x < 1\}$

    Solution: Since this compound inequality uses the word "and", we may write the inequalities as a continued inequality:

    $\{-3 < x < 1\}$

    See the graph on page A15 in your textbook.

25. Graph $\{x|x \leq -1 \text{ and } x \geq 3\}$

    Solution: Since this compound inequality uses the word "and", we graph their intersection--the part they have in common. The answer is Ø (null set) because there are no numbers that are less than or equal to -1 and greater than or equal to 3 at the same time. See the graph on page A15 in the textbook.

29. Graph $\{x|-1 \leq x \leq 2\}$

    Solution: This continued inequality is equivalent to $-1 \leq x$ and $x \leq 2$. Therefore, we graph all the numbers between -1 and 2 on the number line, including -1 and 2. See the graph on page A15 in the textbook.

33. Graph $\{x|-3 < x < 1\}$

    Solution: This continued inequality is equivalent to $-3 < x$ and $x < 1$. Therefore, we graph all the numbers between -3 and 1 on the number line, but not including -3 and 1. See the graph on page A15 in your textbook.

37. Graph $\{x|-4 < x \quad 1\}$

    Solution: This continued inequality is equivalent to $-4 < x$ and $x \leq 1$. Therefore, we graph all the numbers between -4 and 1 on the number line, including 1 but not including -4. See the graph on page A15 in your textbook.

41. Graph $\{x|x \leq -5 \text{ or } 0 \leq x \leq 3\}$

    Solution: Here we have a combination of compound and continued inequalities. We want to graph all real numbers that are either less than -5 or between 0 and 3 including -5, 0 and 3. See the graph on page A15 in your textbook.

45. $x \geq 5$

49. $x \leq 4$

53. $-4 \leq x \leq 4$

4

Section 1.2 continued

57. $-6, -5.2, 0, 1, 2, 2.3, \frac{9}{2}$ are rational numbers.

61. $0, 1, 2$ are nonnegative integers.

65. True because all whole numbers are a part of the set of integers.

69. True because integers are all positive and negative whole numbers plus zero.

73. $266 = 2 \cdot 133 = 2 \cdot 7 \cdot 19$

77. $369 = 3 \cdot 123 = 3 \cdot 3 \cdot 41 = 3^2 \cdot 41$

81. $\dfrac{385}{735} = \dfrac{5 \cdot 7 \cdot 11}{5 \cdot 7 \cdot 21} = \dfrac{11}{21}$

85. $\dfrac{525}{630} = \dfrac{3 \cdot 5 \cdot 5 \cdot 7}{2 \cdot 3 \cdot 3 \cdot 5 \cdot 7} = \dfrac{5}{6}$

89.
$$\frac{1}{3} = 0.333\ldots$$
$$\frac{1}{3} = 0.333\ldots$$
$$\frac{\frac{1}{3} = 0.333\ldots}{1 = 0.999\ldots}$$

93. $20 \leq t < 40$

97. $x < 3$ or $x > 13$ because 5 units to the left of 8 is 3 and 5 units to the right of 8 is 13.

101. $6! = 6 \cdot 5 \cdot 4 \cdot 3 \cdot 2 \cdot 1 = 6 \cdot 5!$

105. $\dfrac{49!}{50!} = \dfrac{49 \cdot 48 \cdot 47 \ldots 1}{50 \cdot 49 \cdot 48 \ldots 1} = \dfrac{1}{50}$

Section 1.3

1. $4 + (2 + x) = (4 + 2) + x$ Associative property
$ = 6 + x$ Addition

5. $5(3y) = (5 \cdot 3)y$ Associative property
$ = 15y$ Multiplication

9. $4(\frac{1}{4}a) = (4 \cdot \frac{1}{4})a$ Associative property
$\phantom{4(\frac{1}{4}a)} = 1a$ Multiplication
$\phantom{4(\frac{1}{4}a)} = a$

5

13. $3(x + 6) = 3x + 3(6)$ Distributive property

$\qquad\qquad\quad = 3x + 18$ Multiplication

17. $5(3a + 2b) = 5(3a) + 5(2b)$ Distributive property

$\qquad\qquad\qquad = 15a + 10b$ Multiplication

21 $\frac{1}{3}(4x + 6) = \frac{1}{3}(4x) + \frac{1}{3}(6)$ Distributive property

$\qquad\qquad\quad = \frac{4}{3}x + 2$ Multiplication

25. $\frac{1}{5}(10 + 5y) = \frac{1}{5}(10) + \frac{1}{5}(5y)$ Distributive property

$\qquad\qquad\quad = 2 + y$ Multiplication

29. $3(5x + 2) + 4 = 3(5x) + 3(2) + 4$ Distributive property

$\qquad\qquad\qquad = 15x + 6 + 4$ Multiplication

$\qquad\qquad\qquad = 15x + 10$ Addition

33. $5(1 + 3t) + 4 = 5(1) + 5(3t) + 4$ Distributive property

$\qquad\qquad\qquad = 5 + 15t + 4$ Multiplication

$\qquad\qquad\qquad = 15t + 9$ Commutative property and addition

37. $9(3x + 5y + 7) + 10$

$\qquad = 9(3x) + 9(5y) + 9(7) + 10$ Distributive property

$\qquad = 27x + 45y + 63 + 10$ Multiplication

$\qquad = 27x + 45y + 73$ Addition

41. $\dfrac{4}{\sqrt{3}} + \dfrac{5}{\sqrt{3}} = 4 \cdot \dfrac{1}{\sqrt{3}} + 5 \cdot \dfrac{1}{\sqrt{3}}$

$\qquad\qquad = (4 + 5)\dfrac{1}{\sqrt{3}}$ Distributive property

$\qquad\qquad = 9 \cdot \dfrac{1}{\sqrt{3}}$ Addition

$\qquad\qquad = \dfrac{9}{\sqrt{3}}$ Multiplication

45. $\dfrac{2}{5} + \dfrac{1}{15} = \dfrac{2}{5} \cdot \dfrac{3}{3} + \dfrac{1}{15} \cdot \dfrac{1}{1}$ LCD = 15

$\qquad\qquad = \dfrac{6}{15} + \dfrac{1}{15}$

$\qquad\qquad = \dfrac{6 + 1}{15}$

$\qquad\qquad = \dfrac{7}{15}$

49. $48 = 2 \cdot 2 \cdot 2 \cdot 2 \cdot 3 = 2^4 \cdot 3$
$54 = 2 \cdot 3 \cdot 3 \cdot 3 \quad = 2 \cdot 3^3$ LCD $= 2^4 \cdot 3^3 = 16 \cdot 27 = 432$

$$\frac{9}{48} + \frac{3}{54}$$

$$= \frac{9}{48} \cdot \frac{9}{9} + \frac{3}{54} \cdot \frac{8}{8}$$

$$= \frac{81}{432} + \frac{24}{432}$$

$$= \frac{105}{432}$$

$$= \frac{35}{144}$$

53. $3x + 5 + 4x + 2$
$= (3x + 4x) + (5 + 2)$ Commutative and associative properties
$= (3 + 4)x + (5 + 2)$ Distributive property
$= 7x + 7$ Addition

57. $5a + 7 + 8a + a$
$= (5a + 8a + a) + 7$ Commutative and associative properties
$= (5 + 8 + 1)a + 7$ Distributive property
$= 14a + 7$ Addition

61. $x + 1 + x + 2 + x + 3$
$= (x + x + x) + (1 + 2 + 3)$ Commutative and associative properties
$= (1 + 1 + 1)x + (1 + 2 + 3)$ Distributive property
$= 3x + 6$ Addition

65. $7 + 2(4y + 2)$
$= 7 + 2(4y) + 2(2)$ Distributive property
$= 7 + 8y + 4$ Multiply
$= 8y + (7 + 4)$ Commutative and associative properties
$= 8y + 11$ Addition

69. $5x + 2(3x + 8) + 4$
$= 5x + 2(3x) + 2(8) + 4$ Distributive property
$= 5x + 6x + 16 + 4$ Multiply
$= (5x + 6x) + (16 + 4)$ Associative property
$= (5 + 6)x + (16 + 4)$ Distributive property
$= 11x + 20$ Addition

73. $5(2x + 4) = 5(2x) + 5(4)$ (Right-hand side)
$= 10x + 20$

77. $\frac{3}{5} + \frac{1}{5} = \frac{4}{5}$ (Right-hand side)

81. Commutative property

7

Section 1.3 continued

85. Commutative property

89. Associative and commutative properties

93. $|-2| = 2$

97. $|\pi| = \pi$

101. $-|-2| = -(2) = -2$

105. $\frac{1}{3} \cdot 6 = \frac{1}{3} \cdot \frac{6}{1} = \frac{6}{3} = 2$

109. $\left(\frac{1}{10}\right)^4 = \left(\frac{1}{10}\right)\left(\frac{1}{10}\right)\left(\frac{1}{10}\right)\left(\frac{1}{10}\right) = \frac{1}{10,000}$

113. $9\left(\frac{1}{3}\right)^2 = 9\left(\frac{1}{3}\right)\left(\frac{1}{3}\right) = 9\left(\frac{1}{9}\right) = \frac{9}{1}\left(\frac{1}{9}\right) = \frac{9}{9} = 1$

117. 0

121. $15 - (8 - 2) = (15 - 8) - 2$
$15 - (6) = 7 - 2$
$9 = 5$ False statement

125. $6 * (-2) = 6(-2) + 6 = -12 + 6 = -6$

129. $3 \triangledown 5 = 3 \cdot 3 + 5 \cdot 5 = 9 + 25 = 34$

133. $a \triangledown b = b \triangledown a$
$aa + bb = bb + aa$
$aa + bb = aa + bb$

The operation $\triangledown$ is commutative.

Section 1.4

1. $6 + (-2) = 4$

5. $-\frac{1}{2} + \left(-\frac{1}{6}\right) + \left(-\frac{1}{18}\right)$ LCD = 18

      $= -\frac{1}{2} \cdot \frac{9}{9} + \left(-\frac{1}{6}\right)\frac{3}{3} + \left(-\frac{1}{18}\right)\frac{1}{1}$

      $= -\frac{9}{18} + \left(-\frac{3}{18}\right) + \left(-\frac{1}{18}\right)$

      $= \frac{-9 + (-3) + (-1)}{18}$

      $= -\frac{13}{18}$

9. $-7 - (-3) = -7 + 3$ Subtracting -3 is equivalent to adding 3
    $ = -4$

13. $\dfrac{11}{42} - \dfrac{17}{30}$ $\begin{aligned} 42 &= 2 \cdot 3 \cdot 7 \\ 30 &= 2 \cdot 3 \cdot 5 \end{aligned}$ LCD $= 2 \cdot 3 \cdot 5 \cdot 7 = 210$

    $= \dfrac{11}{42} \cdot \dfrac{5}{5} - \dfrac{17}{30} \cdot \dfrac{7}{7}$

    $= \dfrac{55}{210} - \dfrac{119}{210}$

    $= \dfrac{55 - 119}{210}$

    $= -\dfrac{64}{210}$

    $= -\dfrac{32}{105}$

17. $-\dfrac{4}{3} - \left(-\dfrac{1}{2}\right) - \dfrac{3}{2} = -\dfrac{4}{3} \cdot \dfrac{2}{2} + \dfrac{1}{2} \cdot \dfrac{3}{3} - \dfrac{3}{2} \cdot \dfrac{3}{3}$

    $= -\dfrac{8}{6} + \dfrac{3}{6} - \dfrac{9}{6}$

    $= -\dfrac{15}{6}$

    $= -\dfrac{5}{2}$

21. $-5 - (2 - 6) - 3 = -5 - (-4) - 3$ Subtract
    $ = -5 + 4 - 3$ Write subtraction as addition
    of the opposite
    $ = -1 - 3$ Addition
    $ = -1 + (-3)$ Write subtraction as addition
    of the opposite
    $ = -4$

25. $-3 - 5 = -3 + (-5)$ Write subtraction as addition of the opposite
    $ = -8$ Addition

29. $-3x - 4x = -3x + (-4x)$ Write subtraction as addition of the opposite
    $ = -7x$ Addition

33. $(2 - 9) + (-7) = -7 + (-7)$ Simplify the parentheses
    $ = -14$ Addition

37. $3(-5) = -15$

41. $-8(3) = -24$

45. $2(-3)(4) = -6(4) = -24$ Multiply left to right

49. $-\frac{1}{3}(-3x) = [-\frac{1}{3}(-3)]x$ Associative property

$$= 1x$$
$$= x$$

53. $-2(4x - 3) = -2(4x) - (-2)3$ Distributive property
$$= -8x + 6$$ Multiply

57. $-\frac{1}{2}(6a - 8) = -\frac{1}{2}(6a) - (-\frac{1}{2})8$ Distributive property

$$= -\frac{6a}{2} + \frac{8}{2}$$ Multiply

$$= -3a + 4$$ Simplify

61. $3(-4) - 2 = -12 - 2$ Multiply
$$= -14$$ Simplify

65. $2 - 5(-4) - 6 = 2 + 20 - 6$ Multiply
$$= 22 - 6$$ Add
$$= 16$$ Subtract

69. $(2 - 5)(-4 - 6) = (-3)(-10)$ Simplify the parentheses
$$= 30$$ Multiply

73. $2(-3)^2 - 4(-2)^3 = 2(9) - 4(-8)$ Exponents
$$= 18 + 32$$ Multiply
$$= 50$$ Add

77. $7(3 - 5)^3 - 2(4 - 7)^3 = 7(-2)^3 - 2(-3)^3$ Simplify the parentheses
$$= 7(-8) - 2(-27)$$ Exponents
$$= -56 + 54$$ Multiply
$$= -2$$ Add

81. $-5|-8 - 2|-3|-2 - 8|$

$$= -5|-10|-3|-10|$$ Simplify inside absolute value

$$= -5(10) - 3(10)$$ Evaluate absolute values
$$= -50 - 30$$ Multiply
$$= -80$$ Add

85. $(8 - 7)[4 - 7(-2)]$
$$= (8 - 7)[4 + 14]$$ Multiply
$$= (1)(18)$$ Simplify
$$= 18$$ Multiply

89. $5 - 6[-3(2 - 9) - 4(8 - 6)]$
$= 5 - 6[-3(-7) - 4(2)]$ Subtract
$= 5 - 6[21 - 8]$ Multiply
$= 5 - 6[13]$ Add
$= 5 - 78$ Multiply
$= -73$ Subtract

93. $6 - 7(3 - m)$
$= 6 - 7(3) - (-7)m$ Distributive property
$= 6 - 21 + 7m$ Multiply
$= 7m - 15$ Subtract

97. $5(3y + 1) - (8y - 5)$
$= 5(3y) + 5(1) - (8y) - (-5)$ Distributive property
$= 15y + 5 - 8y + 5$ Multiply
$= 7y + 10$ Associative and combine

101. $10 - 4(2x + 1) - (3x - 4)$
$= 10 - 4(2x) + (-4)1 - 3x - (-4)$ Distributive property
$= 10 - 8x - 4 - 3x + 4$ Multiply
$= -8x - 3x + 10 - 4 + 4$ Associative property
$= -11x + 10$ Simplify

105. $\dfrac{-8}{-4} = -8\left(-\dfrac{1}{4}\right) = \dfrac{-8}{1}\left(-\dfrac{1}{4}\right) = 2$

109. $\dfrac{0}{-3} = 0\left(-\dfrac{1}{3}\right) = 0$

113. $-8 \div \left(-\dfrac{1}{4}\right) = -8 \cdot \left(-\dfrac{4}{1}\right) = -8(-4) = 32$

117. $\dfrac{4}{9} \div (-8) = \dfrac{4}{9} \cdot \left(-\dfrac{1}{8}\right) = -\dfrac{4}{72} = -\dfrac{1}{18}$

121. $\dfrac{3(-1) - 4(-2)}{8 - 5} = \dfrac{-3 + 4}{8 - 5}$ Multiply

             $= \dfrac{1}{3}$ Add, subtract

125. $6 - (-3)[\dfrac{2 - 4(3 - 8)}{1 - 5(1 - 3)}]$

$\qquad = 6 - (-3)[\dfrac{2 - 4(-5)}{1 - 5(-2)}]$ Simplify the parentheses

$\qquad = 6 - (-3)[\dfrac{2 + 20}{1 + 10}]$ Multiply

$\qquad = 6 - (-3)[\dfrac{22}{11}]$ Add

$\qquad = 6 - (-3)(2)$ Simplify
$\qquad = 6 - (-6)$ Multiply
$\qquad = 6 + 6$ Simplify
$\qquad = 12$ Add

129. $[-3 \div (\tfrac{1}{2})] + (-5)$

$\qquad = [-3 \cdot (\tfrac{2}{1})] + (-5)$ Change division to multiplication

$\qquad = (-6) + (-5)$ Multiply
$\qquad = -11$ Add

1. $x + 2$

5. $2(x + y)$

9. $8^2 = 8 \cdot 8 = 64$

13. $2 + 3 \cdot 5 = 2 + 15$ Multiply
$\qquad\qquad\quad = 17$

17. $3 + 2(5 - 2) = 3 + 2(3)$ Parentheses first
$\qquad\qquad\qquad\;\; = 3 + 6$ Multiply
$\qquad\qquad\qquad\;\; = 9$

21. Let $A = \{1,3,5\}$ and $B = \{2,4,6\}$, find $A \cup B$.

    Solution: The union of A and B means all the elements in A **or** in B.

$\qquad A \cup B = \{1,2,3,4,5,6\}$

25. See the graph on page A16 of the textbook.

29. $|-3| = 3$

33. $-7, 0, 5$

37. $328 = 2 \cdot 164$
$\qquad\quad = 2 \cdot 2 \cdot 82$
$\qquad\quad = 2 \cdot 2 \cdot 2 \cdot 41$
$\qquad\quad = 2^3 \cdot 41$

41. See the graph on page A16 in the textbook.

45. $x \geq 4$

49. $-2y + 4y = (-2 + 4)y$ Distributive property
$\qquad\qquad\;\; = 2y$ Add

53. $x + 3 = 3 + x$ a. (Commutative property of addition)

57. $(x + 2) + y = (x + y) + 2$

$\qquad\qquad\qquad$ a (Commutative property of addition)
$\qquad\qquad\qquad$ c (Associative property of addition)

61. $5 - 3 = 2$

65. $|-4| - |-3| + |-2| = 4 - 3 + 2$ Absolute value
$\qquad\qquad\qquad\qquad\; = 1 + 2$ Subtract
$\qquad\qquad\qquad\qquad\; = 3$

69. $-\dfrac{1}{12} - \dfrac{1}{6} - \dfrac{1}{4} - \dfrac{1}{3}$ $\qquad\qquad\qquad$ LCD = 12

$\qquad = -\dfrac{1}{12} \cdot \dfrac{1}{1} - \dfrac{1}{6} \cdot \dfrac{2}{2} - \dfrac{1}{4} \cdot \dfrac{3}{3} - \dfrac{1}{3} \cdot \dfrac{4}{4}$

$\qquad = -\dfrac{1}{12} - \dfrac{2}{12} - \dfrac{3}{12} - \dfrac{4}{12}$

$\qquad = \dfrac{-1 - 2 - 3 - 4}{12}$

$\qquad = -\dfrac{10}{12}$

$\qquad = -\dfrac{5}{6}$

73. $7(3x) = (7 \cdot 3)x = 21x$

77. $-\dfrac{1}{2}(2x - 6)$

$\qquad = -\dfrac{1}{2}(2x) - (-\dfrac{1}{2})6$ $\qquad$ Distributive property

$\qquad = (-\dfrac{1}{2} \cdot \dfrac{2}{1})x - (-\dfrac{1}{2})(\dfrac{6}{1})$

$\qquad = -\dfrac{2}{2}x + \dfrac{6}{2}$

$\qquad = -1x + 3$

$\qquad = -x + 3$

81. $\dfrac{3}{5} \div 6 = \dfrac{3}{5} \div \dfrac{6}{1}$

$\qquad = \dfrac{3}{5} \cdot \dfrac{1}{6}$

$\qquad = \dfrac{3}{30}$

$\qquad = \dfrac{1}{10}$

85. $6 + 3(-2) = 6 + (-6) = 0$

89. $8 - 2(6 - 10) = 8 - 2(-4)$

$\qquad\qquad\qquad\quad = 8 + 8$

$\qquad\qquad\qquad\quad = 16$

93. $\dfrac{2(-3) - 5(4)}{6 - 8} = \dfrac{-6 - 20}{6 - 8}$ Multiply

$\qquad\qquad\qquad = \dfrac{-26}{-2}$ Add, Subtract

$\qquad\qquad\qquad = 13$ Simplify

97. $2(3x + 1) - 5 = 2(3x) + 2(1) - 5$ Distributive property
$\qquad\qquad\qquad = 6x + 2 - 5$ Multiply
$\qquad\qquad\qquad = 6x - 3$ Subtract

101. $4(3x - 1) - 5(6x + 2)$
$\qquad = 4(3x) - 4(1) - 5(6x) + (-5)2$ Distributive property
$\qquad = 12x - 4 - 30x - 10$ Multiply
$\qquad = -18x - 14$ Simplify

1. $2(3x + 4y)$

5.
$$
\begin{aligned}
12 - 8 \div 4 + 2 \cdot 3 &= 12 - 2 + 2 \cdot 3 && \text{Divide} \\
&= 12 - 2 + 6 && \text{Multiply} \\
&= 10 + 6 && \text{Subtract} \\
&= 16 && \text{Add}
\end{aligned}
$$

9. -3 opposite 3, reciprocal $-\dfrac{1}{3}$

13. $\{-5, 0, 1, 4\}$

17. See the graph on page A16 in the textbook.

21. $3(x \cdot y) = (3y) \cdot x$ Associative and commutative property of multiplication

25.
$$
\begin{aligned}
&-3(5)^2 - 4(-2)^5 \\
&= -3(25) - 4(-32) && \text{Exponents} \\
&= -75 + 128 && \text{Multiply} \\
&= 53
\end{aligned}
$$

29.
$$
\begin{aligned}
\frac{-4(-1) - (-10)}{5 - (-2)} &= \frac{4 + 10}{5 - (-2)} && \text{Multiply} \\[2mm]
&= \frac{4 + 10}{7} && \text{Simplify} \\[2mm]
&= \frac{14}{7} && \text{Add} \\[2mm]
&= 2 && \text{Reduce}
\end{aligned}
$$

33.
$$
\begin{aligned}
-\frac{1}{2}(8x) &= \left[-\frac{1}{2}(8)\right]x \\[2mm]
&= \left[-\frac{1}{2}\left(\frac{8}{1}\right)\right]x \\[2mm]
&= -\frac{8}{2}x \\[2mm]
&= -4x
\end{aligned}
$$

37.
$$
\begin{aligned}
&-4(3x + 2) + 7x \\
&= -4(3x) + (-4)2 + 7x && \text{Distributive property} \\
&= -12x - 8 + 7x && \text{Multiply} \\
&= -5x - 8
\end{aligned}
$$

41.
$$
\begin{aligned}
&3 + 4(2x - 5) - 5x \\
&= 3 + 4(2x) - 4(5) - 5x && \text{Distributive property} \\
&= 3 + 8x - 20 - 5x && \text{Multiply} \\
&= 8x - 5x + 3 - 20 && \text{Associative property} \\
&= 3x - 17 && \text{Simplify}
\end{aligned}
$$

45. $[-4 \div (-\frac{1}{3})] - (-4)$

$$= [-\frac{4}{1} \cdot -\frac{3}{1}] - (-4)$$

$$= 12 - (-4)$$
$$= 12 + 4$$
$$= 16$$

Section 2.1

1.
$$x - 5 = 3$$
$$x - 5 + 5 = 3 + 5 \qquad \text{Add } \mathbf{5} \text{ to both sides}$$
$$x = 8$$

The solution set is {8}.

5.
$$7 = 4a - 1$$
$$7 + 1 = 4a - 1 + 1 \qquad \text{Add } \mathbf{1} \text{ to both sides}$$
$$8 = 4a \qquad \text{Add}$$

$$\frac{1}{4}(8) = \frac{1}{4}(4a) \qquad \text{Multiply by } \frac{1}{4}$$

$$2 = a$$

The solution set is {2}.

9.
$$-3 - 4x = 15$$
$$-3 + 3 - 4x = 15 + 3 \qquad \text{Add } \mathbf{3} \text{ to both sides}$$
$$-4x = 18$$

$$-\frac{1}{4}(-4x) = -\frac{1}{4}(18) \qquad \text{Multiply by } -\frac{1}{4}$$

$$x = -\frac{18}{4}$$

$$x = -\frac{9}{2}$$

The solution set is $\{-\frac{9}{2}\}$.

13.
$$-300y + 100 = 500$$
$$-300y + 100 + (\mathbf{-100}) = 500 + (\mathbf{-100}) \qquad \text{Add } \mathbf{-100} \text{ to both sides}$$
$$-300y = 400$$

$$-\frac{1}{300}(-300y) = -\frac{1}{300}(400) \qquad \text{Multiply by } -\frac{1}{300}$$

$$y = -\frac{4}{3}$$

The solution set is $\{-\frac{4}{3}\}$.

17.
$$3 - 4a = -11$$
$$3 + (-3) - 4a = -11 + (-3) \qquad \text{Add } -3 \text{ to both sides}$$
$$-4a = -14$$

$$-\frac{1}{4}(-4a) = -\frac{1}{4}(-14) \qquad \text{Multiply by } -\frac{1}{4}$$

$$a = \frac{14}{4}$$

$$a = \frac{7}{2}$$

The solution set is $\{\frac{7}{2}\}$.

21.
$$\frac{2}{3}x = 8$$

$$\frac{3}{2}(\frac{2}{3}x) = \frac{3}{2}(8) \qquad \text{Multiply by } \frac{3}{2}$$

$$x = \frac{24}{2} \qquad \text{Multiply}$$

$$x = 12$$

The solution set is $\{12\}$.

25.
$$8 = 6 + \frac{2}{7}y$$

$$8 + (-6) = 6 + (-6) + \frac{2}{7}y \qquad \text{Add } -6 \text{ to both sides}$$

$$2 = \frac{2}{7}y$$

$$\frac{7}{2}(2) = \frac{7}{2}(\frac{2}{7}y) \qquad \text{Multiply by } \frac{7}{2}$$

$$\frac{14}{2} = y$$

$$7 = y$$

The solution set is $\{7\}$.

29.
$$-x = 2$$
$$(-1)(-x) = 2(-1)$$
$$x = 2$$

33.
$$7y + 4 = 2y + 11$$
$$7y + (-2y) - 4 = 2y + (-2y) + 11 \qquad \text{Add } -2y \text{ to both sides}$$
$$5y - 4 = 11$$
$$5y - 4 + 4 = 11 + 4 \qquad \text{Add } 4 \text{ to both sides}$$
$$5y = 15$$

$$\left(\frac{1}{5}\right)5y = \left(\frac{1}{5}\right)15 \qquad \text{Multiply } \frac{1}{5} \text{ to both sides}$$

$$y = 3$$

37.
$$5y - 2 + 4y = 2y + 12$$
$$9y - 2 = 2y + 12 \qquad \text{Simplify the left side}$$
$$9y + (-2y) - 2 = 2y + (-2y) + 12 \qquad \text{Add } -2y \text{ to both sides}$$
$$7y - 2 = 12$$
$$7y - 2 + 2 = 12 + 2 \qquad \text{Add } 2 \text{ to both sides}$$
$$7y = 14$$

$$\frac{1}{7}(7y) = \frac{1}{7}(14) \qquad \text{Multiply both sides by } \frac{1}{7}$$

$$y = 2$$

41.
$$k = 2(3k - 5)$$
$$k = 6k - 10 \qquad \text{Distributive property}$$
$$k + (-6k) = 6k + (-6k) - 10 \qquad \text{Add } -6k \text{ to both sides}$$
$$-5k = -10$$

$$-\frac{1}{5}(-5k) = -\frac{1}{5}(-10) \qquad \text{Multiply by } -\frac{1}{5}$$
$$k = 2$$

The solution set is {2}.

45.
$$5(y + 2) - 4(y + 1) = 3$$
$$5y + 10 - 4y - 4 = 3 \qquad \text{Distributive property}$$
$$y + 6 = 3 \qquad \text{Simplify the left side}$$
$$y + 6 + (-6) = 3 + (-6) \qquad \text{Add } -6 \text{ to both sides}$$
$$y = -3$$

The solution set is {-3}.

49.
$$4(a - 3) + 5 = 7(3a - 1)$$
$$4a - 12 + 5 = 21a - 7 \qquad \text{Distributive property}$$
$$4a - 7 = 21a - 7 \qquad \text{Simplify the left side}$$
$$4a + (-4a) - 7 = 21a + (-4a) - 7 \qquad \text{Add } -4a \text{ to both sides}$$
$$-7 = 17a - 7$$
$$-7 + 7 = 17a - 7 + 7 \qquad \text{Add } 7 \text{ to both sides}$$
$$0 = 17a$$

$$\frac{1}{17}(0) = \frac{1}{17}(17a) \qquad \text{Multiply by } \frac{1}{17}$$

$$0 = a$$

The solution set is {0}.

53.
$$5 = 7 - 2(3x - 1) + 4x$$
$$5 = 7 - 6x + 2 + 4x \qquad \text{Distributive property}$$
$$5 = -2x + 9 \qquad \text{Simplify the left side}$$
$$5 + \mathbf{(-9)} = -2x + 9 + \mathbf{(-9)} \qquad \text{Add } \mathbf{-9} \text{ to both sides}$$
$$-4 = -2x$$

$$-\frac{1}{2}(-4) = -\frac{1}{2}(-2x) \qquad \text{Multiply by } -\frac{1}{2}$$

$$2 = x$$

The solution set is {2}.

57. Method 1

$$\frac{1}{2}x + \frac{1}{4} = \frac{1}{3}x + \frac{5}{4}$$

$$\frac{1}{2}x + \mathbf{\left(-\frac{1}{3}x\right)} + \frac{1}{4} = \frac{1}{3}x + \mathbf{\left(-\frac{1}{3}x\right)} + \frac{5}{4} \qquad \text{Add } -\frac{1}{3}\,\mathbf{x} \text{ to both sides}$$

$$\frac{1}{2}x + \left(-\frac{1}{3}x\right) + \frac{1}{4} = \frac{5}{5}$$

$$\frac{1}{6}x + \frac{1}{4} = \frac{5}{4} \qquad \qquad \frac{1}{2} \cdot \frac{3}{3} + \left(-\frac{1}{3}\right) \cdot \frac{2}{2} = \frac{3}{6} - \frac{2}{6} = \frac{1}{6}$$

$$\frac{1}{6}x + \frac{1}{4} + \left(-\frac{1}{4}\right) = \frac{5}{4} + \left(-\frac{1}{4}\right) \qquad \text{Add } -\frac{1}{4} \text{ to both sides}$$

$$\frac{1}{6}x = \frac{4}{4}$$

$$\frac{6}{1}\left(\frac{1}{6}x\right) = \frac{6}{1}(1) \qquad \text{Multiply both sides by } \frac{6}{1}$$

$$x = 6$$

Method 2

$$\frac{1}{2}x + \frac{1}{4} = \frac{1}{3}x + \frac{5}{4} \qquad \text{LCD} = 12$$

$$\mathbf{12}\left(\frac{1}{2}x\right) + \mathbf{12}\left(\frac{1}{4}\right) = \mathbf{12}\left(\frac{1}{3}x\right) + \mathbf{12}\left(\frac{5}{4}\right) \qquad \text{Multiply both sides by } \mathbf{12}$$

$$6x + 3 = 4x + 15$$
$$6x + \mathbf{(-4x)} + 3 = 4x + \mathbf{(-4x)} + 15 \qquad \text{Add } \mathbf{-4x} \text{ to both sides}$$
$$2x + 3 = 15$$
$$2x + 3 + \mathbf{(-3)} = 15 + \mathbf{(-3)} \qquad \text{Add } \mathbf{-3} \text{ to both sides}$$
$$2x = 12$$
$$\frac{1}{2}(2x) = \frac{1}{2}(12) \qquad \text{Multiply both sides by } \frac{1}{2}$$
$$x = 6$$

61. Method 1

$$\frac{1}{2}y - \frac{2}{7} = \frac{1}{7}y + \frac{11}{14}$$

$$\frac{1}{2}y + (-\frac{1}{7}y) - \frac{2}{7} = \frac{1}{7}y + (-\frac{1}{7}y) + \frac{11}{14} \qquad \text{Add } -\frac{1}{7}y \text{ to both sides}$$

$$\frac{1}{2}y - \frac{1}{7}y - \frac{2}{7} = \frac{11}{14}$$

$$\frac{5}{14}y - \frac{2}{7} = \frac{11}{14} \qquad\qquad \frac{1}{2} \cdot \frac{7}{7} - \frac{1}{7} \cdot \frac{2}{2} = \frac{7}{14} - \frac{2}{14} = \frac{5}{14}$$

$$\frac{5}{14}y - \frac{2}{7} + \frac{2}{7} = \frac{11}{14} + \frac{2}{7} \qquad \text{Add } \frac{2}{7} \text{ to both sides}$$

$$\frac{5}{14}y = \frac{15}{14} \qquad\qquad \frac{11}{14} + \frac{2}{7} \cdot \frac{2}{2} = \frac{11}{14} + \frac{4}{14} = \frac{15}{14}$$

$$\frac{14}{5}(\frac{5}{14}y) = \frac{14}{5}(\frac{15}{14}) \qquad \text{Multiply both sides by } \frac{14}{5}$$

$$y = 3$$

Method 2

$$\frac{1}{2}y - \frac{2}{7} = \frac{1}{7}y + \frac{11}{14} \qquad\qquad \text{LCD} = 14$$

$$14(\frac{1}{2}y) - 14(\frac{2}{7}) = 14(\frac{1}{7}y) + 14(\frac{11}{14}) \qquad \text{Multiply both sides by } 14$$

$$7y - 4 = 2y + 11$$

$$7y + (-2y) - 4 = 2y + (-2y) + 11 \qquad \text{Add } -2y \text{ to both sides}$$

$$5y - 4 = 11$$

$$5y - 4 + 4 = 11 + 4 \qquad \text{Add } 4 \text{ to both sides}$$

$$5y = 15$$

$$\frac{1}{5}(5y) = \frac{1}{5}(15) \qquad \text{Multiply both sides by } \frac{1}{5}$$

$$y = 3$$

22

65. Method 1

$$\frac{1}{4}(12a + 1) - \frac{1}{4} = 5$$

$$\frac{1}{4}(12a) + \frac{1}{4}(1) - \frac{1}{4} = 5 \qquad \text{Distributive property}$$

$$3a + \frac{1}{4} - \frac{1}{4} = 5 \qquad \text{Multiply}$$

$$3a = 5 \qquad \text{Simplify}$$

$$\frac{1}{3}(3a) = \frac{1}{3}(5) \qquad \text{Multiply both sides by } \frac{1}{3}$$

$$a = \frac{5}{3}$$

Method 2

$$\frac{1}{4}(12a + 1) - \frac{1}{4} = 5 \qquad \text{LCD = 4}$$

$$4 \cdot \frac{1}{4}(12a + 1) - 4 \cdot \frac{1}{4} = 4 \cdot 5 \qquad \text{Multiply both sides by } \mathbf{4}$$

$$1(12a + 1) - 1 = 20$$
$$12a + 1 - 1 = 20$$
$$12a = 20$$

$$\frac{1}{12}(12a) = \frac{1}{12}(20) \qquad \text{Multiply both sides by } \frac{1}{12}$$

$$a = \frac{20}{12}$$

$$a = \frac{5}{3}$$

69. Method 1

$$.08x + .09(9000 - x) = 750$$
$$.08x + .09(9000) - .09x = 750 \qquad \text{Distributive property}$$
$$810 - .01x = 750 \qquad \text{Simplify left side}$$
$$810 + (\mathbf{-810}) - .01x = 750 + (\mathbf{-810}) \qquad \text{Add } \mathbf{-810} \text{ to both sides}$$
$$-.01x = -60$$

$$\frac{-.01x}{-.01} = \frac{-60}{-.01} \qquad \text{Divide}$$

$$x = 6000$$

Method 2

$$.08x + .09(9000 - x) = 750$$
$$100(.08x) + 100[.09(9000 - x)] = \mathbf{100}(750) \qquad \text{Multiply each side by } \mathbf{100}$$
$$8x + 9(9000 - x) = 75000 \qquad \text{Multiply}$$
$$8x + 81000 - 9x = 75000 \qquad \text{Distributive property}$$
$$81000 - x = 75000 \qquad \text{Simplify the left side}$$
$$-x = -6000 \qquad \text{Add } \mathbf{-81000} \text{ to each side}$$
$$x = 6000 \qquad \text{Multiply each side by } \mathbf{-1}$$

23

73. Method 1

$$.35y - .2 = .15x + .1$$
$$.35y + (-.15x) - .2 = .15x + (-.15x) + .1 \qquad \text{Add } -.15x \text{ to both sides}$$
$$.20x - .2 = .1 \qquad \text{Simplify left side}$$
$$.20x - .2 + .2 = .1 + .2 \qquad \text{Add } .2 \text{ to both sides}$$
$$.20x = .3$$

$$\frac{.20x}{.20} = \frac{.3}{.20} \qquad \text{Divide}$$

$$x = \frac{30}{20}$$

$$x = \frac{3}{2}$$

Method 2

$$.35x - .2 = .15x + .1$$
$$100(.35x) - 100(.2) = 100(.15x) + 100(.1) \qquad \text{Multiply each side by } 100$$
$$35x - 20 = 15x + 10 \qquad \text{Multiply}$$
$$35x + (-15x) - 20 = 15x + (-15x) + 10 \qquad \text{Add } -15x \text{ to both sides}$$
$$20x - 20 = 10 \qquad \text{Simplify}$$
$$20x - 20 + 20 = 10 + 20 \qquad \text{Add } 20 \text{ to both sides}$$
$$20x = 30$$

$$\frac{20x}{20} = \frac{30}{20} \qquad \text{Divide}$$

$$x = \frac{3}{2}$$

77.
$$6x - 2(x - 5) = 4x + 3$$
$$6x - 2x + 10 = 4x + 3 \qquad \text{Distributive property}$$
$$4x + 10 = 4x + 3$$
$$4x + (-4x) + 10 = 4x + (-4x) + 3 \qquad \text{Add } -4x \text{ to both sides}$$
$$10 = 3$$

This equation has no solution.

81.
$$3x - 6 = 3(x + 4)$$
$$3x - 6 = 3x + 12 \qquad \text{Distributive property}$$
$$3x + (-3x) - 6 = 3x + (-3x) + 12 \qquad \text{Add } -3x \text{ to both sides}$$
$$-6 = 12 \qquad \text{(A false statement)}$$

This equation has no solution.

85.
$$2(4t - 1) + 3 = 5t + 4 + 3t$$
$$8t - 2 + 3 = 5t + 4 + 3t \qquad \text{Distributive property}$$
$$8t + 1 = 8t + 4 \qquad \text{Simplify the left and right sides}$$
$$8t + (-8t) + 1 = 8t + (-8t) + 4 \qquad \text{Add } -8t \text{ to both sides}$$
$$1 = 4 \qquad \text{A false statement}$$

This equation has no solution.

Section 2.1 continued

89. $3 + (x + y) = (3 + x) + y$ Associative property of addition

93. $4(xy) = 4(yx)$ Commutative property of multiplication

97. $2 + 0 = 2$ Additive identity of addition

101.

$$\frac{x + 4}{5} - \frac{x + 3}{3} = -\frac{7}{15} \qquad \text{LCD = 15}$$

$$15\left(\frac{x + 4}{5}\right) - 15\left(\frac{x + 3}{3}\right) = 15\left(-\frac{7}{15}\right) \qquad \text{Multiply both sides by 15}$$

$$3(x + 4) - 5(x + 3) = -7$$
$$3x + 12 - 5x - 15 = -7 \qquad \text{Distributive property}$$
$$-2x - 3 = -7 \qquad \text{Simplify left side}$$
$$-2x - 3 + 3 = -7 + 3 \qquad \text{Add 3 to both sides}$$

$$\frac{-2x}{-2} = \frac{-4}{-2} \qquad \text{Divide}$$

$$x = 2$$

Section 2.2

1. When $x = 0$
the formula $3x - 4y = 12$
becomes $3(0) - 4y = 12$
$$-4y = 12$$
$$y = -3 \qquad \text{Divide each side by -4}$$

5. When $y = 0$
the formula $y = 2x - 3$
becomes $0 = 2x - 3$
$$3 = 2x \qquad \text{Add 3 to each side}$$
$$\frac{3}{2} = x \qquad \text{Divide each side by 2}$$

9. When $x = 800$
the formula $x = 1300 - 100p$
becomes $800 = 1300 - 100p$
$$-500 = -100p \qquad \text{Add -1300 to each side}$$
$$5 = p \qquad \text{Divide each side by -100}$$

The company should charge $5 per ribbon.

25

13. When $V = 308$, $r = 7$ and $\pi = \dfrac{22}{7}$

    the formula $V = \pi r^2 h$

    becomes $308 = \dfrac{22}{7} \cdot 7^2 \cdot h$

    $308 = \dfrac{22}{7} \cdot 49 \cdot h$ Exponent

    $308 = 154h$ Multiply $\dfrac{22}{7}$ and 49

    $2 = h$ Divide each side by 154

    The height is 2 centimeters.

17. When $S = 942$, $r = 10$ and $\pi = 3.14$
    the formula $S = 2\pi r^2 + 2\pi rh$
    becomes $942 = 2 \cdot 3.14 \cdot 10^2 + 2 \cdot 3.14 \cdot 10h$
    $942 = 2 \cdot 3.14 \cdot 100 + 2 \cdot 31.4h$
    $942 = 2 \cdot 314 + 62.8h$
    $942 = 628 + 62.8h$
    $314 = 62.8h$
    $5 = h$

    The height is 5 feet.

21. $I = prt$

    $\dfrac{I}{pr} = t$ Divide each side by pr

25. $y = mx + b$
    $y - mx = b$ Add -mx to each side

29. $A = P + Prt$
    $A - P = Prt$ Add -P to each side
    $\dfrac{A - P}{Pt} = r$ Divide each side by Pt

33. $h = vt + 16t^2$
    $h - 16t^2 = vt$ Add $-16t^2$ to each side
    $\dfrac{h - 16t^2}{t} = v$ Divide each side by t

37. $2x + 3y = 6$
    $3y = 6 - 2x$ Add -2x to each side
    $y = \dfrac{6 - 2x}{3}$ Divide each side by 3
    $y = -\dfrac{3}{2}x + 2$

41. $9x - 3y = 6$
$-3y = 6 - 9x$ Add -9x to each side

$y = \dfrac{6 - 9x}{-3}$ Divide each side by -3

$y = -2 + 3x$
$y = 3x - 2$

45. $8x - 10y - 16 = 0$
$8x - 10y = 16$ Add 16 to each side
$8x = 10y + 6$ Add 10y to each side

$x = \dfrac{10y + 6}{8}$ Divide each side by 8

$x = \dfrac{5}{4}y + 2$

49. $ax + 4 = bx + 9$
$ax - bx + 4 = 9$ Add -bx to each side
$ax - bx = 5$ Add -4 to each side
$(a - b)x = 5$ Distributive property

$x = \dfrac{5}{a - b}$ Divide each side by a - b

53. $ax + b = cx + d$
$ax - cx + b = d$ Add -cx to each side
$ax - cx = d - b$ Add -b to each side
$(a - c)x = d - b$ Distributive property

$x = \dfrac{d - b}{a - c}$ Divide each side by a - c

57. $\dfrac{x}{5} + \dfrac{y}{-3} = 1$ LCD = 15

$15\left(\dfrac{x}{5}\right) + 15\left(\dfrac{y}{-3}\right) = 15(1)$ Multiply both sides by **15**

$3x - 5y = 15$
$3x + \mathbf{(-3x)} - 5y = 15 + \mathbf{(-3x)}$ Add **-3x** to both sides
$-5y = -3x + 15$

$\dfrac{-5y}{-5} = \dfrac{-3x}{-5} + \dfrac{15}{-5}$ Divide

$y = \dfrac{3}{5}x - 3$

61. What number is 5% of 10,000?
$\quad\quad\downarrow\quad\quad\downarrow\ \downarrow\quad\downarrow\quad\quad\downarrow$
$\quad x \quad = .05 \ \cdot \ 10,000$
$\quad\quad\quad = 500$

The number 500 is 5% of 10,000.

65.   37 is 4% of what number?
$$\downarrow \quad \downarrow \ \downarrow \quad \downarrow \qquad \downarrow$$
37 = .04 ·   x
925 = x   Divide each side by .04

37 is 4% of 925.

69. See the graph on page A17 in the textbook.

73. See the graph on page A17 in the textbook.

77.
$$\frac{1}{a} + \frac{1}{b} = \frac{1}{c} \qquad \text{LCD} = abc$$

$$abc\left(\frac{1}{a}\right) + abc\left(\frac{1}{b}\right) = abc\left(\frac{1}{c}\right) \qquad \text{Multiply both sides by } \mathbf{abc}$$

$$bc + ac = ab$$
$$bc + ac + (-ac) = ab + (-ac) \qquad \text{Add } -ac \text{ to both sides}$$
$$bc = ab - ac$$
$$bc = a(b - c) \qquad \text{Distributive property}$$

$$\frac{bc}{b - c} = \frac{a(b - c)}{b - c} \qquad \text{Divide}$$

$$\frac{bc}{b - c} = a$$

Section 2.3

1. $2x \le 3$

$$\frac{1}{2}(2x) \le \frac{1}{2}(3) \qquad \text{Multiply by } \frac{1}{2}$$

$$x \le \frac{3}{2}$$

The solution set is $\{x \mid x \le \frac{3}{2}\}$ and the graph is on page A17 in the textbook.

5. $-5x \ge 25$

$$-\frac{1}{5}(-5x) \ge -\frac{1}{5}(25) \qquad \text{Multiply both sides by } -\frac{1}{5} \text{ and reverse the direction of the inequality symbol.}$$

$$x \ge -5$$

The solution set is $\{x \mid x \ge -5\}$ and the graph is on page A17 in the textbook.

9. $-12 \leq 2x$

$\frac{1}{2}(-12) \leq \frac{1}{2}(2x)$ Multiply by $\frac{1}{2}$

$-6 \leq x$

The solution set is $\{x \mid x \geq -6\}$ and the graph is on page A17 in the textbook.

13. $-3x + 1 > 10$

$-3x + 1 + (-1) > 10 + (-1)$ Add **-1** to both sides

$-3x > 9$

$-\frac{1}{3}(-3x) < -\frac{1}{3}(9)$ Multiply both sides by $-\frac{1}{3}$ and reverse the direction of the inequality symbol.

$x < -3$

The solution set is $\{x \mid x < -3\}$ and the graph is on page A17 in the textbook.

17. $\frac{1}{2} \geq -\frac{1}{6} - \frac{2}{9}x$ LCD = 18

$18(\frac{1}{2}) \geq 18(-\frac{1}{6}) - 18(\frac{2}{9}x)$ Multiply both sides by **18**

$9 \geq -3 - 4x$

$9 + 3 \geq -3 + 3 - 4x$ Add **3** to both sides

$\frac{12}{-4} \geq \frac{-4x}{-4}$ Divide, change signs

$-3 \leq x$

$x \geq -3$

See the graph on page A17 in the textbook.

21. $\frac{2}{3}x - 3 < 1$

$\frac{2}{3}x - 3 + 3 < 1 + 3$ Add **3** to both sides

$\frac{2}{3}x < 4$

$\frac{3}{2}(\frac{2}{3}x) < \frac{3}{2}(4)$ Multiply by $\frac{3}{2}$

$x < 6$

The solution set is $\{x \mid x < 6\}$ and the graph on page A17 in the textbook.

25.
$$2(3y + 1) \leq -10$$
$$6y + 2 \leq -10 \qquad \text{Distributive property}$$
$$6y + 2 + (-2) \leq -10 + (-2) \qquad \text{Add } (-2) \text{ to both sides}$$
$$6y \leq -12$$
$$\frac{1}{6}(6y) \leq \frac{1}{6}(-12) \qquad \text{Multiply by } \frac{1}{6}$$
$$y \leq -2$$

The solution set is $\{y | y \leq -2\}$ and the graph is on page A18 in the textbook.

29.
$$\frac{1}{3}t - \frac{1}{2}(5 - t) \leq 0 \qquad \text{LCD} = 6$$
$$6(\frac{1}{3}t) - 6(\frac{1}{2})(5 - t) \leq (0)6 \qquad \text{Multiply both sides by } 6$$
$$2t - 3(5 - t) \leq 0$$
$$2t - 15 + 3t \leq 0 \qquad \text{Distributive property}$$
$$5t - 15 + 15 \leq 0 + 15 \qquad \text{Add } 15 \text{ to both sides}$$
$$\frac{5t}{5} \leq \frac{15}{5} \qquad \text{Divide}$$
$$t \leq 3$$

See the graph on page A18 in the textbook.

33.
$$-\frac{1}{3}(x + 5) \leq -\frac{2}{9}(x - 1) \qquad \text{LCD} = 9$$
$$9(-\frac{1}{3})(x + 5) \leq 9(-\frac{2}{9})(x - 1) \qquad \text{Multiply both sides by } 9$$
$$-3(x + 5) \leq -2(x - 1)$$
$$-3x - 15 \leq -2x + 2 \qquad \text{Distributive property}$$
$$-3x + 3x - 15 \leq -2x + 3x + 2 \qquad \text{Add } 3x \text{ to both sides}$$
$$-15 \leq x + 2$$
$$-15 + (-2) \leq x + 2 + (-2) \qquad \text{Add } -2 \text{ to both sides}$$
$$-17 \leq x$$
$$x \geq -17$$

See the graph on page A18 in the textbook.

37. $-2 \leq m - 5 \leq 2$

$\qquad 3 \leq \quad m \quad \leq 7 \qquad$ Add **5** to all three members

The graph is on page A18 in the textbook.

41. $5 \leq 3a - 7 \leq 11$

$\qquad 12 \leq \quad 3a \quad \leq 18 \qquad$ Add **7** to all three members

$\qquad 4 \leq \quad a \quad \leq 6 \qquad$ Multiply through by $\frac{1}{3}$

The graph is on page A18 in the textbook.

45. $4 < 6 + \frac{2}{3}x < 8$

$\qquad -2 < \quad \frac{2}{3}x \quad < 2 \qquad$ Add **-6** to all three members

$\qquad -3 < \quad x \quad < 3 \qquad$ Multiply through by $\frac{3}{2}$

The graph is on page A18 in the textbook.

49. $5y + 1 \leq -4 \quad$ or $\quad 5y + 1 > 4$

$\qquad 5y \leq -5 \quad$ or $\qquad 5y > 3 \qquad$ Add **-1** to all three members

$\qquad y \leq -1 \quad$ or $\qquad y > \frac{3}{5} \qquad$ Multiply through by $\frac{1}{5}$

The graph is on page A18 in the textbook.

53. Let $x = 300$,

$\qquad\qquad x \leq 900 - 300p$

$\qquad\qquad 300 \leq 900 - 300p \qquad$ Substitution for x

$\qquad 300 - \mathbf{900} \leq 900 - \mathbf{900} - 300p \qquad$ Add **900** to both sides

$\qquad\qquad \dfrac{-600}{-300} \leq \dfrac{-300p}{-300} \qquad$ Divide

$\qquad\qquad 2 \geq p$

$\qquad\qquad p \leq 2$

Set the price at \$2.00 or less a pad.

57. $\qquad 3x + 2y < 6$

$\qquad 3x + \mathbf{(-3x)} + 2y < \mathbf{-3x} + 6 \qquad$ Add **-3x** to both sides

$\qquad\qquad 2y < -3x + 6$

$\qquad\qquad \dfrac{2y}{2} < \dfrac{-3x}{2} + \dfrac{6}{2} \qquad$ Divide

$\qquad\qquad y < -\dfrac{3}{2}x + 3$

31

61. $\qquad 95° \quad < \quad \dfrac{9}{5}c + 32 \quad < 113°$

$95 + (-32) < \dfrac{9}{5}c + 32 + (-32) < 113 + (-32) \qquad$ Add **-32** to both sides

$\qquad \dfrac{5}{9}(63) < \qquad \dfrac{5}{9}\left(\dfrac{9}{5}c\right) \qquad < \dfrac{5}{9}(81) \qquad$ Multiply both sides by $\dfrac{5}{9}$

$\qquad 35° \quad < \qquad c \qquad < 45°$

35° to 45° Celsius

65. $\quad 2 \quad < \quad -x \quad < 5$

$(-1)2 < -x(-1) < 5(-1) \qquad$ Multiply both sides by **-1**, change signs

$\qquad -2 \quad > \quad x \quad > -5$

69. $2(x + 3) = 16$

73. $3x + 2 = x - 4$

79. $\qquad -1.96 \quad < \quad \dfrac{x - \mu}{s} \quad < 1.96$

$\quad s(-1.96) < s\left(\dfrac{x - \mu}{s}\right) < s(1.96) \qquad$ Multiply both sides by **s**

$\quad -1.96s \quad < \quad x - \mu \quad < 1.96s$

$-1.96s + \mu < x - u + \mu < 1.96s + \mu \qquad$ Add μ to both sides

$-1.96s + \mu < \qquad x \qquad < 1.96s + \mu$

Section 2.4

1. Step 1: Let x = the number asked for.

Step 2: Three times the sum of a number and 4 is 3(x + 4).

Step 3: $\quad 3(x + 4) = 3$

Step 4: Solve

$$3x + 12 = 3$$
$$3x = -9$$
$$x = -3$$

Step 5: $3(-3 + 4) = 3$

$$3(1) = 3$$
$$3 = 3$$

The solution is {-3}.

5. Step 1: Let x = the number asked for.

    Step 2: 5x + 2, 3x + 8

    Step 3:   5x + 2 = 3x + 8

    Step 4: Solve

    $$2x + 2 = 8$$
    $$2x = 6$$
    $$x = 3$$

    Step 5: 5(3) + 2 = 3(3) + 8
    $$15 + 2 = 9 + 8$$
    $$17 = 17$$

    The solution is {3}.

9. Step 1: Let   x = first consecutive number,
             Let x + 1 = second consecutive number

    Step 2: x, (x + 1), 3x - 1

    Step 3: x + (x + 1) = 3x - 1

    Step 4: Solve

    $$2x + 1 = 3x - 1$$
    $$2 = x - 1$$
    $$2 = x$$
    $$3 = x + 1$$

    Step 5:   2 + 2 + 1 = 3(2) - 1
    $$5 = 5$$

    The solutions are {2,3}.

13. Step 1: Let   x = first consecutive number,
             Let x + 2 = second consecutive number

    Step 2: 2x, (x + 2), 5

    Step 3:   2x - (x + 2) = 5

    Step 4: Solve

    $$2x - x - 2 = 5$$
    $$x - 2 = 5$$
    $$x = 7$$
    $$x + 2 = 9$$

    Step 5: 2(7) - (7 + 2) = 5
    $$14 - 9 = 5$$

    The solutions are {7,9}.

17. Step 1: Let S = side of square

    Step 2: 4S

    Step 3: 4S = 28

    Step 4: Solve

    $$S = 7$$

    Step 5: 4(7) = 28

    The solution is 7 feet.

21. Step 1: Let x = width and 2x - 3 = length

    Step 2: 2x, 2(2x - 3)

    Step 3: 2x + 2(2x - 3) = 18

    Step 4: Solve

    $$2x + 4x - 6 = 18$$
    $$6x - 6 = 18$$
    $$6x = 24$$
    $$x = 4 \text{ meters (width)}$$
    $$2x - 3 = 5 \text{ meters (length)}$$

    Step 5: 2(4) + 2(8 - 3) = 18

    $$2(4) + 2(5) = 18$$
    $$8 + 10 = 18$$

    The solution is 4 meters.

25. Step 1: Let x = Stacey's age

    Step 2:

	Now	Four years ago
Travis	x + 5	x + 5 - 4
Stacey	x	x - 4

    Step 3: x + 5 - 4 + x - 4 = 15

    Step 4: Solve

    $$2x - 3 = 15$$
    $$2x = 18$$
    $$x = 9 \qquad \text{Stacey's age}$$
    $$x + 5 = 14 \qquad \text{Travis' age}$$

    Step 5: 9 + 5 - 4 + 9 - 4 = 15

29. Step 1: Let x = the first number

Step 2: $2x$, x

Step 3: $x + 2x = 24$

Step 4: Solve

$$3x = 24$$
$$x = 8 \qquad \text{first number}$$
$$2x = 16 \qquad \text{second number}$$

Step 5: $3(8) = 24$

33. Step 1: Let x = dimes, and $36 - x$ = nickels

Step 2:

		dimes	nickels	total
Number of		x	$36 - x$	36
Value of		$10x$	$5(36 - x)$	280

Step 3: $10x + 5(36 - x) = 280$

Step 4: Solve

$$10x + 180 - 5x = 280$$
$$5x + 180 = 280$$
$$5x = 100$$
$$x = 20 \quad \text{dimes}$$
$$36 - x = 16 \quad \text{nickels}$$

Step 5: $10(20) + 180 - 5(20) = 280$
$$200 + 180 - 100 = 280$$

37. Step 1: Let x = 8%, let $\$9,000 - x$ = 9%

Step 2:

		Dollars at 8%	Dollars at 9%	Total
Number of		x	$9,000 - x$	9,000
Interest on		$.08x$	$.09(9,000 - x)$	750

Step 3: $.08x + .09(9,000 - x) = 750$

Step 4: Solve

$$.08x + 810 - .09x = 750$$
$$810 - .01x = 750$$
$$-.01x = -60$$
$$x = \$6,000 \text{ at } 8\%$$
$$9,000 - x = \$3,000 \text{ at } 9\%$$

Step 5: $.08(6,000) + .09(9,000 - 6,000) = 750$
$$480 + .09(3,000) = 750$$
$$480 + 270 = 750$$

41. Step 1: Let x = 8%, let 6,000 - x = 9%

Step 2:

	Dollars at 8%	Dollars at 9%	Total
Number of	x	6,000 - x	6,000
Interest on	.08x	.09(6,000 - x)	500

Step 3: $\quad$ .08x + .09(6,000 - x) = 500

Step 4: Solve

$$.08x + 540 - .09x = 500$$
$$540 - .01x = 500$$
$$-.01x = -40$$
$$x = \$4,000 \text{ at } 8\%$$
$$6,000 - x = \$2,000 \text{ at } 9\%$$

Step 5: $\quad$ .08(4,000) + .09(6,000 - 4,000) = 500
$$320 + .09(2,000) = 500$$
$$320 + 180 = 500$$

45. Step 1: Let x = amount of money spent at the business

Step 2: x, .06x

Step 3: 1,204 = x + .06x + 250

Step 4: $\quad$ 954 = 1.06x
$$900 = x \quad \text{(amount of money spent)}$$
$$54 = .06x \text{ (amount of 6\% sales tax money)}$$

Step 5: 1,204 = 900 + 900(.06) + 250
$$1,204 = 900 + 54 + 250$$
$$1,204 = 1,204$$

She sent $54 for state taxes collected.

49. -4, 0, 2, 3

53. |7| - |9| = 7 - 9 = -2

57. |-4| - |-7| = 4 - 7 = -3

Section 2.5

1. |x| = 4 means the distance between x and 0 on the number line is 4. If x is 4 units from 0, the x can be 4 or -4.

5. |x| = -3 The solution set is Ø because |x| will always be a positive number or zero.

9.　　　　　　$|y| + 4 = 3$

　　　$y + 4 + (-4) = 3 + (-4)$　　　Add **-4** to both sides

　　　　　　　　$|y| = -1$

The solution set is Ø because $|x|$ will always be a positive number or zero.

13.　　$|x - 2| = 5$

　　　　$x - 2 = 5$　　　or　　　　$x - 2 = -5$
　$x　2 + 2 = 5 + 2$　　　$x - 2 + 2 = -5 + 2$　　　Add **2** to both sides
　　　　　$x = 7$　　　　　　　　$x = -3$

The solution set is $\{-3,7\}$.

17.　　　　　$1 = |3 - x|$

　　　　　　$1 = 3 - x$　　　or　　　　$-1 = 3 - x$
　$1 + (-3) = 3 + (-3) - x$　　　$-1 + (-3) = 3 + (-3) - x$　　　Add **-3** to both sides
　　　　$-2 = -x$　　　　　　　　　$-4 = -x$
　　　$-2(-1) = -x(-1)$　　　　　　$-4(-1) = -x(-1)$　　　　Multiply by **-1**
　　　　　$2 = x$　　　　　　　　　$4 = x$

The solution set is $\{2,4\}$

21.　　　　$60 = 20x - 40$

　　　　　$60 = 20x - 40$　　　or　　　　$-60 = 20x - 40$
　$60 + 40 = 20x - 40 + 40$　　　$-60 + 40 = 20x - 40 + 40$　　　Add **40** to both sides
　　　$100 = 20x$　　　　　　　　$-20 = 20x$

$\frac{1}{20}(100) = \frac{1}{20}(20x)$　　　　$\frac{1}{20}(-20) = \frac{1}{20}(20x)$　　　Multiply by $\frac{1}{20}$

　　　　　$5 = x$　　　　　　　　　$-1 = x$

The solution set is $\{-1,5\}$.

25.　　$\left|\frac{3}{4}x - 6\right| = 9$

　　　$\frac{3}{4}x - 6 = 9$　　　　or　　　$\frac{3}{4}x - 6 = -9$

　$\frac{3}{4}x - 6 + 6 = 9 + 6$　　　$\frac{3}{4}x - 6 + 6 = -9 + 6$　　　Add **6** to both sides

　　　　$\frac{3}{4}x = 15$　　　　　　　$\frac{3}{4}x = -3$

　$\frac{4}{3}(\frac{3}{4}x) = (\frac{4}{3})(15)$　　　$\frac{4}{3}(\frac{3}{4}x) = \frac{4}{3}(-3)$　　　Multiply by $\frac{4}{3}$

　　　　　$x = 20$　　　　　　　$x = -4$

The solution set is $\{-4,20\}$.

29. $|3x + 4| + 1 = 7$

$|3x + 4| + 1 + \mathbf{(-1)} = 7 + \mathbf{(-1)}$ Add **-1** to both sides

$|3x + 4| = 6$

$3x + 4 = 6$ or $3x + 4 = -6$

$3x + 4 + \mathbf{(-4)} = 6 + \mathbf{(-4)}$ $3x + 4 + \mathbf{(-4)} = 6 + \mathbf{(-4)}$ Add **-4** to both sides

$3x = 2$ $3x = -10$

$\frac{1}{3}(3x) = \frac{1}{3}(2)$ $\frac{1}{3}(3x) = \frac{1}{3}(-10)$ Multiply by $\frac{1}{3}$

$x = \frac{2}{3}$ $x = -\frac{10}{3}$

The solution set is $\{-\frac{10}{3}, \frac{2}{3}\}$.

33. $3 + |4t - 1| = 8$

$3 + \mathbf{(-3)} + |4t - 1| = 8 + \mathbf{(-3)}$ Add **-3** to both sides

$|4t - 1| = 5$

$4t - 1 = 5$ or $4t - 1 = -5$

$4t - 1 + \mathbf{1} = 5 + \mathbf{1}$ $4t - 1 + \mathbf{1} = -5 + \mathbf{1}$ Add **1** to both sides

$4t = 6$ $4t = -4$

$\frac{1}{4}(4t) = \frac{1}{4}(6)$ $\frac{1}{4}(4t) = \frac{1}{4}(-4)$ Multiply by $\frac{1}{4}$

$t = \frac{6}{4}$ $t = -1$

$t = \frac{3}{2}$

The solution set is $\{-1, \frac{3}{2}\}$.

37. $5 = |\frac{2x}{7} + \frac{4}{7}| - 3$

$5 + \mathbf{3} = |\frac{2x}{7} + \frac{4}{7}| - 3 + 3$ Add **3** to both sides

$8 = |\frac{2x}{7} + \frac{4}{7}|$

$8 = \frac{2x}{7} + \frac{4}{7}$ or $-8 = \frac{2x}{7} + \frac{4}{7}$ LCD = 7

$7(8) = 7(\frac{2x}{7}) + 7(\frac{4}{7})$ $7(-8) = 7(\frac{2x}{7}) + 7(\frac{4}{7})$ Multiply both sides by 7

$56 = 2x + 4$ $-56 = 2x + 4$

$56 + \mathbf{(-4)} = 2x + 4 + \mathbf{(-4)}$ $-56 + \mathbf{(-4)} = 2x + 4 + \mathbf{(-4)}$ Add **-4** to both sides

$52 = 2x$ $-60 = 2x$

$\frac{1}{2}(52) = \frac{1}{2}(2x)$ $\frac{1}{2}(-60) = \frac{1}{2}(2x)$ Multiply by $\frac{1}{2}$

$26 = x$ $-30 = x$

The solution set is $\{-30, 26\}$.

41. $|3(x + 5) + 2| = 1$

$$3(x + 5) + 2 = 1 \qquad \text{or} \qquad 3(x + 5) + 2 = -1$$
$$3x + 15 + 2 = 1 \qquad\qquad\qquad 3x + 15 + 2 = -1 \qquad \text{Distributive property}$$

$$3x + 17 = 1 \qquad\qquad\qquad 3x + 17 = -1 \qquad \text{Simplify the left side}$$

$$3x + 17 + \mathbf{(-17)} = 1 + \mathbf{(-17)} \qquad 3x + 17 + \mathbf{(-17)} = -1 + \mathbf{(-17)} \qquad \text{Add } \mathbf{-17} \text{ to both sides}$$

$$3x = -16 \qquad\qquad\qquad 3x = -18$$
$$\tfrac{1}{3}(3x) = \tfrac{1}{3}(-16) \qquad\qquad \tfrac{1}{3}(3x) = \tfrac{1}{3}(-18) \qquad \text{Multiply by } \tfrac{1}{3}$$
$$x = -\tfrac{16}{3} \qquad\qquad\qquad x = -6$$

The solution set is $\{-\tfrac{16}{3}, -6\}$.

45. $1 = |2(k + 4) - 3|$

$$1 = 2(k + 4) - 3 \quad \text{or} \qquad -1 = 2(k + 4) - 3$$
$$1 = 2k + 8 - 3 \qquad\qquad\qquad -1 = 2k + 8 - 3 \qquad \text{Distributive property}$$
$$1 = 2k + 5 \qquad\qquad\qquad -1 = 2k + 5 \qquad \text{Simplify the right side}$$
$$1 + \mathbf{(-5)} = 2k + 5 + \mathbf{(-5)} \qquad -1 + \mathbf{(-5)} = 2k + 5 + \mathbf{(-5)} \qquad \text{Add } \mathbf{-5} \text{ to both sides}$$
$$-4 = 2k \qquad\qquad\qquad -6 = 2k$$
$$\tfrac{1}{2}(-4) = \tfrac{1}{2}(2k) \qquad\qquad \tfrac{1}{2}(-6) = \tfrac{1}{2}(2k) \qquad \text{Multiply by } \tfrac{1}{2}$$
$$-2 = k \qquad\qquad\qquad -3 = k$$

The solution set is $\{-3, -2\}$.

49. $\left| x - \tfrac{1}{3} \right| = \left| \tfrac{1}{2}x + \tfrac{1}{6} \right|$

$$\text{Equals} \qquad\qquad\qquad\qquad \text{Opposites}$$

$$x - \tfrac{1}{3} = \tfrac{1}{2}x + \tfrac{1}{6} \qquad \text{or} \qquad x - \tfrac{1}{3} = -\left(\tfrac{1}{2}x + \tfrac{1}{6} \right)$$

$$\mathbf{6}(x) - \mathbf{6}\left(\tfrac{1}{3}\right) = \mathbf{6}\left(\tfrac{1}{2}x\right) + \mathbf{6}\left(\tfrac{1}{6}\right) \qquad\qquad x - \tfrac{1}{3} = -\tfrac{1}{2}x - \tfrac{1}{6}$$

$$6x - 2 = 3x + 1 \qquad\qquad \mathbf{6}(x) - \mathbf{6}\left(\tfrac{1}{3}\right) = \mathbf{6}\left(-\tfrac{1}{2}x\right) - \mathbf{6}\left(\tfrac{1}{6}\right)$$

$$3x - 2 = 1 \qquad\qquad\qquad 6x - 2 = -3x - 1$$
$$3x = 3 \qquad\qquad\qquad 9x - 2 = -1$$
$$x = 1 \qquad\qquad\qquad 9x = 1$$
$$x = \tfrac{1}{9}$$

The solution set is $\{\tfrac{1}{9}, 1\}$.

53. $|3x - 1| = |3x + 1|$

Equals   Opposites

$3x - 1 = 3x + 1$ or $3x - 1 = -(3x + 1)$

     $-1 = +1$ $3x - 1 = -3x - 1$

No solution here $6x - 1 = -1$

                           $6x = 0$

                           $x = 0$

The solution set is $\{0\}$.

57. $|.03 - .01x| = |.04 + .05x|$

Equals   Opposites

$.03 - .01x$ or $.03 - .01x$

     $= .04 + .05x$ $= -(.04 + .05x)$

$100(.03) - 100(.01x)$ $.03 - .01x$

     $= 100(.04) + 100(.05x)$ $= -.04 - .05x$

$3 - x = 4 + 5x$ $100(.03) - 100(.01x)$

                          $= 100(-.04) - 100(.05x)$

     $3 = 4 + 6x$ $3 - x = -4 - 5x$

   $-1 = 6x$ $3 + 4x = -4$

  $-\dfrac{1}{6} = x$ $4x = -7$

                          $x = -\dfrac{7}{4}$

The solution set is $\{-\dfrac{1}{6}, -\dfrac{7}{4}\}$.

61. $|\dfrac{x}{5} - 1| = |1 - \dfrac{x}{5}|$

           Equals   Opposites

    $\dfrac{x}{5} - 1 = 1 - \dfrac{x}{5}$ or $\dfrac{x}{5} - 1 = -(1 - \dfrac{x}{5})$

$5(\dfrac{x}{5}) - 5(1) = 5(1) - 5(\dfrac{x}{5})$ $\dfrac{x}{5} - 1 = -1 + \dfrac{x}{5}$

     $x - 5 = 5 - x$ $5(\dfrac{x}{5}) - 5(1) = 5(-1) + 5(\dfrac{x}{5})$

   $2x - 5 = 5$ $x - 5 = -5 + x$

     $2x = 10$ $-5 = -5$

      $x = 5$ All real numbers

The solution set is all real numbers.

65. $|x + 2| = x + 2$

$x + 2 = x + 2$ or $x + 2 = -(x + 2)$
    $+2 = +2$ $x + 2 = -x - 2$
All real numbers $2x + 2 = -2$
                              $2x = -4$
                                $x = -2$

The solution set is $\{-2,-1,-0,1,2,3,4\}$. When you substitute -4 or -3 the absolute value becomes negative.

69. $x < -2$ or $x > 8$. Since this compound inequality uses the word **or**, we graph their union. That is, we graph all points on either graph.

See the graph on page A19 in the textbook.

73. $4t - 3 \leq -9$

$4t \leq -6$

$t \leq -\dfrac{6}{4}$

$t \leq -\dfrac{3}{2}$

77. $\dfrac{1}{2} \quad < \dfrac{3}{4}a \quad < \dfrac{3}{5}$

$\dfrac{4}{3}\left(\dfrac{1}{2}\right) < \dfrac{4}{3}\left(\dfrac{3}{4}a\right) < \dfrac{4}{3}\left(\dfrac{3}{5}\right)$

$\dfrac{2}{3} \quad < a \quad < \dfrac{4}{5}$

81. $|ax + b| = c$

      Equals   or   Opposites
$ax + b = c$ or $ax + b = -c$
    $ax = -b + c$ $ax = -b - c$

$x = \dfrac{-b + c}{a}$ $x = \dfrac{-b - c}{a}$

Section 2.6

1. $|x| < 3$. The inequality without absolute values that described the situation is $-3 < x < 3$. See the graph on page A19 in the textbook.

5. $|x| + 2 < 5$

$|x| + 2 + (-2) < 5 + (-2)$ Add **-2** to both sides

            $|x| < 3$

This problem is now graphed like problem 1 in this section.

9. $|y| < -5$. The solution set is $\emptyset$ because $|y|$ will never be less than -5.

13. $\quad |x - 3| < 7 \quad$ becomes:

$-7 < x - 3 < 7$

$\quad -4 < x < 10 \qquad$ Add **3** to each member

See the graph on page A19 in the textbook.

17. $|a - 1| < -3$. The solution set is $\emptyset$ because $|a - 1|$ will never be less than -3.

21. $|3y + 9| \geq 6$ becomes:

$$3y + 9 \leq -6 \qquad \text{or} \qquad 3y + 9 \geq 6$$

$$3y + 9 + \textbf{(-9)} \leq 6 + \textbf{(-9)} \qquad 3y + 9 + \textbf{(-9)} \geq 6 + \textbf{(-9)} \qquad \text{Add } \textbf{-9}$$

$$3y \leq -15 \qquad\qquad\qquad 3y \geq -3$$

$$\frac{1}{3}(3y) \leq \frac{1}{3}(-15) \qquad\qquad \frac{1}{3}(3y) \geq \frac{1}{3}(-3) \qquad \text{Multiply by } \frac{1}{3}$$

$$y \leq -5 \qquad\qquad\qquad y \geq -1$$

See the graph on page A19 in the textbook.

25. $|x - 3| + 2 < 6 \qquad$ Add 2

$\quad |x - 3| < 4 \qquad$ becomes:

$-4 < x - 3 < 4$

$\quad -1 < x < 7 \qquad$ Add 3 to each member

See the graph on page A19 in the textbook.

29. $\quad |3x + 5| - 8 < 5 \qquad\qquad$ Add 8

$\qquad |3x + 5| < 13 \qquad\qquad$ becomes:

$\quad -13 < 3x + 5 < 13$

$\qquad -18 < 3x < 8 \qquad\qquad$ Add -5 to each member

$\frac{1}{3}(-18) < \frac{1}{3}(3x) < \frac{1}{3}(8) \qquad$ Multiply $\frac{1}{3}$ to each member

$\qquad -6 < x < \frac{8}{3}$

See the graph on page A19 in the textbook.

33. $|3 - \frac{2}{3}x| \geq 5$ becomes:

$$3 - \frac{2}{3}x \leq -5 \qquad \text{or} \qquad 3 - \frac{2}{3}x \geq 5$$

$$-\frac{2}{3}x \leq -8 \qquad\qquad\qquad -\frac{2}{3}x \geq 2 \qquad\qquad \text{Add } -3$$

$$-\frac{3}{2}(-\frac{2}{3}x) \leq -\frac{3}{2}(-8) \qquad -\frac{3}{2}(-\frac{2}{3}x) \geq -\frac{3}{2}(2) \qquad \text{Multiply by } -\frac{3}{2}$$

$$x \geq 12 \qquad\qquad\qquad x \leq -3$$

See the graph on page A20 in the textbook.

37. $|x - 1| < .01$ becomes:

$$x - 1 > -.01 \quad \text{or} \quad x - 1 < .01$$
$$x > .99 \qquad\qquad x < 1.01$$
$$.99 < x < 1.01$$

41. $|\frac{3x - 2}{5}| \leq \frac{1}{2}$ becomes:

$$\frac{3x - 2}{5} \geq -\frac{1}{2} \qquad \text{or} \qquad \frac{3x - 2}{5} \leq \frac{1}{2}$$

$$10(\frac{3x - 2}{5}) \geq 10(-\frac{1}{2}) \qquad 10(\frac{3x - 2}{5}) \leq 10(\frac{1}{2})$$

$$2(3x - 2) \geq -5 \qquad\qquad 2(3x - 2) \leq 5$$

$$6x - 4 \geq -5 \qquad\qquad 6x - 4 \leq 5$$

$$6x \geq -1 \qquad\qquad\qquad 6x \leq 9$$

$$x \geq -\frac{1}{6} \qquad\qquad\qquad x \leq \frac{9}{6}$$

$$x \leq \frac{3}{2}$$

$$-\frac{1}{6} \leq x \leq \frac{3}{2}$$

45. The continued inequality $-4 \leq x \leq 4$ as an absolute value becomes $|x| \leq 4$.

49. $-9 \div \frac{3}{2} = -\frac{9}{1} \cdot \frac{2}{3}$

$$= -\frac{18}{3}$$

$$= -6$$

53. $-4(-2)^3 - 5(-3)^2$

$$= -4(-8) - 5(9) \qquad \text{Exponents}$$
$$= 32 - 45 \qquad \text{Multiply}$$
$$= -13$$

57. $\dfrac{2(-3) - 5(-6)}{-1 - 2 - 3} = \dfrac{-6 + 30}{-1 - 2 - 3} \qquad \text{Multiply}$

$$= -\frac{24}{6} \qquad \text{Addition and subtraction}$$

$$= -4 \qquad \text{Division}$$

61. $|ax - b| > c$ becomes:

$$ax - b < -c \qquad \text{or} \quad ax - b > c$$
$$ax < b - c \qquad\qquad ax > b + c$$
$$x < \frac{b - c}{a} \qquad\qquad x > \frac{b + c}{a}$$

65. $\left| \dfrac{x}{a} + \dfrac{y}{b} \right| < c$ becomes:

$$\frac{x}{a} + \frac{y}{b} > -c \qquad\qquad \text{or} \qquad\qquad \frac{x}{a} + \frac{y}{b} < c$$

$$ab\left(\frac{x}{a}\right) + ab\left(\frac{y}{b}\right) > ab(-c) \qquad ab\left(\frac{x}{a}\right) + ab\left(\frac{y}{b}\right) < ab(c)$$

$$bx + ay > -abc \qquad\qquad bx + ay < abc$$
$$bx > -ay - abc \qquad\qquad bx < -ay + abc$$
$$x > \frac{-ay - abc}{b} \qquad\qquad x < \frac{-ay + abc}{b}$$
$$x > -ac - \frac{ay}{b} \qquad\qquad x < ac - \frac{ay}{b}$$

$$-ac - \frac{ay}{b} < x < ac - \frac{ay}{b}$$

Chapter 2 Review

1.
$$x - 3 = 7$$
$$x - 3 + 3 = 7 + 3$$
$$x = 10$$

5.
$$400 - 100a = 200$$
$$400 + (-400) - 100a = 200 + (-400)$$
$$\frac{-100a}{-100} = \frac{-200}{-100}$$
$$a = 2$$

9.
$$4x - 2 = 7x + 7$$
$$4x + (-4x) - 2 = 7x + (-4x) + 7$$
$$-2 = 3x + 7$$
$$-2 + (-7) = 3x + 7 + (-7)$$
$$\frac{-9}{3} = \frac{3x}{3}$$
$$-3 = x$$

13.
$$7y - 5 - 2y = 2y - 3$$
$$5y - 5 = 2y - 3$$
$$5y + (-2y) - 5 = 2y + (-2y) - 3$$
$$3y - 5 = -3$$
$$3y - 5 + 5 = -3 + 5$$
$$\frac{3y}{3} = \frac{2}{3}$$
$$y = \frac{2}{3}$$

17.
$$3(2x + 1) = 18$$
$$6x + 3 = 18$$
$$6x + 3 + (-3) = 18 + (-3)$$
$$\frac{6x}{6} = \frac{15}{6}$$
$$x = \frac{5}{2}$$

21.
$$8 - 3(2t + 1) = 5(t + 2)$$
$$8 - 6t - 3 = 5t + 10$$
$$-6t + 5 = 5t + 10$$
$$-6t + \mathbf{6t} + 5 = 5t + \mathbf{6t} + 10$$
$$5 = 11t + 10$$
$$5 + \mathbf{(-10)} = 11t + 10 + \mathbf{(-10)}$$

$$\frac{-5}{11} = \frac{11t}{11}$$

$$\frac{-5}{11} = t$$

25.
$$.08x + .07(900 - x) = 67$$
$$\mathbf{100}(.08)x + \mathbf{100}(.07)(900 - x) = \mathbf{100}(67)$$
$$8x + 7(900 - x) = 6700$$
$$8x + 6300 - 7x = 6700$$
$$x + 6300 = 6700$$
$$x + 6300 + \mathbf{(-6300)} = 6700 + \mathbf{(-6300)}$$
$$x = 400$$

29. $A = P + Prt$: $A = 2{,}000$, $P = 1{,}000$, $r = 0.05$
$$2000 = 1000 + (1000)(0.05)(t)$$
$$2000 = 1000 + 50t$$
$$1000 = 50t$$
$$20 = t$$

33.
$$\frac{I}{rt} = \frac{Prt}{rt}$$

$$\frac{I}{rt} = P$$

37. $4x - 3y = 12$
$$-3y = -4x + 12$$

$$\frac{-3y}{-3} = \frac{-4x}{-3} + \frac{12}{-3}$$

$$y = \frac{4}{3}x - 4$$

41.
$$C = \frac{5}{9}(F - 32)$$

$$\frac{9}{5}C = F - 32$$

$$\frac{9}{5}C + 32 = F$$

45. $6 - a \geq -2$

$-a \geq -8$

$\dfrac{-a}{-1} \geq \dfrac{-8}{-1}$

$a \leq 8$

49. $800 - 200x < 1000$

$-200x < 200$

$\dfrac{-200x}{-200} < \dfrac{200}{-200}$

$x > -1$

53. $-.01 \leq .02x - .01 < .01$

$-1 \leq 2x - 1 \leq 1$

$0 \leq 2x \leq 2$

$0 \leq x \leq 1$

57. $3(x + 1) < 2(x + 2)$ or $2(x - 1) \geq x + 2$

$3x + 3 < 2x + 4$ $\qquad$ $2x - 2 \geq x + 2$

$x + 3 < 4$ $\qquad\qquad$ $x - 2 \geq 2$

$x < 1$ $\qquad\qquad\qquad$ $x \geq 4$

61. Step 1: Let $\quad x$ = first consecutive integer
Let $x + 1$ = second consecutive integer

Step 2: $4(x + x + 1)$, $14(x + 1) + 2$

Step 3: $4(x + x + 1) = 14(x + 1) + 2$

Step 4: Solve

$4(x + x + 1) = 14(x + 1) + 2$
$4x + 4x + 4 = 14x + 14 + 2$
$8x + 4 = 14x + 16$
$4 = 6x + 16$
$-12 = 6x$
$-2 = x$
$-1 = x + 1$

Step 5: $4[-2 + (-1)] = 14(-1) + 2$
$-12 = -14 + 2$
$-12 = -12$

The solutions are {-2,-1}.

65. Step 1: Let   x = first consecutive integer
Let x + 1 = second consecutive integer
Let x + 2 = third consecutive integer

Step 2: x, x + 1, x + 2, 12

Step 3: x + (x + 1) + (x + 2) = 12

Step 4: Solve

$$x + (x + 1) + (x + 2) = 12$$
$$3x + 3 = 12$$
$$3x = 9$$
$$x = 3$$
$$x + 1 = 4$$
$$x + 2 = 5$$

Step 5:   3 + 4 + 5 = 12
12 = 12

The sides of the triangle are 3m, 4m and 5m.

69. Step 1: Let   x = one number
Let 16 - x = other number

Step 2: x, 16 - x, 16

Step 3:   x = 3(16 - x)

Step 4: Solve

$$x = 3(16 - x)$$
$$x = 48 - 3x$$
$$4x = 48$$
$$x = 12$$
$$16 - x = 4$$

Step 5:   12 = 3(4)
12 = 12

The two numbers are 4 and 12.

73. Step 1: Let x = 8%, let 900 - x = 7%

Step 2:

	Dollars at 8%	Dollars at 7%	Total
Number of	x	900 - x	900
Interest on	.08x	.07(900 - x)	67

Step 3: $.08x + .07(900 - x) = 67$

$$8x + 7(900 - x) = 6700$$
$$8x + 6300 - 7x = 6700$$
$$x + 6300 = 6700$$
$$x = 400$$
$$900 - x = 500$$

Step 5: $.08(400) + .07(500) = 67$

$$32 + 35 = 67$$
$$67 = 67$$

$400 at 8%, $500 at 7%

77. $|a| - 3 = 1$

$|a| = 4$ becomes

$a = 4$ or $a = -4$

81. $|2y - 3| = 5$ becomes

$2y - 3 = 5$ or	$2y - 3 = -5$
$2y = 8$	$2y = -2$
$y = 4$	$y = -1$

85. $|\frac{7}{3} - \frac{x}{3}| + \frac{4}{3} = 2$

$|\frac{7}{3} - \frac{x}{3}| = \frac{2}{3}$ becomes: $(-\frac{4}{3} + 2 = -\frac{4}{3} + \frac{6}{3} = \frac{2}{3})$

$\frac{7}{3} - \frac{x}{3} = \frac{2}{3}$ or	$\frac{7}{3} - \frac{x}{3} = -\frac{2}{3}$
$7 - x = 2$	$7 - x = -2$
$-x = -5$	$-x = -9$
$x = 5$	$x = 9$

89. $\left| \frac{1}{2} - x \right| = \left| x + \frac{1}{2} \right|$

Equals	Opposites
$\frac{1}{2} - x = x + \frac{1}{2}$	$\frac{1}{2} - x = -(x + \frac{1}{2})$
$1 - 2x = 2x + 1$	$\frac{1}{2} - x = -x - \frac{1}{2}$
$1 = 4x + 1$	$1 - 2x = -2x - 1$
$0 = 4x$	$1 = -1$
$0 = x$	No solution

There is no solution to the equation.

93. $|.01a| \geq 5$

    $.01a \leq -5$ or $.01a \geq 5$

      $a \leq -500$ $a \geq 500$

See the graph on page A20 in the textbook.

97. $|y + 5| \geq .02$

    $y + 5 \leq -.02$ or $y + 5 \geq .02$

      $y \leq -5.02$ $y \geq -4.98$

See the graph on page A20 in the textbook.

101. $2x - 3 = 2(x - 3)$
$2x - 3 = 2x - 6$
    $-3 = -6$
      $\emptyset$

105. $|4y + 8| = -1$. The absolute value will never be negative so the solution is $\emptyset$.

109. $|5 - 8t| + 4 \leq 1$

    $|5 - 8t| \leq -3$ The absolute value will never be negative so the solution is $\emptyset$.

1. $x - 5 = 7$
$x - 5 + 5 = 7 + 5$
$x = 12$

5. $5(x - 1) - 2(2x + 3) = 5x - 4$
$5x - 5 - 4x - 6 = 5x - 4$
$x - 11 = 5x - 4$
$-11 = 4x - 4$
$-7 = 4x$
$-\dfrac{7}{4} = x$

9. $-5t \leq 30$

$\dfrac{-5t}{-5} \leq \dfrac{30}{-5}$

$t \geq -6$

See the graph on page A21 in the textbook.

13. $\left| \dfrac{1}{4}x - 1 \right| = \dfrac{1}{2}$ becomes:

$\dfrac{1}{4}x - 1 = \dfrac{1}{2}$ or $\dfrac{1}{4}x - 1 = -\dfrac{1}{2}$

$x - 4 = 2$ $\qquad$ $x - 4 = -2$
$x = 6$ $\qquad\qquad$ $x = 2$

17. $|6x - 1| > 7$ becomes:

$6x - 1 < -7$ or $6x - 1 > 7$

$6x < -6$ $\qquad\quad$ $6x > 8$

$x < -1$ $\qquad\qquad$ $x > \dfrac{8}{6}$

$\qquad\qquad\qquad\quad x > \dfrac{4}{3}$

See the graph on page A21 in the textbook.

21. Step 1: Let   x = first consecutive integer
             Let x + 2 = second consecutive integer

Step 2: x, x + 2, 18

Step 3: x + (x + 2) = 18

Step 4: Solve

$$x + (x + 2) = 18$$
$$2x + 2 = 18$$
$$2x = 16$$
$$x = 8$$
$$x + 2 = 10$$

Step 5:   8 + 10 = 18
            18 = 18

The two numbers are 8 and 10.

25. Step 1: Let x = dimes and 14-x = nickels

Step 2:

	Dimes	Nickels	Total
Number of	x	14 - x	14
Value of	10x	5(14 - x)	110

Step 3: 10x + 5(14 - x) = 110

Step 4: Solve

$$10x + 5(14 - x) = 110$$
$$10x + 70 - 5x = 110$$
$$5x + 70 = 110$$
$$5x = 40$$
$$x = 8$$
$$14 - x = 6$$

Step 5:   10(8) + 5(6) = 110
             80 + 30 = 110
                110 = 110

There are 6 nickels and 8 dimes.

Chapter 3

Section 3.1

1. $4^2 = 4 \cdot 4$ Base 4, exponent 2
$\quad = 16$

5. $-.3^3 = -.3 \cdot .3 \cdot .3$ Base 3, exponent 3
$\quad = -.027$

9. $(\frac{1}{2})^3 = \frac{1}{2} \cdot \frac{1}{2} \cdot \frac{1}{2}$ Base $\frac{1}{2}$, exponent 3

$\quad = \frac{1}{8}$

13. $x^5 \cdot x^4 = x^{5+4}$ Property 1
$\quad = x^9$

17. $(-\frac{2}{3}x^2)^3 = (-\frac{2}{3})^3(x^2)^3$ Property 3

$\quad = -\frac{8}{27}x^6$ Property 2

21. $6x^2(-3x^4)(2x^5) = [6(-3)2](x^2 \cdot x^4 \cdot x^5)$ Commutative and associative
$\quad = -36x^{11}$ Property 1

25. $3^{-2} = \frac{1}{3^2}$ Property 4

$\quad = \frac{1}{9}$

29. $(-3)^{-2} = \frac{1}{(-3)^2}$ Property 4

$\quad = \frac{1}{9}$

33. $(\frac{1}{3})^{-2} + (\frac{1}{2})^{-3} = \frac{1}{(\frac{1}{3})^2} + \frac{1}{(\frac{1}{2})^3}$ Property 4

$\quad = \frac{1}{\frac{1}{9}} + \frac{1}{\frac{1}{8}}$

$\quad = \frac{9}{1} + \frac{8}{1}$

$\quad = 17$

37. $x^{-4}x^7 = x^3$ Property 1

41. $(\frac{1}{2}x^{-3})^3(6x^4) = \frac{1}{8}x^{-9}(6x^4)$ Properties 2 and 3

$= (\frac{1}{8} \cdot 6)(x^{-9}x^4)$ Commutative and Associative

$= \frac{6}{8}x^{-5}$ Property 1

$= \frac{6}{8} \cdot \frac{1}{x^5}$ Property 4

$= \frac{6}{8x^5}$ Multiplication

45. $(\frac{1}{2}x^3)(\frac{2}{3}x^4)(\frac{3}{5}x^{-7}) = (\frac{1}{2} \cdot \frac{2}{3} \cdot \frac{3}{5})(x^3 \cdot x^4 \cdot x^{-7})$ Commutative and associative

$= \frac{1}{5}x^0$ Multiplication

$= \frac{1}{5}$

49. $(2x^4y^{-3})(7x^{-8}y^5) = (2 \cdot 7)(x^4x^{-8})(y^{-3}y^5)$ Associative and commutative

$= 14x^{-4}y^2$ Property 1

$= 14 \cdot \frac{1}{x^4} \cdot y^2$ Property 4

$= \frac{14y^2}{x^4}$ Multiplication

53. $(4a^5b^2)(2b^{-5}c^2)(3a^7c^4)$

$= (4 \cdot 2 \cdot 3)(a^5a^7)(b^2b^{-5})(c^2c^4)$ Associative and commutative

$= 24a^{12}b^{-3}c^6$ Property 1

$= 24a^{12} \cdot \frac{1}{b^3} \cdot c^6$ Property 4

$= \frac{24a^{12}c^6}{b^3}$ Multiplication

57. $(3r^2s^{-1})^4(9r^{-6}s^4)^{-3}$

$= (81r^8s^{-4})(\frac{1}{729}r^{18}s^{-12})$ Properties 2 and 3

$= (81 \cdot \frac{1}{729})(r^8r^{18})(s^{-4}s^{-12})$ Commutative and associative

$= \frac{1}{9}r^{26}s^{-16}$ Property 1

$= \frac{1}{9} \cdot r^{26} \cdot \frac{1}{s^{16}}$ Property 4

$= \frac{r^{26}}{9s^{16}}$ Multiplication

61. Remember to rewrite 4,900 to a number between 1 and 10 and a power of 10.

$4,900 = 4,900 \times 10^3 = 4.9 \times 10^3$

65. Remember to write 0.00495 to a number between 1 and 10 and a power of 10.

$0.00495 = 0.00495 \times 10^{-3} = 4.95 \times 10^{-3}$

A number that is less than 1 will have a negative exponent when written in scientific notation.

69. $5.34 \times 10^3 = 5.34 \times 1,000$ $(10^3 = 1,000)$
$= 5,340$

The exponent 3 indicates the number of places we need to move the decimal point in order to write our number in expanded form.

73. $3.44 \times 10^{-3} = 0.00344$

The negative exponent when written in scientific notation indicates a number less than 1.

77. $\frac{1}{2^{-3}} = \frac{1}{\frac{1}{2^3}}$

$= 1 \cdot \frac{2^3}{1}$

$= 2^3$
$= 8$

81. $(2^2)^3 = (4)^3 = 64$
$\quad 2^{2^3} = 2^8 = 256$

85. When $\quad\quad$ a = 2 and b = 4

$\quad\quad$ the equation $\quad (a + b)^{-1} = a^{-1} + b^{-1}$

$\quad\quad$ becomes $\quad\quad (2 + 4)^{-1} = 2^{-1} + 4^{-1}$

$$(6)^{-1} = \frac{1}{2} + \frac{1}{4}$$

$$\frac{1}{6} = \frac{3}{4} \quad\quad\quad \text{False}$$

89. $630,000,000 = 6.3 \times 10^8$

93. $6 - (-8) = 6 + 8 = 14$

97. $-4 - (-3) = -4 + 3 = -1$

101. $x^{m+2} \cdot x^{-2m} \cdot x^{m-5} = x^{m+2-2m+m-5} \quad\quad$ Property 1

$$= x^{-3}$$

$$= \frac{1}{x^3} \quad\quad\quad \text{Property 4}$$

105. $(2x^m y^n)^2 (3x^{-m} y^{-n})^2$

$$= 2^2 x^{2m} y^{2n} 3^2 x^{-2m} y^{-2n} \quad\quad\quad \text{Property 2}$$

$$= (4 \cdot 9)(x^{2m} \cdot x^{-2m})(y^{2n} \cdot y^{-2n}) \quad\quad \text{Commutative and associative}$$

$$= 36 x^0 y^0$$

$$= 36$$

Section 3.2

1. $\left(\dfrac{x^3}{y^2}\right)^2 = \dfrac{(x^3)^2}{(y^2)^2} \quad\quad$ Property 5

$$= \frac{x^6}{y^4} \quad\quad\quad \text{Property 2}$$

5. $\dfrac{3^4}{3^6} = 3^{4-6} \quad\quad$ Property 6

$$= 3^{-2}$$

$$= \frac{1}{3^2} \quad\quad\quad \text{Property 4}$$

$$= \frac{1}{9} \quad\quad\quad \text{Definition of exponents}$$

9. $\dfrac{x^{-1}}{x^9} = x^{-1-9}$ Property 6

       $= x^{-10}$

       $= \dfrac{1}{x^{10}}$ Property 4

13. $\dfrac{t^{-10}}{t^{-4}} = t^{-10-(-4)}$ Property 6

       $= t^{-6}$

       $= \dfrac{1}{t^6}$ Property 4

17. $\dfrac{(a^3)^4}{a^7} = \dfrac{a^{12}}{a^7}$ Property 2

       $= a^{12-7}$ Property 6

       $= a^5$

21. $\dfrac{(x^5)^6}{(x^3)^4} = \dfrac{x^{30}}{x^{12}}$ Property 2

       $= x^{30-12}$ Property 6

       $= x^{18}$

25. $\dfrac{a^3}{a^5 a^6} = \dfrac{a^3}{a^{11}}$ Property 1

       $= a^{3-11}$ Property 6

       $= a^{-8}$

       $= \dfrac{1}{a^8}$ Property 4

29. $\dfrac{(x^{-2})^3 (x^3)^{-2}}{x^{10}} = \dfrac{x^{-6} x^{-6}}{x^{10}}$ Property 2

       $= \dfrac{x^{-12}}{x^{10}}$ Property 1

       $= x^{-12-10}$ Property 6

       $= x^{-22}$

       $= \dfrac{1}{x^{22}}$ Property 4

33. $\dfrac{5a^8b^3}{20a^5b^{-4}} = \dfrac{1}{4} \cdot a^{8-5}b^{3-(-4)}$ Property 6

$\qquad\qquad = \dfrac{1}{4} \cdot a^3 \cdot b^7$

$\qquad\qquad = \dfrac{a^3b^7}{4}$ Multiplication

37. $\dfrac{12r^{-6}s^0t^{-3}}{3r^{-4}s^{-3}t^{-5}}$

$\qquad = 4 \cdot r^{-6-(-4)}s^{0-(-3)}t^{-3-(-5)}$ Property 6

$\qquad = 4 \cdot r^{-2}s^3t^2$

$\qquad = 4 \cdot \dfrac{1}{r^2}s^3t^2$ Property 4

$\qquad = \dfrac{4s^3t^2}{r^2}$ Multiplication

41. $\dfrac{(3x^{-2}y^8)^4}{(9x^4y^{-3})^2}$

$\qquad = \dfrac{3^4x^{-8}6^{32}}{9^2x^8y^{-6}}$ Property 2

$\qquad = \dfrac{81}{81} \cdot x^{-8-8}6^{32-(-6)}$ Property 6 and definition of exponents

$\qquad = x^{-16}y^{38}$

$\qquad = \dfrac{1}{x^{16}} \cdot y^{38}$ Property 4

$\qquad = \dfrac{y^{38}}{x^{16}}$ Multiplication

45. $\left(\dfrac{8x^2y}{4x^4y^{-3}}\right)^4$

$\qquad = (2x^{2-4}y^{1-(-3)})^4$ Property 6

$\qquad = (2x^{-2}y^4)^4$

$\qquad = 2^4x^{-8}y^{16}$ Property 2

$\qquad = 16 \cdot \dfrac{1}{x^8} \cdot y^{16}$ Property 4

$\qquad = \dfrac{16y^{16}}{x^8}$ Multiplication

NOTE: In problem 45, we could have obtained the same result by applying Property 5 first:

$$\left(\frac{8x^2y}{4x^4y^{-3}} \right)^4$$

$$= \frac{8^4x^8y^4}{4^4x^{16}y^{-2}} \qquad \text{Property 5}$$

$$= \frac{4096}{256} \cdot x^{8-16}y^{4-(-12)} \qquad \text{Property 6 and definition of exponents}$$

$$= 16x^{-8}y^{16}$$

$$= 16 \cdot \frac{1}{x^8} \cdot y^{16} \qquad \text{Property 4}$$

$$= \frac{16y^{16}}{x^8} \qquad \text{Multiplication}$$

49. $\left(\frac{2x^{-3}y^0}{4x^6y^{-5}} \right)^{-2}$

$$= \left(\frac{1}{2} \cdot x^{-3-6}y^{0-(-5)} \right)^{-2} \qquad \text{Property 6}$$

$$= \left(\frac{1}{2} \cdot x^{-9}y^5 \right)^{-2}$$

$$= \left(\frac{1}{2} \right)^{-2} x^{18}y^{-10} \qquad \text{Property 2}$$

$$= 4x^{18} \cdot \frac{1}{y^{10}} \qquad \text{Property 4:} \quad \frac{1}{2^{-2}} = 2^2 = 4$$

$$= \frac{4x^{18}}{y^{10}} \qquad \text{Multiplication}$$

Note: In problem 49, we could have obtained the same result by applying Property 5 first:

$$\left(\frac{2x^{-3}y^0}{4x^6y^{-5}} \right)^{-2}$$

$$= \frac{2^{-2}x^6y^0}{4^{-2}x^{-12}y^{10}} \qquad \text{Property 5}$$

$$= \frac{4^2}{2^2} \cdot x^{6-(-12)}y^{0-10} \qquad \text{Properties 2 and 6}$$

$$= 4x^{18}y^{-10}$$

$$= 4x^{18} \cdot \frac{1}{y^{10}} \qquad \text{Property 4}$$

$$= \frac{4x^{18}}{y^{10}} \qquad \text{Multiplication}$$

53. $\left(\dfrac{x^{-3}y^2}{x^4y^{-5}}\right)^{-2}\left(\dfrac{x^{-4}y}{x^0y^2}\right)$

$\quad\quad = (x^{-3-4}y^{2-(-5)})^{-2}(x^{-4-0}y^{1-2})$ $\quad\quad$ Property 6

$\quad\quad = (x^{-7}y^7)^{-2}(x^{-4}y^{-1})$

$\quad\quad = (x^{14}y^{-14})(x^{-4}y^{-1})$ $\quad\quad$ Property 3

$\quad\quad = x^{10}y^{-15}$

$\quad\quad = x^{10}\cdot\dfrac{1}{y^{15}}$ $\quad\quad$ Property 4

$\quad\quad = \dfrac{x^{10}}{y^{15}}$ $\quad\quad$ Multiplication

Note: In problem 53, we could have obtained the same result by applying Property 5 first:

$\left(\dfrac{x^{-3}y^2}{x^4y^{-5}}\right)^{-2}\left(\dfrac{x^{-4}y}{x^0y^2}\right)$

$\quad\quad = \left(\dfrac{x^6y^{-4}}{x^{-8}y^{10}}\right)\left(\dfrac{x^{-4}y}{x^0y^2}\right)$ $\quad\quad$ Property 5

$\quad\quad = (x^{6-(-8)}y^{-4-10})(x^{-4-0}y^{1-2})$ $\quad\quad$ Property 6

$\quad\quad = (x^{14}y^{-14})(x^{-4}y^{-1})$

$\quad\quad = x^{10}y^{-15}$

$\quad\quad = x^{10}\cdot\dfrac{1}{y^{15}}$ $\quad\quad$ Property 4

$\quad\quad = \dfrac{x^{10}}{y^{15}}$ $\quad\quad$ Multiplication

57. $7x^{-3}y^4\left(\dfrac{3x^{-1}y^5}{9x^3y^{-2}}\right)^{-3}$

$= 7x^{-3}y^4\left(\dfrac{1}{3}\cdot x^{-1-3}y^{5-(-2)}\right)^{-3}$ Property 6

$= 7x^{-3}y^4\left(\dfrac{1}{3}x^{-4}y^7\right)^{-3}$

$= 7x^{-3}y^4\left[\left(\dfrac{1}{3}\right)^{-3}x^{12}y^{-21}\right]$ Property 3

$= 7\cdot x^{-3+12}y^{4+(-21)}\cdot 27$ Properties 6 and 4:
$\left(\dfrac{1}{3}\right)^{-3}=\dfrac{1}{3^{-3}}=3^3=27$

$= 189\cdot x^9y^{-17}$

$= 189\cdot x^9\cdot \dfrac{1}{y^{17}}$ Property 4

$= \dfrac{189x^9}{y^{17}}$ Multiplication

Note: In problem 57, we could have obtained the same result by applying Property 5 first:

$7x^{-3}y^4\left(\dfrac{3x^{-1}y^5}{9x^3y^{-2}}\right)^{-3}$

$= 7x^{-3}y^4\left(\dfrac{x^3y^{-15}}{3^{-3}x^{-9}y^6}\right)$ Property 5

$= 7x^{-3}y^4\cdot 27\cdot x^{3-(-9)}y^{-15-6}$ Properties 6 and 4:
$\dfrac{1}{3^{-3}}=3^3=27$

$= 189x^9y^{-17}$

$= 189x^9\cdot \dfrac{1}{y^{17}}$ Property 4

$= \dfrac{189x^9}{y^{17}}$ Multiplication

61. $\dfrac{6.8 \times 10^6}{3.4 \times 10^{10}} = \dfrac{6.8}{3.4} \times \dfrac{10^6}{10^{10}}$

$\qquad\qquad\quad = 2 \times 10^{6-10}$

$\qquad\qquad\quad = 2 \times 10^{-4}$

65. $\dfrac{(2.4 \times 10^{-3})(3.6 \times 10^{-7})}{(4.8 \times 10^6)(1 \times 10^{-9})}$

$\qquad = \left(\dfrac{2.4 \times 3.6}{4.8 \times 1}\right)\left(\dfrac{10^{-3} \times 10^{-7}}{10^6 \times 10^{-9}}\right)$

$\qquad = \left(\dfrac{3.6}{2}\right)\left(\dfrac{10^{-10}}{10^{-3}}\right)$

$\qquad = 1.8 \times 10^{-10-(-3)}$

$\qquad = 1.8 \times 10^{-7}$

69. $\dfrac{69{,}800}{0.000349} = \dfrac{6.98 \times 10^4}{3.49 \times 10^{-4}}$

$\qquad\qquad = \dfrac{6.98}{3.49} \times \dfrac{10^4}{10^{-4}}$

$\qquad\qquad = 2 \times 10^{4-(-4)}$

$\qquad\qquad = 2 \times 10^8$

73. $4^3 \cdot 4^x = 4^9$

$\qquad 4^{3+x} = 4^9 \qquad$ The exponents become:

$\qquad 3 + x = 9$

$\qquad\quad\ x = 6$

77. $2^x \cdot 2^3 = \dfrac{1}{16}$

$\qquad 2^{x+3} = \dfrac{1}{2^4}$

$\qquad 2^{x+3} = 2^{-4} \qquad$ The exponents become:

$\qquad x + 3 = -4$

$\qquad\quad\ x = -7$

81. $6 + 2(x + 3) = 6 + 2x + 6$

$\qquad\qquad\qquad\ = 2x + 12$

85. $\frac{1}{2}(4y - 2) - \frac{1}{3}(6y - 3)$

$$= \frac{1}{2}(4y) - \frac{1}{2}(2) = \frac{1}{3}(6y) - (-\frac{1}{3})(3)$$

$$= 2y - 1 - 2y + 1$$
$$= 0$$

89. $-\frac{1}{3} + \frac{7}{12} - \frac{1}{4} = -\frac{1}{3} \cdot \frac{\mathbf{4}}{\mathbf{4}} + \frac{7}{12} \cdot \frac{\mathbf{1}}{\mathbf{1}} - \frac{1}{4} \cdot \frac{\mathbf{3}}{\mathbf{3}}$ LCD = 12

$$= -\frac{4}{12} + \frac{7}{12} - \frac{3}{12}$$

$$= \frac{-4 + 7 - 3}{12}$$

$$= 0$$

93. $\dfrac{(1.98 \times 10^{25})(3.85 \times 10^{35})}{6.93 \times 10^{40}}$

$$= \frac{1.98 \times 3.85}{6.93} \times \frac{10^{25} \times 10^{35}}{10^{40}}$$

$$= 1.1 \times 10^{60} \times 10^{-40}$$

$$= 1.1 \times 10^{20}$$

97. $\dfrac{x^{n+2}}{x^{n-3}} = x^{n+2-(n-3)}$

$$= x^{n+2-n+3}$$
$$= x^{5}$$

101. $\dfrac{(y^{r})^{-2}}{y^{-2r}} = \dfrac{y^{-2r}}{y^{-2r}}$

$$= y^{-2r-(-2r)}$$
$$= y^{0}$$
$$= 1$$

Section 3.3

1. $5x^2 = 3x + 2$ A trinomial of degree 2, leading coefficient 5.

5. $8a^2 = 3a - 5$ A trinomial of degree 2, leading coefficient 8.

9. $-\frac{3}{4}$ A monomial of degree 0, leading coefficient $-\frac{3}{4}$

13. $(4x + 2) + (3x - 1) = (4x + 3x) + (2 - 1)$ Distributive property
    $$= 7x + 1 \qquad\qquad\qquad \text{Addition}$$

17. $12a^2 + 8ab - 15ab - 10b^2$

    $= 12a^2 + (8 - 15)ab - 10b^2$ Distributive property

    $= 12a^2 + (-7)ab - 10b^2$

    $= 12a^2 - 7ab - 10b^2$

21. $\left(\frac{1}{2}x^2 - \frac{1}{3}x - \frac{1}{6}\right) - \left(\frac{1}{4}x^2 + \frac{7}{12}x\right) + \left(\frac{1}{3}x - \frac{1}{12}\right)$

    $= \frac{1}{2}x^2 - \frac{1}{3}x - \frac{1}{6} - \frac{1}{4}x^2 - \frac{7}{12}x + \frac{1}{3}x - \frac{1}{12}$ The opposite of a sum is the sum of the opposites

    $= \left(\frac{1}{2}x^2 - \frac{1}{4}x^2\right) + \left(-\frac{1}{3}x - \frac{7}{12}x + \frac{1}{3}x\right) + \left(-\frac{1}{6} - \frac{1}{12}\right)$ LCD = 12

    $= \left(\frac{6}{12}x^2 - \frac{3}{12}x^2\right) + \left(-\frac{4}{12}x - \frac{7}{12}x + \frac{4}{12}x\right) + \left(-\frac{2}{12} - \frac{1}{12}\right)$

    $= \frac{3}{12}x^2 - \frac{7}{12}x - \frac{3}{12}$

    $= \frac{1}{4}x^2 - \frac{7}{12}x - \frac{1}{4}$

25. $(5x^3 - 4x^2) - (3x + 4) + (5x^2 - 7) - (3x^3 + 6)$

    $= 5x^3 - 4x^2 - 3x - 4 + 5x^2 - 7 - 3x^3 - 6$

    $= (5x^3 - 3x^3) + (-4x^2 + 5x^2) - 3x + (-4 - 7 - 6)$

    $= 2x^3 + x^2 - 3x - 17$

29. $(3a^3 + 2a^2b + ab^2 - b^3) - (6a^3 - 4a^2b + 6ab^2 - b^3)$

    $= 3a^3 + 2a^2b + ab^2 - b^3 - 6a^3 + 4a^2b - 6ab^2 + b^3$

    $= (3a^3 - 6a^3) + (2a^2b + 4a^2b) + (ab^2 - 6ab^2) + (-b^3 + b^3)$

    $= -3a^3 + 6a^2b - 5ab^2$

33. $(x^2 - 6xy + y^2) + (2x^2 - 6xy - y^2)$

    $= (x^2 + 2x^2) + (-6xy - 6xy) + (y^2 - y^2)$

    $= 3x^2 - 12xy$

37. $(11a^2 + 3ab + 2b^2) + (9a^2 - 2ab + b^2) + (-6a^2 - 3ab + 5b^2)$

    $= (11a^2 + 9a^2 - 6a^2) + (3ab - 2ab - 3ab) + (2b^2 + b^2 + 5b^2)$

    $= 14a^2 - 2ab + 8b^2$

41. $-5[-(x - 3) - (x + 2)]$

$= -5(-x + 3 - x - 2)$

$= -5(-2x + 1)$

$= 10x - 5$

45. $-(3x - 4y) - [(4x + 2y) - (3x + 7y)]$

$= -3x + 4y - (4x + 2y - 3x - 7y)$

$= -3x + 4y - 4x - 2y + 3x + 7y$

$= -4x + 9y$

$= 9y - 4x$

49. When $x = 2$

the polynomial $2x^2 - 3x - 4$

becomes $2 \cdot 2^2 - 3 \cdot 2 - 4$

$2 \cdot 4 - 3 \cdot 2 - 4$

$8 - 6 - 4$

-2

53. When $x = -2$

the polynomial $x^3 - x^2 + x - 1$

becomes $(-2)^3 - (-2)^2 + (-2) - 1$

$-8 - 4 - 2 - 1$

-15

57. Using the equation $h = -16t^2 + 128t$ and the information given in the problem, we have

$h = -16t^2 + 128t$

$h = -16(3)^2 + 128(3)$

$h = -16(9) + 128(3)$

$h = -144 + 384$

$h = 240$ feet for 3 seconds

$h = -16t^2 + 128t$

$h = -16(5)^2 + 128(5)$

$h = -16(25) + 128(5)$

$h = -400 + 640$

$h = 240$ feet for 5 seconds

61. Using the equation $P = R - C$ and the information given in the problem, we have

$$P = R - C$$
$$= 10x - .002x^2 - (800 + 6.5x)$$
$$= 10x - .002x^2 - 800 - 6.5x$$
$$= -800 + 3.5x - .002x^2$$

If they produced 1,000 patterns in May, the profit would be

$$P = -800 + 3.5(1,000) - .002(1,000)^2$$
$$= -800 + 3,500 - 2,000$$
$$= \$700$$

65. $4x^3(5x^2) = (4 \cdot 5)(x^3 x^2)$
$$= 20x^5$$

69. $2x(3x^2) = (2 \cdot 3)(x x^2)$
$$= 6x^3$$

73. $3 - \frac{2}{3}y = -9$

$-\frac{2}{3} = -12$ Add -3 to each side

$y = 18$ Multiply by $-\frac{3}{2}$

Section 3.4

1. $2x(6x^2 - 5x + 4)$
$$= 2x(6x^2) + 2x(-5x) + 2x(4) \qquad \text{Distributive property}$$
$$= 12x^3 - 10x^2 + 8x$$

5. $2a^2b(a^3 - ab + b^3)$
$$= 2a^2b(a^3) + 2a^2b(-ab) + 2a^2b(b^3) \qquad \text{Distributive property}$$
$$= 2a^5b - 2a^3b^2 + 2a^2b^4$$

9. $3r^3s^2(r^3 - 2r^2s + 3rs^2 + s^3)$
$$= 3r^3s^2(r^3) + 3r^3s^2(-2r^2s) + 3r^3s^2(3rs^2) + 3r^3s^2(s^3)$$
$$= 3r^6s^2 - 6r^5s^3 + 9r^4s^4 + 3r^3s^5$$

13.
$$2x^2 - 3$$
$$3x^2 - 5$$

$-10x^2 + 15 \qquad$ Multiply -5 times $2x^2 - 3$

$\underline{6x^4 - 9x^2 \qquad\qquad}$ Multiply $3x^2$ times $2x^2 - 3$

$6x^4 - 19x^2 + 15 \qquad$ Add in columns

17.
$$2a^3 - 3a^2 + a$$
$$3a + 5$$

$10a^3 - 15a^2 + 5a \qquad$ Multiply $(2a^3 - 3a^2 + a)$ by 5

$\underline{6a^4 - 9a^3 + 3a^2 \qquad\qquad}$ Multiply $(2a^3 - 3a^2 + a)$ by $3a$

$6a^4 + \ a^3 - 12a^2 + 5a \qquad$ Add similar terms

21.
$$4x^2 - 2xy + y^2$$
$$2x + y$$

$4x^2y - 2xy^2 + y \qquad$ Multiply $(4x^2 - 2xy + y^2)$ by y

$\underline{8x^3 - 4x^2y + 2xy^2 \qquad\qquad}$ Multiply $(4x^2 - 2xy + y^2)$ by $2x$

$8x^3 \qquad\qquad\quad + y^3 \qquad$ Add similar terms

25.
$$6x^2 + 3xy + 4y^2$$
$$3x - 4y$$

$-24x^2y - 12xy^2 - 16y \qquad$ Multiply $(6x^2 + 3xy + 4y^2)$ by $-4y$

$\underline{18x^3 + 9x^2y + 12xy^2 \qquad\qquad}$ Multiply $(6x^2 + 3xy + 4y^2)$ by $3x$

$18x^3 - 15x^2y \qquad\qquad - 16y^3$

29.
$$x - 2 \qquad\qquad\qquad 2x^2 - \ x - 6$$
$$2x + 3 \qquad\qquad\qquad 3x - 4$$

$3x - 6 \qquad\qquad\qquad -8x^2 + 4x + 24$

$\underline{2x^2 - 4x \qquad}\qquad\qquad \underline{6x^3 - 3x^2 - 18x \qquad\qquad}$

$2x^2 - \ x - 6 \qquad\qquad 6x^3 - 11x^2 - 14x + 24$

$(x - 2)(2x + 3)(3x - 4) = 6x^3 - 11x^2 - 14x + 24$

33. $(x^2 - 2)(x^2 + 3) = x^4 + 3x^2 - 2x^2 - 6$

$\qquad\qquad\qquad$ F $\quad$ O $\quad$ I $\quad$ L

$\qquad\qquad = x^4 + x^2 - 6$

37. $(5 - 3t)(4 + 2t) = 20 + 10t - 12t - 6t^2$

$\qquad\qquad\qquad$ F $\quad$ O $\quad$ I $\quad$ L

$\qquad\qquad = 20 - 2t - 6t^2$

41. $(4a + 1)(5a + 1) = 20a^2 + 4a + 5a + 1$

 F O I L

$ = 20a^2 + 9a + 1$

45. $(3t + \frac{1}{3})(6t - \frac{2}{3}) = 18t^2 - 2t + 2t - \frac{2}{9}$

 F O I L

$ = 18t^2 - \frac{2}{9}$

49. $(5x + 2y)^2 = (5x)^2 + 2(5x)(2y) + (2y)^2$

$ = 25x^2 + 20xy + 4y^2$

53. $(2a + 3b)(2a - 3b) = (2a)^2 - (3b)^2 = 4a^2 - 9b^2$

57. $(\frac{1}{3}x - \frac{2}{5})(\frac{1}{3}x + \frac{2}{5}) = (\frac{1}{3}x)^2 - (\frac{2}{5})^2 = \frac{1}{9}x^2 - \frac{4}{25}$

65. $3(x - 1)(x - 2)(x - 3)$

$ = (3x - 3)(x - 2)(x - 3) $ Distributive property

$ = (3x^2 - 9x + 6)(x - 3) $ Multiply first two binomials

$ = 3x^3 - 18x^2 + 33x - 18 $ Multiply $3x^2 - 9x + 6$ and $x - 3$

69. $(x - 2)(3y^2 + 4) = x(3y^2) + x(4) + (-2)(3y^2) + (-2)(4)$

 F O I L

$ = 3xy^2 + 4x - 6y^2 - 8$

73. $(2x + 3)^2 - (2x - 3)^2$

$ = (2x + 3)(2x + 3) - [(2x - 3)(2x - 3)]$

$ = 4x^2 + 2(2x)(3) + 9 - [4x^2 + 2(2x)(-3) + 9]$

$ = 4x^2 + 12x + 9 - [4x^2 - 12x + 9]$

$ = 4x^2 + 12x + 9 - 4x^2 + 12x - 9$

$ = 24x$

77. $(x + y - 4)(x + y + 5) = [(x + y) - 4][(x + y) + 5]$

$ = (x + y)^2 + (x + y)(5) + (-4)(x + y) - 20$

 F O I L

$ = (x + y)^2 + (x + y) - 20$

$ = x^2 + 2xy + y^2 + x + y - 20$

Section 3.4 continued

81. If we expand $(p + q)^2 = p^2 + 2pq + q^2$, we find the middle term is $2pq$.

$$\text{When} \quad p = \frac{1}{4}, \quad q = \frac{3}{4}$$

the expression $2pq$

$$\text{becomes} \quad 2\left(\frac{1}{4}\right)\left(\frac{3}{4}\right)$$

$$= \frac{3}{8} \text{ of the next generation will have pink flowers.}$$

85. If $R = xp$, where R = revenue, x = tapes and p = price, given $x = 230 - 20p$ and $p = \$6.50$, we have

$$R = xp$$
$$= (230 - 20p)p$$
$$= 230p - 20p^2$$
$$= 230(6.50) - 20(6.50)^2$$
$$= \$650$$

89. If we let x = the first integer, then $x + 1$ is the second integer and $x + 3$ is the third integer. See example 10 on page 115 in your textbook for the drawing of a rectangular box.

Total Surface Area	Area of the two sides	Area of the top and bottom	Area of the front and back
S	$= 2(x + 1)(x + 2)$	$+ \ 2x(x + 2) +$	$2x(x + 1)$
	$= 2x^2 + 6x + 4 + 2x^2 + 4x + 2x^2 + 2x$		
	$= 6x^2 + 12x + 4$		

93. $|3x - 5| = 7$

$$3x - 5 = 7 \qquad \text{or} \qquad 3x - 5 = -7$$
$$3x = 12 \qquad\qquad\qquad 3x = -2$$
$$x = 4 \qquad\qquad\qquad x = -\frac{2}{3}$$

Our solution set is $\{-\frac{2}{3}, 4\}$.

97. $5 + |6t + 2| = 3$

$$|6t + 2| = -2$$

The solution set is $\varnothing$ because the left side cannot be negative and the right side is negative.

101. $(x^n - 2)(x^n - 3) = x^n \cdot x^n + x^n(-3) + (-2)x^n + (-2)(-3)$

        F   O   I   L

      $= x^{2n} - 3x^n - 2x^n + 6$

      $= x^{2n} - 5x^n + 6$

105. $(2x^n + 3)(5x^n - 1) = 2x^n(5x^n) + 2x^n(-1) + 3(5x^n) + 3(-1)$

         F   O   I   L

      $= 10x^{2n} - 2x^n + 15x^n - 3$

      $= 10x^{2n} + 13x^n - 3$

109. $(x^n + 1)^2 = (x^n + 1)(x^n + 1)$

      $= x^n(x^n) + 2(x^n)(1) + 1$

      $= x^{n^2} + 2x^n + 1$

$$x^{n^2} + 2x^n + 1$$

$$\underline{\qquad x^n + 1}$$

$$x^{n^2} + 2x^n + 1$$

$$\underline{x^{n^3} + 2x^{n^2} + x^n \qquad}$$

$$x^{n^3} + 3x^{n^2} + 3x^n + 1$$

$$(x^n + 1)^3 = (x^n + 1)(x^n + 1)^2 = x^{n^3} + 3x^{n^2} + 3x^n + 1$$

Section 3.5

1. $10x^3 - 15x^2 = 5x^2(2x - 3)$

5. $9a^2b - 6ab^2 = 3ab(3a - 2b)$

9. $3a^2 - 21a + 30 = 3(a^2 - 7a + 10)$

13. $10x^4y^2 + 20x^3y^3 - 30x^2y^4 = 10x^2y^2(x^2 + 2xy - 3y^2)$

17. $4x^3y^2z - 8x^2y^2z^2 + 6xy^2z^3 = 2xy^2z(2x^2 - 4xz + 3z^2)$

21. $5x(a - 2b) - 3y(a - 2b) = (a - 2b)(5x - 3y)$

25. $2x^2(x + 5) + 7x(x + 5) + 6(x + 5) = (x + 5)(2x^2 + 7x + 6)$

29. $x^2y + x + 3xy + 3 = x(xy + 1) + 3(xy + 1)$
$$= (xy + 1)(x + 3)$$

33. $x^2 - ax - bx + ab = x(x - a) - b(x - a)$
$$= (x - a)(x - b)$$

37. $a^4b^2 + a^4 - 5b^2 - 5 = a(b^2 + 1) - 5(b^2 + 1)$
$$= (b^2 + 1)(a^4 - 5)$$

41. $x^3 + 2x^2 - 25x - 50 = x^2(x + 2) - 25(x + 2)$
$$= (x + 2)(x^2 - 25)$$

45. $4x^3 + 12x^2 = 9x - 27 = 4x^2(x + 3) - 9(x + 3)$
$$= (x + 3)(4x^2 - 9)$$

49. $P + Pr + (P + Pr)r = P(1 + r) + P(1 + r)r$
$$= (1 + r)(P + Pr)$$
$$= (1 + r)P(1 + r)$$
$$= P(1 + r)^2$$

53. If $R = xp$, where R = revenue, x = programs and p = price, given
$R = 35x - .1x^2$ and $x = 65$, we have
$$R = xp$$
$$35x - .1x^2 = xp$$
$$35 - .1x = p$$
$$\$28.50 = p$$

They should charge $28.50 for each program.

57. $(2y + 5)(3y - 7) = 6x^2 - 14y + 15y - 35$

 F O I L

$$= 6y^2 + y - 35$$

61. $-5x < 30$

$x > -6$ Multiplying both sides of an inequality by a negative number reverses the sense of the inequality.

65. $5t - 4 > 3t - 8$
$$5t > 3t - 4$$
$$2t > -4$$
$$t > -2$$

Section 3.5 continued

69. $x^{n+2} + x^{n+1} + x^n$

$$= x^n \cdot x^2 + x^n \cdot x + x^n$$

$$= x^n(x^2 + x + 1)$$

Section 3.6

1. Factor $x^2 + 7x + 12$

    Solution: The leading coefficient is 1. We need two numbers whose sum is 7 and whose product is 12. The numbers are 3 and 4. (12 and 1, and 2 and 6 do not work because their sum is not 7.)

    $$x^2 + 7x + 12 = (x + 3)(x + 4)$$

5. Factor $y^2 + y - 6$

    Solution: The leading coefficient is 1. We need two numbers whose sum is 1 and whose product is 6. The numbers are 3 and -2.

    $$y^2 + y - 6 = (y + 3)(y - 2)$$

9. Factor $12 + 8x + x^2$

    Solution: The coefficient of the third term is 1. We need two numbers whose sum is 8 and whose product is 12. The numbers are 6 and 2.

    $$12 + 8x + x = (2 + x)(6 + x)$$

13. $4x^3 - 16x^2 - 20x = 4x(x^2 - 4x - 5)$ Sum is -4, product is -5

    $$= 4x(x - 5)(x + 1)$$

17. $a^2 + 3ab - 18b^2 = (a + 6b)(a - 3b)$ Sum is 3, product is -18

21. $x^2 - 12xb + 36b^2 = (x - 6b)(x - 6b)$ Sum is -12, product is 36

    $$= (x - 6b)^2$$

25. $2a^5 + 4a^4b + 4a^3b^2 = 2a^3(a^2 + 2ab + 2b^2)$

    Since there is no pair of integers whose product is 2 and whose sum is 8, $2a^3(a^2 + 2ab + 2b^2)$ is the answer.

29. $2x^2 + 7x - 15 = (2x - 3)(x + 5)$ Middle term is 7x

33. $2x^2 - 13x + 15 = (2x - 3)(x - 5)$ Middle term is -13x

37. $2x^2 + 7x + 15$ Prime

41. $60y^2 - 15y - 45 = 15(4y^2 - y - 3)$

    $$= 15(4y + 3)(y - 1)$$ Middle term is -y

72

45. $40r^3 - 120r^2 + 90r = 10r(4r^2 - 12r + 9)$

$\qquad\qquad\qquad\quad = 10r(2r - 3)(2r - 3)$

$\qquad\qquad\qquad\quad = 10r(2r - 3)^2$ $\qquad$ Middle term is $-12r$

49. $10x^2 - 3xa - 18a^2 = (2x - 3a)(5x + 6a)$ $\quad$ Middle term is $-3xa$

53. $600 + 800t - 800t^2 = 200(3 + 4t - 4t^2)$

$\qquad\qquad\qquad\quad\;\; = 200(1 + 2t)(3 - 2t)$ $\quad$ Middle term is $8t$

57. $24a^2 - 2a^3 - 12a^4 = 2a^2(12 - a - 6a^2)$

$\qquad\qquad\qquad\qquad = 2a^2(3 + 2a)(4 - 3a)$ $\quad$ Middle term is $-a$

61. $300x^4 + 1000x^2 + 300 = (300x^2 + 100)(x^2 + 300)$ $\quad$ Middle term is $10x^2$

65. $9 + 3r^2 - 12r^4 = 3(1 + r^2 - 4r^4)$

$\qquad\qquad\qquad = 3(3 + 4r^2)(1 - r^2)$ $\quad$ Middle term is r^2

69. $x^2(2x + 3) + 7x(2x + 3) + 10(2x + 3)$

$\qquad = (2x + 3)(x^2 + 7x + 10)$

$\qquad = (2x + 3)(x + 5)(x + 2)$ $\qquad$ Sum is $7x$, product is 10

73. $(a^2 + 260a + 2{,}500) = (a + 10)(a + 250)$

$\qquad$ The other factor is $(a + 250)$

77. $\qquad\qquad h = 96 + 80t - 16t^2$

$\qquad\qquad\quad h = 16(6 + 5t - t^2)$

$\qquad\qquad\quad h = 16(6 - t)(1 + t)$

When $t = 6$

$\qquad\quad h = 16(6 - t)(1 + t)$

becomes $h = 16(6 - 6)(1 + 6)$

$\qquad\quad h = 0$

When $t = 3$

$\qquad\quad h = 16(6 - t)(1 + t)$

becomes $h = 16(6 - 3)(1 + 3)$

$\qquad\quad h = 16(3)(4)$

$\qquad\quad h = 192$

When t is 6 seconds, h is 0 feet and when t is 3 seconds, h is 192 feet.

81. $(2x - 3)^2 = (2x - 3)(2x - 3)$

$\qquad = (2x)^2 + 2(2x)(-3) + (-3)^2$

$\qquad = 4x^2 - 12x + 9$

85. $\left|\dfrac{x}{5} + 1\right| \geq \dfrac{4}{5}$

The quantity $\dfrac{x}{5} + 1$ is greater than or equal to $\dfrac{4}{9}$ units from 0. It

must be either above $+\dfrac{4}{9}$ or below $-\dfrac{4}{9}$:

$$\dfrac{x}{5} + 1 \leq -\dfrac{4}{5} \qquad \text{or} \qquad \dfrac{x}{5} + 1 \geq \dfrac{4}{5}$$

$$\dfrac{x}{5} \leq -1\dfrac{4}{5} \qquad\qquad\qquad \dfrac{x}{5} \geq -\dfrac{1}{5}$$

$$\dfrac{x}{5} \leq -\dfrac{9}{5} \qquad\qquad\qquad x \geq -1$$

$$x \leq -9$$

See the graph on page A23 in your textbook.

89. $-8 + |3y + 5| < 5$

$\qquad |3y + 5| < 13$

The absolute value of $3y + 5$ is the distance that $3y + 5$ is from 0
on the number line.

$$-13 < 3y + 5 < 13$$

$$-18 < \quad 3y \quad < 8$$

$$-6 < \quad y \quad < \dfrac{8}{3}$$

See the graph on page A23 in the textbook.

93. $3x^2 + 295x - 500 = (3x - 5)(x + 100)$

97. $2x^2 + 1.5x + .25 = (2x + .5)(x + .5)$

101. $20x^{2n} - 119x^n - 6 = (20x^n + 1)(x^n - 6)$

Section 3.7

1. $x^2 - 6x + 9 = (x - 3)(x - 3)$

$\qquad\qquad = (x - 3)^2$

5. $25 - 10t + t^2 = (5 - t)(5 - t)$
$$= (5 - t)^2$$

9. $4y^4 - 12y^2 + 9 = (2y^2 - 3)(2y^2 - 3)$
$$= (2y^2 - 3)^2$$

13. $\frac{1}{25} + \frac{1}{10}t^2 + \frac{1}{16}t^4 = (\frac{1}{5} + \frac{1}{4}t)(\frac{1}{5} + \frac{1}{4}t)$
$$= (\frac{1}{5} + \frac{1}{4}t)^2$$

17. $75a^3 + 30a^2 + 3a = 3a(25a^2 + 10a + 1)$
$$= 3a(5a + 1)(5a + 1)$$
$$= 3a(5a + 1)^2$$

21. $x^2 - 9 = (x + 3)(x - 3)$

25. $4a^2 - \frac{1}{4} = (2a + \frac{1}{2})(2a - \frac{1}{2})$

29. $9x^2 - 16y^2 = (3x + 4y)(3x - 4y)$

33. $x^4 - 81 = (x^2 + 9)(x^2 - 9)$
$$= (x^2 + 9)(x + 3)(x - 3)$$

37. $16a^4 - 81 = (4a^2 + 9)(4a^2 - 9)$
$$= (4a^2 + 9)(2a + 3)(2a - 3)$$

41. $x^6 - y^6 = (x^3 - y^3)(x^3 + y^3)$
$$= (x - y)(x^2 + xy + y^2)(x + y)(x^2 - xy + y^2)$$
$$= (x - y)(x + y)(x^2 + xy + y^2)(x^2 - xy + y^2)$$

45. $(x - 2)^2 - 9 = (x - 2)^2 - 3^2$ Write 9 as 3^2
$$= [(x - 2) + 3][(x - 2) - 3]$$
$$= (x + 1)(x - 5)$$

49. $x^2 - 10x + 25 - y^2$
$= (x^2 - 10x + 25) - y^2$ Group first 3 terms together
$= (x - 5)^2 - y^2$ This has the form $a^2 - b^2$
$= [(x - 5) + y][(x - 5) - y]$
$= (x - 5 + y)(x - 5 - y)$

53. $x^2 + 2xy + y^2 - a^2$

$\quad = (x^2 + 2xy + y^2 - a^2)$ Group first 3 terms together

$\quad = (x + y)^2 - a^2$ This has the form $a^2 - b^2$

$\quad = [(x + y) + a][(x + y) - a]$

$\quad = (x + y + a)(x + y - a)$

57. $x^3 + 2x^2 - 25x - 50 = x^2(x + 2) - 25(x + 2)$

$\qquad\qquad\qquad\qquad\quad = (x + 2)(x^2 - 25)$

$\qquad\qquad\qquad\qquad\quad = (x + 2)(x + 5)(x - 5)$

61. $4x^3 + 12x^2 - 9x - 27 = 4x^2(x + 3) - 9(x + 3)$

$\qquad\qquad\qquad\qquad\quad = (x + 3)(4x^2 - 9)$

$\qquad\qquad\qquad\qquad\quad = (x + 3)(2x + 3)(2x - 3)$

65. $a^3 + 8 = a^3 + 2^3$ Perfect cubes, $2^3 = 8$

$\quad = (a + 2)(a^2 - 2a + 4)$

69. $y^3 - 1 = y^3 - 1^3$ Perfect cubes, $1^3 = 1$

$\quad = (y - 1)(y^2 + y + 1)$

73. $64 + 27a^3$

$\quad = 4^3 + (3a)^3$ Perfect cubes, $4^3 = 64$ and $(3a)^3 = 27a^3$

$\quad = (4 + 3a)(16 + 12a + 9a^2)$

77. $t^3 + \dfrac{1}{27} = t^3 + \left(\dfrac{1}{3}\right)^3$ Perfect cube, $\left(\dfrac{1}{3}\right)^3 = \dfrac{1}{27}$

$\quad = \left(t + \dfrac{1}{3}\right)\left(t^2 - \dfrac{1}{3}t + \dfrac{1}{9}\right)$

81. $64a^3 + 125b^3$

$\quad = (4a)^3 + (5b)^3$ Perfect cubes, $(4a)^3 = 64a^3$

$\qquad\qquad\qquad\qquad\qquad\qquad\qquad\qquad (5b)^3 = 125b^3$

$\quad = (4a + 5b)(16a^2 - 20ab + 25b^2)$

85.   When $r = 12\%$

the formula $A = 100\left(1 + r + \dfrac{r^2}{4}\right)$

becomes $A = 100\left(1 + .12 + \dfrac{(.12)^2}{4}\right)$

$\qquad\qquad A = 100\left(1 + .12 + \dfrac{.0144}{4}\right)$

$\qquad\qquad A = 100(1 + .12 + .0036)$

$\qquad\qquad A = 100(1.1236)$

$\qquad\qquad A = \$112.36$

89. Solve $3x + 2y = 12$ for y

$$3x + 2y = 12$$
$$2y = -3x + 12$$
$$y = -\frac{3}{2}x + 6$$

93. $x^{2n} - y^{2n} = (x^n - y^n)(x^n + y^n)$

97. $x^{3n} - 8 = x^{3n} - (2)^3$ $\qquad\qquad 2^3 = 8$

$\qquad\qquad = (x^n - 2)(x^{2n} + 2x^n + 4)$

Section 3.8

1. $x^2 - 81 = (x + 9)(x - 9)$

5. $x^2(x + 2) + 6x(x + 2) + 9(x + 2) = (x + 2)(x^2 + 6x + 9)$
$$= (x + 2)(x + 3)(x + 3)$$
$$= (x + 2)(x + 3)^2$$

9. $2a^3b + 6a^2b + 2ab = 2ab(a^2 + 3a + 1)$

13. $12a^2 - 75 = 3(4a^2 - 25)$
$$= 3(2a + 5)(2a - 5)$$

17. $25 - 10t + t^2 = (5 - t)(5 - t)$
$$= (5 - t)^2$$

21. $2y^3 + 20y^2 + 50y = 2y(y^2 + 10y + 25)$
$$= 2y(y + 5)(y + 5)$$
$$= 2y(y + 5)^2$$

25. $t^2 + 6t + 9 - x^2 = (t^2 + 6t + 9) - x^2$
$$= (t + 3)^2 - x^2$$
$$= [(t + 3) + x][(t + 3) - x]$$
$$= (t + 3 + x)(t + 3 - x)$$

29. $5a^2 + 10ab + 5b^2 = 5(a^2 + 2ab + b^2)$
$$= 5(a + b)(a + b)$$
$$= 5(a + b)^2$$

77

33. $3x^2 + 15xy + 18y^2 = 3(x^2 + 5xy + 6y^2)$
$$= 3(x + 2y)(x + 3y)$$

37. $x^2(x - 3) - 14x(x - 3) + 49(x - 3) = (x - 3)(x^2 - 14x + 49)$
$$= (x - 3)(x - 7)(x - 7)$$
$$= (x - 3)(x - 7)^2$$

41. $8 - 14x - 15x^2 = (2 - 5x)(4 + 3x)$

45. $r^2 - \dfrac{1}{25} = (r + \dfrac{1}{5})(r - \dfrac{1}{5})$

49. $100x^2 - 100x - 600 = 100(x^2 - x - 6)$
$$= 100(x - 3)(x + 2)$$

53. $3x^4 - 14x^2 - 5 = (3x^2 + 1)(x^2 - 5)$

57. $64 - r^3 = 4^3 - r^3$
$$= (4 - r)(16 + 4r + r^2)$$

61. $400t^2 - 900 = 100(4t^2 - 9)$
$$= 100(2t + 3)(2t - 3)$$

65. $y^6 - 1 = (y^3 + 1)(y^3 - 1)$
$$= (y + 1)(y^2 - y + 1)(y - 1)(y^2 + y + 1)$$
$$= (y + 1)(y - 1)(y^2 - y + 1)(y^2 + y + 1)$$

69. $12x^4y^2 + 36x^3y^3 + 27x^2y^4$
$$= 3x^2y^2(4x^2 + 12xy + 9y^2)$$
$$= 3x^2y^2(2x + 3y)(2x + 3y)$$
$$= 3x^2y^2(2x + 3y)^2$$

73. Let x = the smaller consecutive even integer and x + 2 = the larger consecutive even integer.

$$(x + 2) + 2x = 26$$
$$3x + 2 = 26$$
$$3x = 24$$
$$x = 8$$

The smaller consecutive even integer is x = 8, and the larger consecutive even integer is x + 2 = 10.

77.

	nickels	quarters	Total
Number of	x	21 - x	21
Value of	5x	25(21 - x)	265

$$5x + 25(21 - x) = 265$$
$$5x + 525 - 25x = 265$$
$$525 - 20x = 265$$
$$-20x = -260$$
$$x = 13$$

David has x = 13 nickels and 21 - x = 8 quarters.

Section 3.9

1. $x^2 - 5x - 6 = 0$

$(x - 6)(x + 1) = 0$ Factor

$x - 6 = 0$ or $x + 1 = 0$ Zero-factor property

$x = 6$ $x = -1$

The two solutions are 6 and -1.

5. $3y^2 - 11y - 4 = 0$

$(y + 4)(3y - 1) = 0$ Factor

$y + 4 = 0$ or $3y - 1 = 0$ Zero-factor property

$y = -4$ $y = \dfrac{1}{3}$

The two solutions are -4 and $\dfrac{1}{3}$.

9. $\dfrac{1}{10}t^2 - \dfrac{5}{2} = 0$

$t^2 - 25 = 0$ Multiply both sides by 10

$(t + 5)(t - 5) = 0$ Factor

$t + 5 = 0$ or $t - 5 = 0$ Zero-factor property

$t = -5$ $t = 5$

The two solutions are -5 and 5.

13. $\dfrac{1}{5}y^2 - 2 = -\dfrac{3}{10}y$

$2y^2 + 3y - 20 = 0$ Multiply both sides by 10, Standard form

$(y + 4)(2y - 5) = 0$ Factor

$y + 4 = 0$ or $2y - 5 = 0$ Zero-factor property

$y = -4$ $y = \dfrac{5}{2}$

The two solutions are -4 and $\dfrac{5}{2}$.

17.
$$.02r + .01 = .15r^2$$
$$-.15r^2 + .02r + .01 = 0$$
$$15r^2 - 2r - 1 = 0 \qquad \text{Multiply both sides by -100, then the equation is in standard form.}$$
$$(5r + 1)(3r - 1) = 0 \qquad \text{Factor}$$
$$5r + 1 = 0 \qquad \text{or} \qquad 3r - 1 = 0 \qquad \text{Zero-factor property}$$
$$r = -\frac{1}{5} \qquad\qquad r = \frac{1}{3}$$

The two solutions are $-\frac{1}{5}$ and $\frac{1}{3}$.

21.
$$-100x = 10x^2$$
$$-10x^2 - 100x = 0$$
$$10x^2 + 100x = 0 \qquad \text{Multiply both sides by -1, then the equation is in the standard form.}$$
$$10x(x + 10) = 0$$
$$10x = 0 \qquad \text{or} \quad x + 10 = 0 \qquad \text{Zero-factor property}$$
$$x = 0 \qquad\qquad x = -10$$

The two solutions are 0 and -10.

25.
$$(y - 4)(y + 1) = -6$$
$$y^2 - 3y - 4 = -6 \qquad \text{Multiply the left side}$$
$$y^2 - 3y + 2 = 0 \qquad \text{Standard form}$$
$$(y - 2)(y - 1) = 0 \qquad \text{Factor}$$
$$y - 2 = 0 \qquad \text{or} \quad y - 1 = 0 \qquad \text{Zero-factor property}$$
$$y = 2 \qquad\qquad y = 1$$

The two solutions are 2 and 1.

29.
$$(2r + 3)(2r - 1) = -(3r - 1)$$
$$4r^2 + 4r - 3 = -3r + 1 \qquad \text{Multiply both sides}$$
$$4r^2 + 7r - 4 = 0 \qquad \text{Standard form}$$
$$(r + 2)(4r - 1) = 0$$
$$r + 2 = 0 \qquad \text{or} \qquad 4r - 1 = 0$$
$$r = -2 \qquad\qquad r = \frac{1}{4}$$

The two solutions are -2 and $\frac{1}{4}$.

33. $x^3 + 2x^2 - 25x - 50 = 0$

$x^2(x + 2) - 25(x + 2) = 0$ Factoring by grouping

$(x + 2)(x^2 - 25) = 0$ Factoring by grouping

$(x + 2)(x + 5)(x - 5) = 0$

$x + 2 = 0$	or	$x + 5 = 0$	or	$x - 5 = 0$
$x = -2$		$x = -5$		$x = 5$

The three solutions are -2, -5 and 5.

37. $4x^3 + 12x^2 - 9x - 27 = 0$

$4x^2(x + 3) - 9(x + 3) = 0$ Factoring by grouping

$(x + 3)(4x^2 - 9) = 0$ Factoring by grouping

$(x + 3)(2x + 3)(2x - 3) = 0$

$x + 3 = 0$	or	$2x + 3 = 0$	or	$2x - 3 = 0$
$x = -3$		$2x = -3$		$2x = 3$
		$x = -\dfrac{3}{2}$		$x = \dfrac{3}{2}$

The three solutions are -3, $-\dfrac{3}{2}$ and $\dfrac{3}{2}$.

41.   Let x = the first integer; then x + 1 = the next consecutive integer. The square of the sum of x and x + 1 is 81.

$$[x + (x + 1)]^2 = 81$$

$$[2x + 1]^2 = 81$$

$$4x^2 + 4x + 1 = 81$$

$$4x^2 + 4x - 80 = 0$$

$$x^2 + x - 20 = 0$$ Divide both sides by 4

$$(x + 5)(x - 4) = 0$$

$x + 5 = 0$	or	$x - 4 = 0$
$x = -5$		$x = 4$
$x + 1 = -4$		$x + 1 = 5$

The two solutions are -5,-4 or 4,5.

45.   Let x = first integer (shortest side)
Then $x + 2$ = next consecutive even integer
      $x + 4$ = last consecutive even integer (longest side)

By the Pythagorean Theorem,
$$c^2 = a^2 + b^2$$
We have $(x + 4)^2 = (x + 2)^2 + x^2$
$$x^2 + 8x + 16 = x^2 + 4x + 4 + x^2$$
$$x^2 - 4x + 12 = 0$$
$$(x + 2)(x - 6) = 0$$

$x + 2 = 0$ or $x - 6 = 0$
$\quad x = -2$ $\qquad\qquad\qquad x = 6$

The shortest side is 6. The other two sides are 8 and 10.

49.   Let x = height and let $4x + 2$ = base

$A = \frac{1}{2} bh$

$36 = \frac{1}{2}(4x + 2)x$

$36 = (2x + 1)x$

$36 = 2x^2 + x$

$0 = 2x^2 + x - 36$

$0 = (2x + 9)(x - 4)$

$2x + 9 = 0$ or $x - 4 = 0$
$\quad x = -9$ $\qquad\qquad\qquad x = 4$
No solution $\qquad\qquad 4x + 2 = 18$

The height is 4 in., base is 18 in.

53.   Let v = 48 ft/sec and height = 32 ft

$h = vt - 16t^2$

$32 = 48t - 16t^2$

$16t^2 - 48t + 32 = 0$

$16(t^2 - 3t + 2) = 0$

$16(t - 1)(t - 2) = 0$

$t - 1 = 0$ or $t - 2 = 0$
$\quad t = 1$ $\qquad\qquad\qquad t = 2$

It will reach the height of 32 ft. at 1 second and 2 seconds.

57. Let v = 80ft/sec and height = 192 ft.

$$h = 96 + 80t - 16t^2$$
$$192 = 96 + 80t - 16t^2$$
$$16t^2 - 80t + 96 = 0$$
$$16(t^2 - 5t + 6) = 0$$
$$16(t - 2)(t - 3) = 0$$

$$t - 2 = 0 \quad\text{or}\quad t - 3 = 0$$
$$t = 2 \qquad\qquad t = 3$$

The bullet will be 192 feet high at 2 seconds and 3 seconds.

61. Let R = $7,000

Substitute R = xp, letting x = 1700 - 100p

$$R = (1700 - 100p)p$$
$$7000 = 1700p - 100p^2$$
$$100p^2 - 1700p + 7000 = 0$$
$$100(p^2 - 17p + 70) = 0$$
$$100(p - 10)(p - 7) = 0$$

$$p - 10 = 0 \quad\text{or}\quad p - 7 = 0$$
$$p = 10 \qquad\qquad p = 7$$

The calculators will cost $7 and $10.

65. $|3a - 5| - 4 = -2$

$$|3a - 5| = 2$$

$$3a - 5 = 2 \quad\text{or}\quad 3a - 5 = -2$$
$$3a = 7 \qquad\qquad 3a = 3$$

$$a = \frac{7}{3} \qquad\qquad a = 1$$

The solution set is $\{1, \frac{7}{3}\}$.

69. $|x| < .01$ without the absolute value symbol becomes

$$-.01 < x < .01$$

73. $|1 - 3t| \geq 5$ without the absolute value symbol becomes:

$$1 - 3t \leq -5 \quad\text{or}\quad 1 - 3t \geq 5$$
$$-3t \leq -6 \qquad\qquad -3t \geq 4$$

$$t \geq 2 \qquad\qquad t \leq \frac{4}{3}$$

1. $x^3 \cdot x^7 = x^{3+7} = x^{10}$

5. $(2x^3y)^2(-2x^4y^2)^3$

$\qquad = (2)^2(x^3)^2(y)^2(-2)^3(x^4)^3(y^2)^3$

$\qquad = (4)(-8)(x^6)(x^{12})(y^2)(y^6)$

$\qquad = -32x^{18}y^8$

9. $\left(\dfrac{2}{3}\right)^{-2} = \dfrac{(2)^{-2}}{(3)^{-2}} = \dfrac{(3)^2}{(2)^2} = \dfrac{9}{4}$

13. $34{,}500{,}000 = 3.45 \times 10^7$

17. $\dfrac{a^{-4}}{a^5} = a^{-4-5} = a^{-9} = \dfrac{1}{a^9}$

21. $\dfrac{x^n x^{3n}}{x^{4n-2}} = \dfrac{x^{4n}}{x^{4n}x^{-2}} = x^{4n-(4n-2)} = x^{4n-4n+2} = x^2$

25. $\dfrac{(600{,}000)(0.000008)}{(4{,}000)(3{,}000{,}000)}$

$\qquad = \dfrac{(6.0 \times 10^5)(8.0 \times 10^{-6})}{(4.0 \times 10^3)(3.0 \times 10^6)}$

$\qquad = \dfrac{(6 \times 8.0)(10^5 \times 10^{-6})}{(4 \times 3)(10^3 \times 10^6)}$

$\qquad = \dfrac{8.0}{2} \times \dfrac{10^{-1}}{10^9}$

$\qquad = 4.0 \times 10^{-10}$

29. $(x^3 - x) - (x^2 + x) + (x^3 - 3) - (x^2 + 1)$

$\qquad = x^3 - x - x^2 - x + x^3 - 3 - x^2 - 1$

$\qquad = x^3 + x^3 - x^2 - x^2 - x - x - 3 - 1$

$\qquad = 2x^3 - 2x^2 - 2x - 4$

33. $-3[2x - 4(3x + 1)]$

$\qquad = -3[2x - 12x - 4]$

$\qquad = -6x + 36x + 12$

$\qquad = 30x + 12$

37. $3x(4x^2 - 2x + 1)$

$\quad\quad = 3x(4x^2) - (3x)(2x) + (3x)(1)$

$\quad\quad = 12x^3 - 6x^2 + 3x$

41. $(6 - y)(3 - y) = (6)(3) - (6)(y) - (y)(3) + y^2$

$\quad\quad\quad\quad\quad\quad\quad$ F $\quad\quad$ O $\quad\quad$ I $\quad\quad$ L

$\quad\quad\quad\quad\quad = 18 - 6y - 3y + y^2$

$\quad\quad\quad\quad\quad = 18 - 9y + y^2$

45. $2t(t + 1)(t - 3) = (2t^2 + 2t)(t - 3)$

$\quad\quad\quad\quad\quad\quad\quad\quad = 2t^2(t) + (2t^2)(-3) + (2t)(t) + (2t)(-3)$

$\quad\quad\quad\quad\quad\quad\quad\quad = 2t^3 - 6t^2 + 2t^2 - 6t$

$\quad\quad\quad\quad\quad\quad\quad\quad = 2t^3 - 4t^2 - 6t$

49.

$$\begin{array}{r} 4x^2 + 6x + 9 \\ 2x - 3 \\ \hline -12x^2 - 18x - 27 \\ 8x^3 + 12x^2 + 18x \\ \hline 8x^3 \quad\quad\quad\quad - 27 \end{array}$$

53. $(3x + 5)^2 = (3x + 5)(3x + 5)$

$\quad\quad\quad\quad\quad = (3x)^2 + 2(3x)(5) + (5)^2$

$\quad\quad\quad\quad\quad = 9x^2 + 30x + 25$

57. $\left(x - \dfrac{1}{3}\right)\left(x + \dfrac{1}{3}\right) = (x)^2 - \left(\dfrac{1}{3}\right)^2 = x^2 - \dfrac{1}{9}$

61. $(x - 1)^2 = (x - 1)(x - 1)$

$\quad\quad\quad\quad = x^2 + 2(x)(-1) + (-1)^2$

$\quad\quad\quad\quad = x^2 - 2x + 1$

$$\begin{array}{r} x^2 - 2x + 1 \\ x - 1 \\ \hline -x^2 + 2x - 1 \\ x^3 - 2x^2 + x \\ \hline x^3 - 3x^2 + 3x - 1 \end{array}$$

$(x - 1)^3 = (x - 1)^2(x - 1) = x^3 - 3x^2 + 3x - 1$

65. $6x^4y - 9xy^4 + 18x^3y^3 = 3xy(2x^3 - 3y^3 + 6x^2y^2)$

69. $8x^2 + 10 - 4x^2y - 5y$

$\quad\quad = (8x^2 + 10) + (-4x^2y - 5y)$

$\quad\quad = 2(4x^2 + 5) - y(4x^2 + 5)$

$\quad\quad = (4x^2 + 5)(2 - y)$

73. $2x^3 + 4x^2 - 30x = 2x(x^2 + 2x - 15)$
$$= 2x(x + 5)(x - 3)$$

77. $6x^4 - 11x^3 - 10x^2 = x^2(6x^2 - 11x - 10)$
$$= x^2(3x + 2)(2x - 5)$$

81. $x^4 - 16 = (x^2 + 4)(x^2 - 4)$
$$= (x^2 + 4)(x + 2)(x - 2)$$

85. $a^3 - 8 = a^3 - 2^3$
$$= (a - 2)(a^2 + 2a + 4)$$

89. $3a^3b - 27ab^3 = 3ab(a^2 - 9b^2)$
$$= 3ab(a + 3b)(a - 3b)$$

93. $36 - 25a^2 = (6 + 5a)(6 - 5a)$

97. $x^2 + 5x + 6 = 0$
$(x + 2)(x + 3) = 0$

$x + 2 = 0 \qquad$ or $\qquad x + 3 = 0$
$x = -2 \qquad\qquad\qquad x = -3$

101. $9x^2 - 25 = 0$
$(3x + 5)(3x - 5) = 0$

$3x + 5 = 0 \qquad$ or $\qquad 3x - 5 = 0$
$3x = -5 \qquad\qquad\qquad 3x = 5$

$x = -\dfrac{5}{3} \qquad\qquad\qquad x = \dfrac{5}{3}$

105. $(x + 2)(x - 5) = 0$
$x^2 - 3x - 10 = 8$
$x^2 - 3x - 18 = 0$
$(x + 3)(x - 6) = 0$

$x + 3 = 0 \qquad$ or $\qquad x - 6 = 0$
$x = -3 \qquad\qquad\qquad x = 6$

109. Let x = first consecutive even integer
       $x + 2$ = second consecutive even integer

$$x(x + 2) = 80$$

$$x^2 + 2x = 80$$

$$x^2 + 2x - 80 = 0$$

$$(x + 10)(x - 8) = 0$$

$x + 10 = 0$	or	$x - 8 = 0$
$x = -10$		$x = 8$
No solution		$x + 2 = 10$

The two consecutive even integers are 8 and 10.

113. Let x = first consecutive integer
       $x + 1$ = second consecutive integer
       $x + 2$ = third consecutive integer

$$a^2 + b^2 = c^2$$

$$x^2 + (x + 1)^2 = (x + 2)^2$$

$$x^2 + x^2 + 2x + 1 = x^2 + 4x + 4$$

$$2x^2 + 2x + 1 = x^2 + 4x + 4$$

$$x^2 - 2x - 3 = 0$$

$$(x + 1)(x - 3) = 0$$

$x + 1 = 0$	or	$x - 3 = 0$
$x = -1$		$x = 3$
No solution		$x + 1 = 4$
		$x + 2 = 5$

The three sides are 3, 4 and 5.

1. $x^4 \cdot x^7 \cdot x^{-3} = x^{4+7-3} = x^8$

5. $\dfrac{a^{-5}}{a^{-7}} = a^{-5-(-7)} = a^2$

9. $0.00087 = 8.7 \times 10^{-4}$

13. $3 - 4[2x - 3(x + 6)]$

   $= 3 - 4[2x - 3x - 18]$

   $= 3 - 8x + 12x + 72$

   $= 4x + 75$

17. $(1 - 6y)(1 + 6y) = (1)^2 - (6y)^2 = 1 - 36y^2$

21. $12x^4 + 26x^2 - 10 = 2(6x^4 + 13x^2 - 5)$

   $= 2(3x^2 - 1)(2x^2 + 5)$

25. $4a^5b - 24a^4b^2 - 64a^3b^3$

   $= 4a^3b(a^2 - 6ab - 16b^2)$

   $= 4a^3b(a - 8b)(a + 2b)$

29. $\qquad 100x^3 = 500x^2$

   $100x^3 - 500x^2 = 0$

   $100x^2(x - 5) = 0$

   $100x^2 = 0 \qquad$ or $\qquad x - 5 = 0$
   $\qquad x = 0 \qquad\qquad\qquad x = 5$

33. Let $\quad x$ = shortest side
   $\quad x + 2$ = third side
   $\quad x + 4$ = longest side

   $$a^2 + b^2 = c^2$$

   $$x^2 + (x + 2)^2 = (x + 4)^2$$

   $$x^2 + x^2 + 4x + 4 = x^2 + 8x + 16$$

   $$2x^2 + 4x + 4 = x^2 + 8x + 16$$

   $$x^2 - 4x - 12 = 0$$

   $$(x + 2)(x - 6) = 0$$

   $x + 2 = 0 \qquad$ or $\qquad x - 6 = 0$
   $\qquad x = -2 \qquad\qquad\qquad x = 6$
   $\quad$ No solution $\qquad\qquad x + 2 = 8$
   $\qquad\qquad\qquad\qquad\qquad x + 4 = 10$

   The three sides are 6 in., 8 in., and 10 in.

37. c = 2x + 100,   letting x = 100

   c = 2(100) + 100
   c = 300

   The cost is $300.

CHAPTER 4

Section 4.1

1. $-\dfrac{12}{36} = -\dfrac{1}{3} \cdot \dfrac{12}{12} - -\dfrac{1}{3}$ Factor numerator and denominator

5. $\dfrac{-24x^3y^5}{16x^4y} = \dfrac{-3y^3 \cdot 8x^3y^2}{2x \cdot 8x^3y^2}$ Factor numerator and denominator

   $= -\dfrac{3y^3}{2x}$ Divide out common factor $8x^3y^2$

9. $\dfrac{x^2 - 16}{6x + 24} = \dfrac{(x + 4)(x - 4)}{6(x + 4)}$ Factor numerator and denominator

   $= \dfrac{x - 4}{6}$ Divide out common factor $(x + 4)$

13. $\dfrac{a^4 - 81}{a - 3} = \dfrac{(a^2 + 9)(a + 3)(a - 3)}{(a - 3)}$ Factor numerator and denominator

   $= (a^2 + 9)(a + 3)$ Divide out common factor $(a - 3)$

17. $\dfrac{20y^2 - 45}{10y^2 - 5y - 15} = \dfrac{4y^2 - 9}{2y^2 - y - 3} \cdot \dfrac{5}{5}$ Factor numerator and denominator

   $= \dfrac{(2y + 3)(2y - 3)}{(y + 1)(2y - 3)}$ Factor numerator and denominator

   $= \dfrac{2y + 3}{y + 1}$

21. $\dfrac{12y - 2xy - 2x^2y}{6y - 4xy - 2x^2y}$

   $= \dfrac{6 - x - x^2}{3 - 2x - x^2} \cdot \dfrac{2y}{2y}$ Factor numerator and denominator

   $= \dfrac{(3 + x)(2 - x)}{(3 + x)(1 - x)}$

   $= \dfrac{2 - x}{1 - x}$

25. $\dfrac{a^3 + b^3}{a^2 - b^2} = \dfrac{(a + b)(a^2 - ab + b^2)}{(a + b)(a - b)}$ Factor numerator and denominator

   $= \dfrac{a^2 - ab + b^2}{a - b}$ Divide out common factor $(a + b)$

29. $\dfrac{6x^2 + 7xy - 3y^2}{6x^2 + xy - y^2}$

$\qquad = \dfrac{(2x + 3y)(3x - y)}{(2x + y)(3x - y)}$ Factor numerator and denominator

$\qquad = \dfrac{2x + 3y}{2x + y}$ Divide out common factor $3x - y$

33. $\dfrac{x^2 + bx - 3x - 3b}{x^2 - 2bx - 3x + 6b} = \dfrac{x(x + b) - 3(x + b)}{x(x - 2b) - 3(x - 2b)}$

$\qquad\qquad = \dfrac{(x - 3)(x + b)}{(x - 3)(x - 2b)}$

$\qquad\qquad = \dfrac{x + b}{x - 2b}$

37. $\dfrac{3x^3 + 21x^2 + 36x}{3x^4 + 12x^3 - 27x^2 - 108x} = \dfrac{3x(x^2 + 7x + 12)}{3x(x^3 + 4x^2 - 9x - 36)}$

$\qquad\qquad = \dfrac{3x(x + 3)(x + 4)}{3x[x^2(x + 4) - 9(x + 4)]}$

$\qquad\qquad = \dfrac{3x(x + 3)(x + 4)}{3x(x + 4)(x^2 - 9)}$

$\qquad\qquad = \dfrac{3x(x + 3)(x + 4)}{3x(x + 4)(x + 3)(x - 3)}$

$\qquad\qquad = \dfrac{1}{x - 3}$

41. $\dfrac{x - 4}{4 - x} = \dfrac{x - 4}{-1(-4 + x)}$ Factor -1 from each term in the denominator

$\qquad = \dfrac{x - 4}{-1(x - 4)}$ Reverse the order of the terms in the denominator

$\qquad = -1$ Divide out common factor $(x - 4)$

45. $\dfrac{1 - 9a^2}{9a^2 - 6a + 1} = \dfrac{(1 + 3a)(1 - 3a)}{(3a - 1)(3a - 1)}$ Factor numerator and denominator

$\qquad = \dfrac{(1 + 3a)(-1)(3a - 1)}{(3a - 1)(3a - 1)}$ Factor -1 from $(1 - 3a)$ in the numerator

$\qquad = \dfrac{-1(3a + 1)}{3a - 1}$

49. $\dfrac{x^3 - 8}{x - 2} - \dfrac{x^3 + 8}{x + 2}$

$= \dfrac{x^3 - 2^3}{x - 2} - \dfrac{x^3 + 2^3}{x + 2}$ Rewrite as cubes

$= \dfrac{(x - 2)(x^2 + 2x + 8)}{x - 2} - \dfrac{(x + 2)(x^2 - 2x + 8)}{x + 2}$ Factor as cube

$= x^2 + 2x + 8 - (x^2 - 2x + 8)$

$= x^2 + 2x + 8 - x^2 + 2x - 8$

$= 4x$

53. $\dfrac{x + y}{x} = y$ Do not factor the terms "x"

57.   When $R_1 = 10$ ohms and $R_2 = 5$ ohms

the equation $R = \dfrac{R_1 + R_2}{R_1 R_2}$

becomes $R = \dfrac{10 + 5}{10(5)}$

$R = \dfrac{15}{50}$

$R = \dfrac{3}{10}$ ohms

61. $(10x - 11) - (10x - 20) = 10x - 11 - 10x + 20 = 9$

65. $\dfrac{\frac{1}{2}x^2 + \frac{5}{12}x - \frac{1}{2}}{\frac{1}{6}x^2 + \frac{1}{2}x + \frac{3}{8}}$

$= \dfrac{12(\frac{1}{2}x^2) + 12(\frac{5}{12}x) - 12(\frac{1}{2})}{24(\frac{1}{6}x^2) + 24(\frac{1}{2}x) + 24(\frac{3}{8})}$ $\dfrac{LCD = 12}{LCD = 24}$

$= \dfrac{6x^2 + 5x - 6}{4x^2 + 12x + 9}$

$= \dfrac{(2x + 3)(3x - 2)}{(2x + 3)(2x + 3)}$

$= \dfrac{3x - 2}{2x + 3}$

Section 41. continued

69. $\dfrac{x^{3n} + 2x^{2n} - 9x^n - 18}{x^{2n} - x^n - 6}$

$\quad = \dfrac{x^{2n}(x^n + 2) - 9(x^n + 2)}{(x^n + 2)(x^n - 3)}$

$\quad = \dfrac{(x^n + 2)(x^{2n} - 9)}{(x^n + 2)(x^n - 3)}$

$\quad = \dfrac{(x^n + 2)(x^n + 3)(x^n - 3)}{(x^n + 2)(x^n - 3)}$

$\quad = x^n + 3$

Section 4.2

1. $\dfrac{4x^3 - 8x^2 + 6x}{2x} = \dfrac{4x^3}{2x} - \dfrac{8x^2}{2x} + \dfrac{6x}{2x}$

$\qquad = 2x^2 - 4x + 3$

5. $\dfrac{8y^5 + 10y^3 - 6y}{4y^3} = \dfrac{8y^5}{4y^3} + \dfrac{10y^3}{4y^3} - \dfrac{6y}{4y^3}$

$\qquad = 2y^2 + \dfrac{5}{2} - \dfrac{3}{2y^2}$

9. $\dfrac{28a^3b^5 + 42a^4b^3}{7a^2b^2} = \dfrac{28a^3b^5}{7a^2b^2} + \dfrac{42a^4b^3}{7a^2b^2}$

$\qquad = 4ab^3 + 6a^2b$

13. $\dfrac{x^2 - x - 6}{x - 3} = \dfrac{(x - 3)(x + 2)}{(x - 3)}$

$\qquad = x + 2$

17. $\dfrac{5x^2 - 14xy - 24y^2}{x - 4y} = \dfrac{(x - 4y)(5x + 6y)}{x - 4y}$

$\qquad = 5x + 6y$

21. $\dfrac{y^4 - 16}{y - 2} = \dfrac{(y^2 + 4)(y^2 - 4)}{(y - 2)}$

$\qquad = \dfrac{(y^2 + 4)(y + 2)(y - 2)}{y - 2}$

$\qquad = (y^2 + 4)(y + 2)$

25. $\dfrac{4x^3 + 12x^2 - 9x - 27}{x + 3} = \dfrac{4x^2(x + 3) - 9(x + 3)}{(x + 3)}$

$\qquad\qquad\qquad\qquad\quad = \dfrac{\cancel{(x + 3)}(4x^2 - 9)}{\cancel{(x + 3)}}$

$\qquad\qquad\qquad\qquad\quad = (2x + 3)(2x - 3)$

29.

$$
\begin{array}{r}
2x + \frac{5}{18} \\
3x - 4 \overline{\smash{\big)}\ 6x^2 + 7x - 18} \\
\ \underset{-\qquad+}{} \\
\underline{6x^2 - 8x} \\
15x - 18 \\
\underset{-\qquad+}{} \\
\underline{15x - 20} \\
2
\end{array}
$$

Change signs

Change signs

$\dfrac{6x^2 + 7x - 18}{3x - 4} = 2x + 5 + \dfrac{2}{3x - 4}$

33.

$$
\begin{array}{r}
y^2 \quad - \quad 5 \\
2y - 3 \overline{\smash{\big)}\ 2y^3 - 3y^2 - 4y + 5} \\
\underset{+\qquad+}{} \\
\underline{2y^3 - 3y^2} \\
0 - 4y + 5 \\
\underset{+\qquad-}{} \\
\underline{- 4y + 6} \\
-1
\end{array}
$$

Change signs

Change signs

$\dfrac{2y^3 - 3y^2 - 4y + 5}{2y - 3} = y^2 - 2 + \dfrac{-1}{2y - 3}$

37.

$$
\begin{array}{r}
3y^2 + 6y + 8 \\
2y - 4 \overline{\smash{\big)}\ 6y^3 + 0y^2 - 8y + 5} \\
\underset{-\qquad+}{} \\
\underline{6y^3 - 12y^2} \\
12y^2 - 8y \\
\underset{-\qquad+}{} \\
\underline{12y^2 - 24y} \\
16y + 5 \\
\underset{-\qquad+}{} \\
\underline{16y + 32} \\
37
\end{array}
$$

Notice: Adding the $0y^2$ term gives us a column in which to write $+12y^2$

$\dfrac{6y^3 - 8y + 5}{2y - 4} = 3y^2 + 6y + 8 + \dfrac{37}{2y - 4}$

41.

$$\begin{array}{r} y^3 + 2y^2 + 4y + 8 \\ y - 2\overline{)\,y^4 + 0y^3 + 0y^2 + 0y - 16} \end{array}$$

$$\begin{array}{r} - \quad - \\ y^4 + 2y^3 \\ \hline - 2y^3 + 0y^2 \\ + \quad - \\ - 2y^3 + 4y^2 \\ \hline - 4y^2 + 0y \\ + \quad - \\ - 4y^2 + 8y \\ \hline - 8y - 16 \\ + \quad - \\ - 8y + 6 \\ \hline 0 \end{array}$$

Notice: Adding the 0 terms gives us a column in which to write $-2y^3$, $-4y^2$ and $-8y$.

$$\frac{y^4 - 16}{y - 2} = y^3 + 2y^2 + 4y + 8$$

45.

$$\begin{array}{r} x^2 + 3x + 2 \\ x + 3\overline{)\,x^3 + 6x^2 + 11x + 6} \end{array}$$

$$\begin{array}{r} - \quad - \\ x^3 + 3x^2 \\ \hline 3x^2 + 11x \\ - \quad - \\ 3x^2 + 9x \\ \hline 2x + 6 \\ - \quad - \\ 2x + 6 \\ \hline 0 \end{array}$$

$$x^3 + 6x^2 + 11x + 6 = (x + 3)(x^2 + 3x + 2)$$
$$= (x + 3)(x + 2)(x + 1)$$

49. The answer to problem 21 is $(y^2 + 4)(y + 2) = y^3 + 2y^2 + 4y + 8$ which is the answer to problem 41.

53. $\dfrac{3}{5} \div \dfrac{2}{7} = \dfrac{3}{5} \cdot \dfrac{7}{2} = \dfrac{21}{10}$

57. $\dfrac{4}{9} \div 8 = \dfrac{4}{9} \cdot \dfrac{1}{8} = \dfrac{4}{72} = \dfrac{1}{18}$

61. $\left(\frac{1}{3}\right)^{-2} + \left(\frac{1}{2}\right)^{-3} - \dfrac{1}{\left(\frac{1}{3}\right)^2} + \dfrac{1}{\left(\frac{1}{2}\right)^3}$

$$= \dfrac{1}{\frac{1}{9}} + \dfrac{1}{\frac{1}{8}}$$

$$= \frac{9}{1} + \frac{8}{1}$$

$$= 17$$

65.

$$
\require{enclose}
\begin{array}{r}
4x^3 - x^2 + 3 \\[2pt]
x^2 - 5 \enclose{longdiv}{4x^5 - x^4 - 20x^3 + 8x^2 - 15} \\
\end{array}
$$

$$
\begin{array}{l}
\quad\;\; - \qquad\qquad + \\
\quad\;\; 4x^5 \qquad - 20x^3 \\
\hline
\quad\;\;\; - x^4 \qquad\quad + 8x^2 \\
\quad\;\; + \qquad\qquad - \\
\quad\;\;\; - x^4 \qquad\quad + 5x^2 \\
\hline
\qquad\qquad\qquad 3x^2 - 15 \\
\qquad\qquad\;\; + \qquad - \\
\qquad\qquad\quad 3x^2 - 15 \\
\hline
\qquad\qquad\qquad\qquad 0
\end{array}
$$

Change signs

Change signs

Change signs

69.

$$
\begin{array}{r}
\frac{3}{2}x - \frac{5}{2} + \dfrac{1}{2x+4} \\[2pt]
2x + 4 \enclose{longdiv}{3x^2 + x - 9} \\
\end{array}
$$

$$
\begin{array}{l}
\quad - \quad\; - \\
\quad 3x^2 + 6x \\
\hline
\qquad\; - 5x - 9 \\
\qquad\; + \quad\; + \\
\qquad\; - 5x - 10 \\
\hline
\qquad\qquad\quad 1
\end{array}
$$

Change signs

Change signs

Section 4.3

1. $\dfrac{2}{9} \cdot \dfrac{3}{4} = \dfrac{2}{9} \cdot \dfrac{3}{4} = \dfrac{1}{6}$ Factor

5. $\dfrac{3}{7} \cdot \dfrac{14}{24} \div \dfrac{1}{2} = \dfrac{3}{7} \cdot \dfrac{14}{24} \cdot \dfrac{2}{1}$ Write division in terms of multiplication

$$= \dfrac{3 \cdot 14 \cdot 2}{7 \cdot 24 \cdot 1}$$ Multiply numerators and denominators

$$= \dfrac{3 \cdot 2 \cdot 7 \cdot 2}{7 \cdot 3 \cdot 2 \cdot 2 \cdot 2}$$ Factor

$$= \dfrac{1}{2}$$

9. $\dfrac{11a^2b}{5ab^2} \div \dfrac{22a^3b^2}{10ab^4}$

       $= \dfrac{11a^2b}{5ab^2} \cdot \dfrac{10ab^4}{22a^3b^2}$ Write division in terms of multiplication

       $= \dfrac{11a^2b \cdot 10ab^4}{5ab^2 \cdot 22a^3b^2}$ Factor and multiply

       $= \dfrac{b}{a}$

13. $\dfrac{x^2 - 9}{x^2 - 4} \cdot \dfrac{x - 2}{x - 3} = \dfrac{(x + 3)\cancel{(x - 3)}\cancel{(x - 2)}}{(x + 2)\cancel{(x - 2)}\cancel{(x - 3)}}$ Factor and multiply

                 $= \dfrac{x + 3}{x + 2}$

17. $\dfrac{3x - 12}{x^2 - 4} \cdot \dfrac{x^2 + 6x + 8}{x - 4}$

       $= \dfrac{3\cancel{(x - 4)}(x + 2)(x + 4)}{\cancel{(x + 2)}(x - 2)\cancel{(x - 4)}}$ Factor and multiply

       $= \dfrac{3(x + 4)}{x - 2}$

21. $\dfrac{a^2 - 5a + 6}{a^2 - 2a - 3} \div \dfrac{a - 5}{a^2 + 3a + 2}$

       $= \dfrac{a^2 - 5a + 6}{a^2 - 2a - 3} \cdot \dfrac{a^2 + 3a + 2}{a - 5}$ Change division to multiplication by the reciprocal.

       $= \dfrac{(a - 2)\cancel{(a - 3)}\cancel{(a + 1)}(a + 2)}{\cancel{(a + 1)}\cancel{(a - 3)}(a - 5)}$ Factor and multiply

       $= \dfrac{(a - 2)(a + 2)}{a - 5}$

25. $\dfrac{2x^2 - 5x - 12}{4x^2 + 8x + 3} \div \dfrac{x^2 - 16}{2x^2 + 7x + 3}$

       $= \dfrac{2x^2 - 5x + 12}{4x^2 + 8x + 3} \cdot \dfrac{2x^2 + 7x + 3}{x^2 - 16}$ Multiply by the reciprocal of the divisor

       $= \dfrac{\cancel{(2x + 3)}\cancel{(x - 4)}\cancel{(2x + 1)}(x + 3)}{\cancel{(2x + 1)}\cancel{(2x + 3)}(x + 4)\cancel{(x - 4)}}$ Factor and multiply

       $= \dfrac{x + 3}{x + 4}$

29. $\dfrac{360x^3 - 490x}{36x^2 + 84x + 49} \cdot \dfrac{30x^2 + 83x + 56}{150x^3 + 65x^2 - 280x}$

$$= \frac{10x(36x^2 - 49)}{(6x + 7)(6x + 7)} \cdot \frac{(6x + 7)(5x + 8)}{5x(30x^2 + 13x - 56)}$$

$$= \frac{\overset{2}{\cancel{10x}}\cancel{(6x + 7)}(6x - 7)}{\cancel{(6x + 7)}\cancel{(6x + 7)}} \cdot \frac{\cancel{(6x + 7)}\cancel{(5x + 8)}}{\cancel{5x}\cancel{(5x + 8)}\cancel{(6x - 7)}}$$

$$= 2$$

33. $\dfrac{a^2 - 16b^2}{a^2 - 8ab + 16b^2} \cdot \dfrac{a^2 - 9ab + 20b^2}{a^2 - 7ab + 12b^2} \div \dfrac{a^2 - 25}{a^2 - 6ab + 9b^2}$

$$= \frac{(a^2 - 16b^2)(a^2 - 9ab + 20b^2)(a^2 - 6ab + 9b^2)}{(a^2 - 8ab + 16b^2)(a^2 - 7ab + 12b^2)(a^2 - 25b^2)}$$

$$= \frac{(a + 4b)\cancel{(a - 4b)}\cancel{(a - 4b)}\cancel{(a - 5b)}\cancel{(a - 3b)}(a - 3b)}{\cancel{(a - 4b)}(a - 4b)\cancel{(a - 3b)}\cancel{(a - 4b)}(a + 5b)\cancel{(a - 5b)}}$$

$$= \frac{(a + 4b)(a - 3b)}{(a - 4b)(a + 5b)}$$

37. $\dfrac{xy - 2x + 3y - 6}{xy + 2x - 4y - 8} \cdot \dfrac{xy + x - 4y - 4}{xy - x + 3y - 3}$

$$= \frac{x(y - 2) + 3(y - 2)}{(y + 2) - 4(y + 2)} \cdot \frac{x(y + 1) - 4(y + 1)}{x(y - 1) + 3(y - 1)}$$

$$= \frac{(y - 2)\cancel{(x + 3)}(y + 1)\cancel{(x - 4)}}{(y + 2)\cancel{(x - 4)}(y - 1)\cancel{(x + 3)}}$$

$$= \frac{(y - 2)(y + 1)}{(y + 2)(y - 1)}$$

41. $\dfrac{2x^3 + 10x^2 - 8x - 40}{x^3 + 4x^2 - 9x - 36} \cdot \dfrac{x^2 + x - 12}{2x^2 + 14x + 20}$

$$= \frac{2x^2(x + 5) - 8(x + 5)}{x^2(x + 4) - 9(x + 4)} \cdot \frac{(x + 4)(x - 3)}{2(x + 2)(x + 5)}$$

$$= \frac{(2x^2 - 8)(x + 5)(x + 4)(x - 3)}{2(x^2 - 9)(x + 4)(x + 2)(x + 5)}$$

$$= \frac{2\cancel{(x + 2)}(x - 2)\cancel{(x + 5)}\cancel{(x + 4)}\cancel{(x - 3)}}{2(x + 3)\cancel{(x - 3)}\cancel{(x + 4)}\cancel{(x + 2)}\cancel{(x + 5)}}$$

$$= \frac{x - 2}{x + 3}$$

45. $(x^2 - 25) \cdot \dfrac{2}{x - 5}$

$\qquad = \dfrac{x^2 - 25}{1} \cdot \dfrac{2}{x - 5}$ Write $x^2 - 25$ with denominator 1

$\qquad = \dfrac{(x - 5)(x - 5)2}{(x - 5)}$ Factor

$\qquad = 2(x + 5)$

49. $(y - 3)(y - 4)(y + 3) \cdot \dfrac{-1}{y^2 - 9}$

$\qquad = \dfrac{(y - 3)(y - 4)(y + 3)}{1} \cdot \dfrac{-1}{y^2 - 9}$ Write $(y - 3)(y - 4)(y + 3)$
$\qquad\qquad\qquad\qquad\qquad\qquad\qquad\qquad$ with denominator 1

$\qquad = \dfrac{(y - 3)(y - 4)(y + 3)(-1)}{(y + 3)(y - 3)}$

$\qquad = -(y - 4)$

53. $2x^2(5x^3 + 4x - 3)$
$\qquad = 2x^2(5x^3) + 2x^2(4x) - (2x^2)(3)$ Distributive property
$\qquad = 10x^5 + 8x^3 - 6x^2$

57. $(x + 3)(x - 3)(x^2 + 9) = (x^2 - 9)(x^2 + 9) = x^4 - 81$

61. $(3x + 7)(4y - 2) = 12xy - 6x + 28y - 14$
$\qquad\qquad\qquad\qquad$ F $\qquad$ O $\quad$ I $\qquad$ L

65. $3(x + 1)(x + 2)(x + 3)$
$\qquad = (3x + 3)(x + 2)(x + 3)$ Distributive property
$\qquad = (3x^2 + 6x + 3x + 6)(x + 3)$
$\qquad\quad$ F $\quad$ O $\quad$ I $\quad$ L
$\qquad = (3x^2 + 9x + 6)(x + 3)$

$$\begin{array}{r} 3x^2 + 9x + 6 \\ x + 3 \\ \hline 9x^2 + 27x + 18 \\ 3x^3 + 9x^2 + 6x \\ \hline 3x^3 + 18x^2 + 33x + 18 \end{array}$$

69. $\dfrac{4}{15} - \dfrac{5}{21}$ $\qquad$ $\begin{aligned}15 &= 3 \cdot 5\\ 21 &= 3 \cdot 7\end{aligned}$ $\qquad$ LCD $= 3 \cdot 5 \cdot 7 = 105$

$$= \dfrac{4}{15} \cdot \dfrac{7}{7} - \dfrac{5}{21} \cdot \dfrac{5}{5}$$

$$= \dfrac{28}{105} - \dfrac{25}{105}$$

$$= \dfrac{3}{105}$$

$$= \dfrac{1}{35}$$

73. $(1 + 2^{-1})(1 + 3^{-1})(1 + 4^{-1})\ldots(1 + 99^{-1})(1 + 100^{-1})$

$$= \left(1 + \dfrac{1}{2}\right)\left(1 + \dfrac{1}{3}\right)\left(1 + \dfrac{1}{4}\right)\ldots\left(1 + \dfrac{1}{99}\right)\left(1 + \dfrac{1}{100}\right)$$

$$= \left(\dfrac{3}{2}\right)\left(\dfrac{4}{3}\right)\left(\dfrac{5}{4}\right)\ldots\left(\dfrac{100}{99}\right)\left(\dfrac{101}{100}\right)$$

$$= \dfrac{101}{2}$$

Section 4.4

1. $\dfrac{3}{4} + \dfrac{1}{2} = \dfrac{3}{4} + \dfrac{1}{2} \cdot \dfrac{2}{2}$ $\qquad$ LCD $= 4$

$$= \dfrac{3}{4} + \dfrac{2}{4}$$

$$= \dfrac{5}{4}$$

5. $\dfrac{5}{6} + \dfrac{7}{8} = \dfrac{5}{6} \cdot \dfrac{4}{4} + \dfrac{7}{8} \cdot \dfrac{3}{3}$ $\qquad$ LCD $= 24$

$$= \dfrac{20}{24} + \dfrac{21}{24}$$

$$= \dfrac{41}{24}$$

9. $\dfrac{3}{4} - \dfrac{1}{8} + \dfrac{2}{3} = \dfrac{3}{4} \cdot \dfrac{6}{6} - \dfrac{1}{8} \cdot \dfrac{3}{3} + \dfrac{2}{3} \cdot \dfrac{8}{8}$ LCD = 24

$\qquad\qquad = \dfrac{18}{24} - \dfrac{3}{24} + \dfrac{16}{24}$

$\qquad\qquad = \dfrac{15}{24} + \dfrac{16}{24}$

$\qquad\qquad = \dfrac{31}{24}$

13. $\dfrac{4}{y-4} - \dfrac{y}{y-4} = \dfrac{4-y}{y-4}$

$\qquad\qquad\quad = \dfrac{-1(\cancel{y-4})}{\cancel{y-4}}$ Factor out -1 in the numerator

$\qquad\qquad\quad = -1$

17. $\dfrac{2x-3}{x-2} - \dfrac{x-1}{x-2} = \dfrac{2x-3-(x-1)}{x-2}$ Subtract numerators

$\qquad\qquad\quad = \dfrac{2x-3-x+1}{x-2}$ Remove parentheses

$\qquad\qquad\quad = \dfrac{x-2}{x-2}$ Combine similar terms in the numerator

$\qquad\qquad\quad = 1$ Reduce (or divide)

21. $\dfrac{7x-2}{2x+1} - \dfrac{5x-3}{2x+1} = \dfrac{7x-2-(5x-3)}{2x+1}$ Subtract numerators

$\qquad\qquad\quad = \dfrac{7x-2-5x+3}{2x+1}$ Remove parentheses

$\qquad\qquad\quad = \dfrac{2x+1}{2x+1}$ Combine similar terms in the numerator

$\qquad\qquad\quad = 1$ Reduce (or divide)

25. $\dfrac{3x+1}{2x-6} - \dfrac{x+2}{x-3} = \dfrac{3x+1}{2(x-3)} - \dfrac{x+2}{x-3}$

The LCD = 2(x - 3). Completing the problem, we have

$\qquad = \dfrac{3x+1}{2(x-3)} - \dfrac{x+2}{x-3} \cdot \dfrac{2}{2}$

$\qquad = \dfrac{3x+1}{2(x-3)} - \dfrac{2x+4}{2(x-3)}$

$\qquad = \dfrac{3x+1-2x-4}{2(x-3)}$

$\qquad = \dfrac{x-3}{2(x-3)}$

$\qquad = \dfrac{1}{2}$

29. $\dfrac{x+1}{2x-2} - \dfrac{2}{x^2-1} = \dfrac{x+1}{2(x-1)} - \dfrac{2}{(x+1)(x-1)}$

The LCD = $2(x-1)(x+1)$. Completing the problem, we have

$$= \dfrac{x+1}{2(x-1)} \cdot \dfrac{x+1}{x+1} - \dfrac{2}{(x+1)(x-1)} \cdot \dfrac{2}{2}$$

$$= \dfrac{x^2+2x+1}{2(x-1)(x+1)} - \dfrac{4}{2(x-1)(x+1)}$$

$$= \dfrac{x^2+2x-3}{2(x-1)(x+1)}$$

$$= \dfrac{(x+3)(x-1)}{2(x-1)(x+1)}$$

$$= \dfrac{x+3}{2(x+1)}$$

33. $\dfrac{1}{2y-3} - \dfrac{18y}{8y^3-27} = \dfrac{1}{2y-3} - \dfrac{18y}{(2y-3)(4y^2+6y+9)}$

The LCD is $(2y-3)(4y^2+6y+9)$. Completing the problem, we have

$$= \dfrac{1}{2y-3} \cdot \dfrac{4y^2+6y+9}{4y^2+6y+9} - \dfrac{18y}{(2y-3)(4y^2+6y+9)}$$

$$= \dfrac{4y^2+6y+9}{(2y-3)(4y^2+6y+9)} - \dfrac{18y}{(2y-3)(4y^2+6y+9)}$$

$$= \dfrac{4y^2-12y+9}{(2y-3)(4y^2+6y+9)}$$

$$= \dfrac{(2y-3)(2y-3)}{(2y-3)(4y^2+6y+9)}$$

$$= \dfrac{2y-3}{4y^2+6y+9}$$

37. $\dfrac{2}{4t - 5} + \dfrac{9}{8t^2 - 38t + 35} = \dfrac{2}{4t - 5} + \dfrac{9}{(4t - 5)(2t - 7)}$

The LCD is $(4t - 5)(2t - 7)$. Completing the problem, we have

$$= \frac{2}{4t - 5} \cdot \frac{2t - 7}{2t - 7} + \frac{9}{(4t - 5)(2t - 7)}$$

$$= \frac{4t - 14}{(4t - 5)(2t - 7)} + \frac{9}{(4t - 5)(2t - 7)}$$

$$= \frac{\cancel{4t - 5}}{\cancel{(4t - 5)}(2t - 7)}$$

$$= \frac{1}{2t - 7}$$

41. $\dfrac{1}{8x^3 - 1} - \dfrac{1}{4x^2 - 1} = \dfrac{1}{(2x - 1)(4x^2 + 2x + 1)} - \dfrac{1}{(2x + 1)(2x - 1)}$

The LCD is $(2x - 1)(4x^2 + 2x + 1)(2x + 1)$. Completing the problem, we have

$$= \frac{1}{(2x - 1)(4x^2 + 2x + 1)} \cdot \frac{2x + 1}{2x + 1} - \frac{1}{(2x + 1)(2x - 1)} \cdot \frac{4x^2 + 2x + 1}{4x^2 + 2x + 1}$$

$$= \frac{2x + 1}{(2x - 1)(4x^2 + 2x + 1)(2x + 1)} - \frac{4x^2 + 2x + 1}{(2x + 1)(2x - 1)(4x^2 + 2x + 1)}$$

$$= \frac{-4x^2}{(2x - 1)(2x + 1)(4x^2 + 2x + 1)}$$

45. $\dfrac{4a}{a^2 + 6a + 5} - \dfrac{3a}{a^2 + 5a + 4} = \dfrac{4a}{(a + 1)(a + 5)} - \dfrac{3a}{(a + 1)(a + 4)}$

The LCD is $(a + 1)(a + 5)(a + 4)$. Completing the problem, we have

$$= \frac{4a}{(a + 1)(a + 5)} \cdot \frac{a + 4}{a + 4} - \frac{3a}{(a + 1)(a + 4)} \cdot \frac{a + 5}{a + 5}$$

$$= \frac{4a^2 + 16a}{(a + 1)(a + 5)(a + 4)} - \frac{3a^2 + 15a}{(a + 1)(a + 5)(a + 4)}$$

$$= \frac{4a^2 + 16a - 3a^2 - 15a}{(a + 1)(a + 5)(a + 4)}$$

$$= \frac{a^2 + a}{(a + 1)(a + 5)(a + 4)}$$

$$= \frac{a\cancel{(a + 1)}}{\cancel{(a + 1)}(a + 5)(a + 4)}$$

$$= \frac{a}{(a + 5)(a + 4)}$$

49. $\dfrac{2x - 8}{3x^2 + 8x + 4} + \dfrac{x + 3}{3x^2 + 5x + 2} = \dfrac{2x - 8}{(x + 2)(3x + 2)} + \dfrac{x + 3}{(3x + 2)(x + 1)}$

The LCD $= (x + 2)(3x + 2)(x + 1)$. Completing the problem, we have

$$= \dfrac{2x - 8}{(x + 2)(3x + 2)} \cdot \dfrac{x + 1}{x + 1} + \dfrac{x + 3}{(3x + 2)(x + 1)} \cdot \dfrac{x + 2}{x + 2}$$

$$= \dfrac{2x^2 - 6x - 8}{(x + 2)(3x + 2)(x + 1)} + \dfrac{x^2 + 5x + 6}{(x + 2)(3x + 2)(x + 1)}$$

$$= \dfrac{3x^2 - x - 2}{(x + 2)(3x + 2)(x + 1)}$$

$$= \dfrac{(3x + 2)(x - 1)}{(x + 2)(3x + 2)(x + 1)}$$

$$= \dfrac{x - 1}{(x + 2)(x + 1)}$$

53. $\dfrac{2x + 8}{x^2 + 5x + 6} - \dfrac{x + 5}{x^2 + 4x + 3} - \dfrac{x - 1}{x^2 + 3x + 2}$

$$= \dfrac{2x + 8}{(x + 2)(x + 3)} - \dfrac{x + 5}{(x + 1)(x + 3)} - \dfrac{x - 1}{(x + 2)(x + 1)}$$

The LCD $= (x + 1)(x + 2)(x + 3)$. Completing the problem, we have

$$= \dfrac{2x + 8}{(x + 2)(x + 3)} \cdot \dfrac{(x + 1)}{(x + 1)} - \dfrac{x + 5}{(x + 1)(x + 3)} \cdot \dfrac{(x + 2)}{(x + 2)} - \dfrac{x - 1}{(x + 2)(x + 1)} \cdot \dfrac{(x + 3)}{(x + 3)}$$

$$= \dfrac{2x^2 + 10x + 8}{(x + 1)(x + 2)(x + 3)} - \dfrac{x^2 + 7x + 10}{(x + 1)(x + 2)(x + 3)} - \dfrac{x^2 + 2x - 3}{(x + 1)(x + 2)(x + 3)}$$

$$= \dfrac{2x^2 + 10x + 8 - x^2 - 7x - 10 - x^2 - 2x + 3}{(x + 1)(x + 2)(x + 3)}$$

$$= \dfrac{x + 1}{(x + 1)(x + 2)(x + 3)}$$

$$= \dfrac{1}{(x + 2)(x + 3)}$$

57. $5 + \dfrac{2}{4 - t} = \dfrac{5}{1} + \dfrac{2}{4 - t}$

The LCD $= 4 - t$. Completing the problem, we have

$$= \dfrac{5}{1} \cdot \dfrac{4 - t}{4 - t} + \dfrac{2}{4 - t}$$

$$= \dfrac{20 - 5t}{4 - t} + \dfrac{2}{4 - t}$$

$$= \dfrac{22 - 5t}{4 - t}$$

61. $\dfrac{x}{x + 2} + \dfrac{1}{2x + 4} - \dfrac{3}{x^2 + 2x} = \dfrac{x}{x + 2} + \dfrac{1}{2(x + 2)} - \dfrac{3}{x(x + 2)}$

The LCD = $2x(x + 2)$. Completing the problem, we have

$$= \dfrac{x}{x + 2} \cdot \dfrac{2x}{2x} + \dfrac{1}{2(x + 2)} \cdot \dfrac{x}{x} - \dfrac{3}{x(x + 2)} \cdot \dfrac{2}{2}$$

$$= \dfrac{2x^2}{2x(x + 2)} + \dfrac{x}{2x(x + 2)} - \dfrac{6}{2x(x + 2)}$$

$$= \dfrac{2x^2 + x - 6}{2x(x + 2)}$$

$$= \dfrac{(2x - 3)(x + 2)}{2x(x + 2)}$$

$$= \dfrac{2x - 3}{2x}$$

65. Letting a = 10 and b = 0.2

the equation $P = \dfrac{1}{a} + \dfrac{1}{b}$

becomes $P = \dfrac{1}{10} + \dfrac{1}{0.2}$

$$P = \dfrac{1}{10} + \dfrac{1}{.2} \cdot \dfrac{50}{50}$$

$$P = \dfrac{1}{10} + \dfrac{50}{10}$$

$$P = \dfrac{51}{10}$$

69. $(1 - 3^{-2}) \div (1 - 3^{-1}) = \left(1 - \dfrac{1}{3^2}\right) \div \left(1 - \dfrac{1}{3}\right)$

$$= \left(1 - \dfrac{1}{9}\right) \div \left(1 - \dfrac{1}{3}\right)$$

$$= \dfrac{8}{9} \div \dfrac{2}{3}$$

$$= \dfrac{8}{9} \cdot \dfrac{3}{2}$$

$$= \dfrac{4}{3}$$

73. Let x = a number, then the reciprocal of a number would be $\frac{1}{x}$.

The sum of a number and four times its reciprocal.

$$x + 4\left(\frac{1}{x}\right) = \frac{x}{1} + \frac{4}{x} \qquad LCD = x$$

$$= \frac{x}{1} \cdot \frac{x}{x} + \frac{4}{x}$$

$$= \frac{x^2}{x} + \frac{4}{x}$$

$$= \frac{x^2 + 4}{x}$$

77. $54{,}000 = 5.4 \times 10^4$

Remember, in scientific notation the number must first be changed to a number between 1 and 10, times a power of ten.

81. $6.44 \times 10^3 = 6.44 \times 1{,}000 = 6{,}440$

85. $(3 \times 10^8)(4 \times 10^{-5}) = (3 \times 4)(10^8 \times 10^{-5})$

$$= 12 \times 10^{8 + (-5)}$$

$$= 12 \times 10^3$$

$$= 1.2 \times 10^4 \qquad \text{Changed to scientific notation}$$

89. $\left(1 - \frac{1}{x}\right)\left(1 - \frac{1}{x+1}\right)\left(1 - \frac{1}{x+2}\right)\left(1 - \frac{1}{x+3}\right)$

$$= \left(\frac{x}{x} - \frac{1}{x}\right)\left(\frac{x+1}{x+1} - \frac{1}{x+1}\right)\left(\frac{x+2}{x+2} - \frac{1}{x+2}\right)\left(\frac{x+3}{x+3} - \frac{1}{x+3}\right)$$

$$= \left(\frac{x-1}{\cancel{x}}\right)\left(\frac{\cancel{x}}{x+1}\right)\left(\frac{\cancel{x+1}}{\cancel{x+2}}\right)\left(\frac{\cancel{x+2}}{x+3}\right)$$

$$= \frac{x-1}{x+3}$$

93. $(1 + \frac{1}{x})(1 + \frac{1}{x-1})(1 + \frac{1}{x-2})\ldots(1 + \frac{1}{x-99})(1 + \frac{1}{x-100})$

$$= (\frac{x}{x} + \frac{1}{x})(\frac{x-1}{x-1} + \frac{1}{x-1})(\frac{x-2}{x-2} + \frac{1}{x-2})\ldots$$

$$(\frac{x-99}{x-99} + \frac{1}{x-99})(\frac{x-100}{x-100} + \frac{1}{x-100})$$

$$= (\frac{x+1}{x})(\frac{x}{x-1})(\frac{x-1}{x-2})\ldots(\frac{x-98}{x-99})(\frac{x-99}{x-100})$$

$$= \frac{x+1}{x-100}$$

Section 4.5

1. **Method 1.** First find the LCD for both fractions, which in this case is 12:

$$\frac{\frac{3}{4}}{\frac{2}{3}} = \frac{\frac{3}{4} \cdot 12}{\frac{2}{3} \cdot 12} = \frac{9}{8}$$

Method 2. Instead of dividing by $\frac{2}{3}$, we can multiply by $\frac{3}{2}$:

$$\frac{\frac{3}{4}}{\frac{2}{3}} = \frac{3}{4} \cdot \frac{3}{2} = \frac{9}{8}$$

5. Using Method 1, we find the LCD for both fractions, which in this case is 35:

$$\frac{3 + \frac{2}{5}}{1 - \frac{3}{7}} \qquad \frac{(3 + \frac{2}{5}) \cdot 35}{(1 - \frac{3}{7}) \cdot 35}$$

$$= \frac{3(35) + \frac{2}{5}(35)}{1(35) - \frac{3}{7}(35)}$$

Apply the distributive property to distribute 35 over both terms in the numerator and denominator

$$= \frac{105 + 14}{35 - 15}$$

$$= \frac{119}{20}$$

9. Applying Method 1, we find the LCD for both fractions, which in this case is a:

$$\frac{1 + \frac{1}{a}}{1 - \frac{1}{a}} = \frac{a(1 + \frac{1}{a})}{a(a - \frac{1}{a})}$$ Multiply numerator and denominator by a

$$= \frac{a \cdot 1 + a \cdot \frac{1}{a}}{a \cdot 1 - a \cdot \frac{1}{a}}$$ Distributive property

$$= \frac{a + 1}{a - 1}$$ Simplify

13. Applying Method 2, we have

$$\frac{\frac{x - 5}{x^2 - 4}}{\frac{x^2 - 25}{x + 2}} = \frac{x - 5}{x^2 - 4} \cdot \frac{x + 2}{x^2 - 25}$$

$$= \frac{\cancel{(x - 5)}(x + 2)}{\cancel{(x + 2)}(x - 2)(x + 5)\cancel{(x - 5)}}$$ Factor

$$= \frac{1}{(x - 2)(x + 5)}$$ Reduce

17. Applying Method 1, we find the LCD for both fractions, which in this case is x^2:

$$\frac{1 - \frac{9}{x^2}}{1 - \frac{1}{x} - \frac{6}{x^2}} = \frac{x^2(1 - \frac{9}{x^2})}{x^2(1 - \frac{1}{x} - \frac{6}{x^2})}$$ Multiply numerator and denominator by x^2

$$= \frac{x^2 \cdot 1 - x^2 \cdot \frac{9}{x^2}}{x^2 \cdot 1 - x^2 \cdot \frac{1}{x} - x^2 \cdot \frac{6}{x^2}}$$ Distributive property

$$= \frac{x^2 - 9}{x^2 - x - 6}$$ Simplify

$$= \frac{(x + 3)\cancel{(x - 3)}}{(x + 2)\cancel{(x - 3)}}$$ Factor

$$= \frac{x + 3}{x + 2}$$ Reduce

21. Applying Method 1, we find the LCD for both fractions, which in this case is x^3:

$$\frac{2 + \frac{3}{x} - \frac{18}{x^2} - \frac{27}{x^3}}{2 + \frac{9}{x} + \frac{9}{x^2}}$$

$$= \frac{x^3(2) + x^3\left(\frac{3}{x}\right) - x^3\left(\frac{18}{x^2}\right) - x^3\left(\frac{27}{x^3}\right)}{x^3(2) + x^3\left(\frac{9}{x}\right) + x^3\left(\frac{9}{x^2}\right)}$$
  Multiply numerator and denominator by x^3

$$= \frac{2x^3 + 3x^2 - 18x - 27}{2x^3 + 9x^2 + 9x}$$
  Distributive property

$$= \frac{x^2(2x + 3) - 9(2x + 3)}{x(x + 3)(2x + 3)}$$

$$= \frac{(x^2 - 9)(2x + 3)}{x(x + 3)(2x + 3)}$$

$$= \frac{(x + 3)(x - 3)(2x + 3)}{x(x + 3)(2x + 3)}$$

$$= \frac{x - 3}{x}$$

25. First, we simplify the numerator and denominator separately:

Numerator

$$1 + \frac{1}{x + 3} = 1 \cdot \frac{x + 3}{x + 3} + \frac{1}{x + 3}$$

$$= \frac{x + 3 + 1}{x + 3}$$

$$= \frac{x + 4}{x + 3}$$

Denominator

$$1 + \frac{7}{x - 3} = 1 \cdot \frac{x - 3}{x - 3} + \frac{7}{x - 3}$$

$$= \frac{x - 3 + 7}{x - 3}$$

$$= \frac{x + 4}{x - 3}$$

Therefore:

$$\frac{1 + \frac{1}{x + 3}}{1 + \frac{7}{x - 3}} = \frac{\frac{x + 4}{x + 3}}{\frac{x + 4}{x - 3}}$$

$$= \frac{x + 4}{x + 3} \cdot \frac{x - 3}{x + 4}$$
  Method 2, reduce

$$= \frac{x - 3}{x + 3}$$

29. Applying Method 1, we find the LCD for both fractions, which in this case is $(x + 3)(x - 3)$:

$$\frac{\dfrac{1}{x + 3} + \dfrac{1}{x - 3}}{\dfrac{1}{x + 3} - \dfrac{1}{x - 3}}$$

$$= \frac{(x + 3)(x - 3)\left(\dfrac{1}{x + 3}\right) + (x + 3)(x - 3)\left(\dfrac{1}{x - 3}\right)}{(x + 3)(x - 3)\left(\dfrac{1}{x + 3}\right) - (x + 3)(x - 3)\left(\dfrac{1}{x - 3}\right)}$$

Multiply numerator and denominator by $(x + 3)(x - 3)$

$$= \frac{x - 3 + x + 3}{x - 3 - (x + 3)}$$

$$= \frac{2x}{x - 3 - x - 3}$$

$$= \frac{2x}{-6}$$

$$= -\frac{x}{3}$$

33. First, we simplify the expression that follows the subtraction sign:

$$1 - \frac{x}{1 - \dfrac{1}{x}} = 1 - \frac{x \cdot x}{x\left(1 - \dfrac{1}{x}\right)} = 1 - \frac{x^2}{x - 1}$$

Now, we subtract by rewriting the first term, 1, with the LCD, $x - 1$:

$$1 - \frac{x^2}{x - 1} = \frac{1}{1} \cdot \frac{x - 1}{x - 1} - \frac{x^2}{x - 1} = \frac{x - 1 - x^2}{x - 1} = \frac{-x^2 + x - 1}{x - 1}$$

37. First, we simplify the numerator and denominator separately:

Numerator   Denominator

$$1 - \cfrac{1}{x + \frac{1}{2}} = 1 - \cfrac{2 \cdot 1}{2(x + \frac{1}{2})} \qquad\qquad 1 + \cfrac{1}{x + \frac{1}{2}} = 1 + \cfrac{2 \cdot 1}{2(x + \frac{1}{2})}$$

$$= 1 - \frac{2}{2x + 1} \qquad\qquad\qquad = 1 + \frac{2}{2x + 1}$$

$$= \frac{1}{1} \cdot \frac{\mathbf{2x + 1}}{\mathbf{2x + 1}} - \frac{2}{2x + 1} \qquad = \frac{1}{1} \cdot \frac{\mathbf{2x + 1}}{\mathbf{2x + 1}} + \frac{2}{2x + 1}$$

$$= \frac{2x + 1 - 2}{2x + 1} \qquad\qquad\quad = \frac{2x + 1 + 2}{2x + 1}$$

$$= \frac{2x - 1}{2x + 1} \qquad\qquad\qquad = \frac{2x + 3}{2x + 1}$$

Therefore:

$$\cfrac{1 - \cfrac{1}{x + 2}}{1 + \cfrac{1}{x + 2}} = \cfrac{\frac{2x - 1}{2x + 1}}{\frac{2x + 3}{2x + 1}} = \frac{2x - 1}{\cancel{2x + 1}} \cdot \frac{\cancel{2x + 1}}{2x + 3} = \frac{2x - 1}{2x + 3}$$

41. $\cfrac{1 - x^{-1}}{1 + x^{-1}} = \cfrac{1 - \frac{1}{x}}{1 + \frac{1}{x}}$ Remember $x^{-1} = \frac{1}{x}$

$$= \cfrac{\mathbf{x}(1 - \frac{1}{x})}{\mathbf{x}(1 + \frac{1}{x})}$$ LCD is x

$$= \cfrac{x \cdot 1 - x \cdot \frac{1}{x}}{x \cdot 1 + x \cdot \frac{1}{x}}$$ Distributive property

$$= \frac{x - 1}{x + 1}$$ Simplify

45.
$$3x + 60 = 15$$
$$3x + 60 - \mathbf{60} = 15 - \mathbf{60}$$
$$3x = -45$$

$$\frac{1}{3}(3x) = \frac{1}{3}(-45)$$

$$x = -15$$

The solution set is {-15}.

49. $10 - 2(x + 3) = x + 1$

$\qquad 10 - 2x - 6 = x + 1$

$\qquad\quad -2x + 4 = x + 1$

$\quad -2x + \mathbf{2x} + 4 = x + \mathbf{2x} + 1$

$\qquad\qquad\qquad 4 = 3x + 1$

$\qquad\qquad 4 - \mathbf{1} = 3x + 1 - \mathbf{1}$

$\qquad\qquad\qquad 3 = 3x$

$$\frac{1}{3}(3) = \frac{1}{3}(3x)$$

$$1 = x$$

The solution set is {1}.

53.
$$3x^2 + x - 10 = 0$$
$$(x + 2)(3x - 5) = 0$$

$x + 2 = 0 \qquad$ or $\qquad 3x - 5 = 0$

$\qquad x = -2 \qquad\qquad\qquad 3x = 5$

$$x = \frac{5}{3}$$

The solution set is $\{-2, \frac{5}{3}\}$.

57. Method 1

$$\frac{(x + 3)(x - 3)(1) + (x + 3)(x - 3)(\frac{1}{x - 3})}{(x + 3)(x - 3)(x) - (x + 3)(x - 3)(\frac{10}{x + 3})}$$

$$= \frac{(x + 3)(x - 3) + (x + 3)}{(x + 3)(x - 3)x - (x - 3)10}$$

$$= \frac{x^2 - 9 + x + 3}{x^3 - 9x - 10x + 30}$$

$$= \frac{x^2 + x - 6}{x^3 - 19x + 30}$$

$$= \frac{(x + 3)(x - 2)}{(x + 5)(x - 3)(x - 2)}$$

$$= \frac{x + 3}{(x + 5)(x - 3)}$$

Problem number 57 continued on next page.

Section 4.5 continued

57. continued

Method 2

$$\frac{\dfrac{(x - 3)}{(x - 3)}(1) + \left(\dfrac{1}{x - 3}\right)}{\dfrac{(x + 3)}{(x + 3)}(x) - \left(\dfrac{10}{x + 3}\right)}$$

$$= \frac{\dfrac{x - 3 + 1}{x - 3}}{\dfrac{x^2 + 3x - 10}{x + 3}}$$

$$= \frac{\dfrac{x - 2}{x - 3}}{\dfrac{(x + 5)(x - 2)}{x + 3}}$$

$$= \frac{x - 2}{x - 3} \cdot \frac{x + 3}{(x + 5)(x - 2)}$$

$$= \frac{x + 3}{(x - 3)(x + 5)}$$

As you can see, Method 2 is easier.

Section 4.6

1. $\dfrac{x}{5} + 4 = \dfrac{5}{3}$

Solution: The LCD for 5 and 3 is 15. Multiplying both sides by 15, we have

$$15\left(\frac{x}{5} + 4\right) = 15\left(\frac{5}{3}\right)$$

$$15\left(\frac{x}{5}\right) + 15(4) = 15\left(\frac{5}{3}\right)$$

$$3x + 60 = 25$$

$$3x = -35$$

$$x = -\frac{35}{3}$$

The solution set is $\left\{-\dfrac{35}{3}\right\}$, which checks in the original equation.

5. $\frac{y}{2} + \frac{y}{4} + \frac{y}{6} = 3$

Solution: The LCD for 2, 4 and 6 is 12. Multiplying both sides by 12, we have

$$12\left(\frac{y}{2} + \frac{y}{4} + \frac{y}{6}\right) = 12(3)$$
$$6y + 3y + 2y = 36$$
$$11y = 36$$
$$y = \frac{36}{11}$$

The solution set is $\{\frac{36}{11}\}$, which checks in the original equation.

9. $\frac{1}{x} = \frac{1}{3} - \frac{2}{3x}$

Solution: The LCD is 3x. We are assuming $x \neq 0$ when we multiply both sides of the equation by 3x.

$$3x\left(\frac{1}{x}\right) = 3x\left(\frac{1}{3} - \frac{2}{3x}\right)$$
$$3 = x - 2$$
$$5 = x$$

The solution set is $\{5\}$, which checks in the original equation.

13. $1 - \frac{1}{x} = \frac{12}{x^2}$

Solution: To clear the equation of denominators, we multiply both sides by x^2:

$$x^2\left(1 - \frac{1}{x}\right) = x^2\left(\frac{12}{x^2}\right)$$
$$x^2 - x = 12$$

Rewrite in standard form and solve:

$$x^2 - x - 12 = 0$$
$$(x + 3)(x - 4) = 0$$

$$x + 3 = 0 \qquad \text{or} \qquad x - 4 = 1$$
$$x = -3 \qquad\qquad\qquad x = 4$$

The solution set is $\{-3,4\}$. Remember: We must check all solutions any time we multiply both sides of the equation by an expression that contains the variable.

17. $\dfrac{x + 2}{x + 1} = \dfrac{1}{x + 1} + 2$

Solution: To clear the equation of deonominators, we multiply both sides by x + 1:

$$(x + 1)\left(\dfrac{x + 2}{x + 2}\right) = x + 1\left(\dfrac{1}{x + 1} + 2\right)$$

$$x + 2 = 1 + 2(x + 1)$$
$$x + 2 = 1 + 2x + 2$$
$$x + 2 = 2x + 3$$
$$2 = x + 3$$
$$-1 = x$$

The first and second terms are undefined. The proposed solution, x = -1, does not check in the original solution. The solution set is the empty set.

21. $6 - \dfrac{5}{x^2} = \dfrac{7}{x}$

Solution: The LCD is x^2. To clear the denominators, we multiply both sides of the equation by x^2:

$$x^2\left(6 - \dfrac{5}{x^2}\right) = x^2\left(\dfrac{7}{x}\right)$$

$$6x^2 - 5 = 7x$$

Rewrite in standard form and solve:

$$6x^2 - 7x - 5 = 0$$
$$(2x + 1)(3x - 5) = 1$$

$$2x + 1 = 0 \qquad \text{or} \qquad 3x - 5 = 0$$
$$2x = -1 \qquad\qquad\qquad 3x = 5$$
$$x = -\dfrac{1}{2} \qquad\qquad\qquad x = \dfrac{5}{3}$$

The solution set $\{-\dfrac{1}{2}, \dfrac{5}{3}\}$ does check in the original equation.

25. $\dfrac{x}{x - 3} + \dfrac{x}{x^2 - 9} = \dfrac{4}{x + 3}$

Solution: Factoring each denominator, we find the LCD is $(x - 3)(x + 3)$. Multiplying each side of the equation by the LCD clears the equation of denominators and leads us to our possible solutions:

$$(x - 3)(x + 3) \cdot \dfrac{2}{x - 3} + (x - 3)(x + 3) \cdot \dfrac{x}{x^2 - 9}$$

$$= (x - 3)(x + 3)\dfrac{4}{x + 3}$$

$$2(x + 3) + x = 4(x - 3)$$
$$2x + 6 + x = 4x - 12$$
$$3x + 6 = 4x - 12$$
$$6 = x - 12$$
$$18 = x$$

The solution set is {18} which checks in the original equation.

29. $\dfrac{t - 4}{t^2 - 3t} = \dfrac{-2}{t^2 - 9}$

Solution: Writing the equation again with the denominators in factored form, we have

$$\dfrac{t - 4}{t(t - 3)} = \dfrac{-2}{(t + 3)(t - 3)}$$

The LCD is $t(t + 3)(t - 3)$. Multiplying through by the LCD, we have

$$t(t + 3)(t - 3) \div \dfrac{t - 4}{t(t - 3)} = t(t + 3)(t - 3) \cdot \dfrac{-2}{(t + 3)(t - 3)}$$

$$(t + 3)(t - 4) = t(-2)$$
$$t^2 - t - 12 = -2t \qquad \text{Multiplying both sides}$$
$$t^2 + t - 12 = 0 \qquad \text{Standard form}$$
$$(t + 4)(t - 3) = 0$$

$$t + 4 = 0 \qquad \text{or} \qquad t - 3 = 0$$
$$t = -4 \qquad\qquad\qquad t = 3$$

The two possible solutions are -4 and 3. If we substitute -4 for t in the original equation, we find that it leads to a true statement. It is therefore a solution. If we substitute 3 for t in the original equation, we find both sides of the equation are undefined. The only solution to our original equation is t = -4. The other possible solution t = 3 is extraneous.

116

33. $\dfrac{2}{1 + a} = \dfrac{3}{1 - a} + \dfrac{5}{a}$

Solution: The LCD is $a(1 + a)(1 - a)$. Multiplying both sides by this quantity yields

$$a(1 + a)(1 - a) \cdot \dfrac{2}{1 + a}$$

$$= a(1 + a)(1 - a) \cdot \dfrac{3}{1 - a} + a(1 + a)(1 - a) \cdot \dfrac{5}{a}$$

$$a(1 - a) \cdot 2 = a(1 + a) \cdot 3 + (1 + a)(1 - a)5$$
$$2a - 2a^2 = 3a + 3a^2 + 5 - 5a^2$$
$$2a - 2a^2 = 5 + 3a - 2a^2$$

$$2a = 5 + 3a \qquad \text{Add } 2a^2 \text{ to both sides}$$
$$-a = 5$$
$$a = -5 \qquad \text{Multiply both sides by -1}$$

The solution set is $\{-5\}$, which checks in the original equation.

37. $\dfrac{y + 2}{y^2 - y} - \dfrac{6}{y^2 - 1} = 0$

Factoring the denominators, we have

$$\dfrac{y + 2}{y(y - 1)} - \dfrac{6}{(y + 1)(y - 1)} = 0$$

The LCD is $y(y + 1)(y - 1)$. Multiplying both sides by this quantity yields

$$\cancel{y}(y + 1)\cancel{(y - 1)} \cdot \dfrac{y + 2}{\cancel{y}\cancel{(y - 1)}} - y\cancel{(y + 1)}\cancel{(y - 1)} \cdot \dfrac{6}{\cancel{(y + 1)}\cancel{(y - 1)}}$$

$$= y(y + 1)(y - 1) \cdot 0$$
$$(y + 1)(y + 2) - y(6) = 0$$
$$y^2 + 3y + 2 - 6y = 0$$
$$y^2 - 3y + 2 = 0 \qquad \text{Standard form}$$
$$(y - 2)(y - 1) = 0$$

$$y - 2 = 0 \quad \text{or} \quad y - 1 = 0$$
$$y = 2 \qquad\qquad y = 1$$

The two possible solutions are 2 and 1. If we substitute 2 for y in the original equation, we find that it leads to a true statement. It is therefore a solution. If we substitute 1 for y in the original equation, we find the left side of the equation becomes undefined. The only solution to our original equation is $y = 2$. The other possible solution $y = 1$ is extraneous.

41. $\dfrac{2}{y^2 - 7y + 12} - \dfrac{1}{y^2 - 9} = \dfrac{4}{y^2 - y - 12}$

Solution: Writing the equation again with the denominators in factored form, we have

$$\frac{2}{(y - 4)(y - 3)} - \frac{1}{(y + 3)(y - 3)} = \frac{4}{(y + 3)(y - 4)}$$

The LCD is $(y - 4)(y - 3)(y + 3)$. Multiplying both sides by this quantity yields

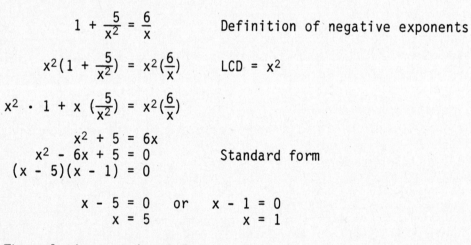

$$(y + 3) \cdot 2 - (y - 4) \cdot 1 = (y - 3) \cdot 4$$
$$2y + 6 - y + 4 = 4y - 12$$
$$y + 10 = 4y - 12$$
$$10 = 3y - 12$$
$$22 = 3y$$

$$\frac{22}{3} = y$$

The solution set is $\{\frac{22}{3}\}$.

45. $\qquad 1 + 5x^{-2} = 6x^{-1}$

$\qquad 1 + \dfrac{5}{x^2} = \dfrac{6}{x}$ $\qquad$ Definition of negative exponents

$\qquad x^2\left(1 + \dfrac{5}{x^2}\right) = x^2\left(\dfrac{6}{x}\right)$ $\qquad$ LCD $= x^2$

$x^2 \cdot 1 + x\left(\dfrac{5}{x^2}\right) = x^2\left(\dfrac{6}{x}\right)$

$\qquad\qquad x^2 + 5 = 6x$
$\qquad x^2 - 6x + 5 = 0$ $\qquad$ Standard form
$(x - 5)(x - 1) = 0$

$\qquad x - 5 = 0 \quad$ or $\quad x - 1 = 0$
$\qquad\qquad x = 5 \qquad\qquad x = 1$

The solution set is $\{1, 5\}$.

49. Solve $\frac{1}{R} = \frac{1}{R_1} + \frac{1}{R_2}$ for R

Solution: Multiplying for sides by the least common denominator RR_1R_2, we have

$$RR_1R_2 \cdot \frac{1}{R} = RR_1R_2 \cdot \frac{1}{R_1} + RR_1R_2 \cdot \frac{1}{R_2}$$

$$R_1R_2 = RR_2 + RR_1$$

$$R_1R_2 = (R_2 + R_1)R \qquad \text{Factor R from the right side}$$

$$\frac{R_1R}{R_1 + R_2} = R$$

We know we are finished because the variable we were solving for is alone on one side of the equation and does not appear on the other side.

53. $x = \frac{2y + 1}{3y + 1}$

Multiply both sides by $3y + 1$.

$$(3y + 1)x = (3y + 1)\left(\frac{2y + 1}{3y + 1}\right)$$

$$3xy + x = 2y + 1$$
$$3xy - 2y = 1 - x \qquad\qquad \text{Combined y values on the left side}$$
$$y(3x - 2) = 1 - x$$

$$\frac{y(3x - 2)}{3x - 2} = \frac{1 - x}{3x - 2}$$

$$y = \frac{1 - x}{3x - 2}$$

57. Step 1: Let x = the width of the rectangle.

Step 2: The length is 3 less than twice the width or 2x - 3.

Step 3: The perimeter is 42 meters. Remember: P = 2L + 2W, thus we have

$$42 = 2(2x - 3) + 2x$$

Step 4: Solve $\quad 42 = 2(2x - 3) + 2x$
$$42 = 4x - 6 + 2x$$
$$42 = 6x - 6$$
$$48 = 6x$$
$$8 = x$$

The width is 8 meters. The length is 13 meters.

Step 5: Since the sum of twice 8 plus twice 13 is 42, the solution set checks in the original problem.

119

Section 4.6 continued

61. Solution: Let x = first integer (shortest side)
Then x + 1 = next consecutive integer
x + 2 = last consecutive integer (longest side)

By the Pythagorean Theorem, we have

$$(x + 2)^2 = (x + 1)^2 + x^2$$
$$x^2 + 4x + 4 = x^2 + 2x + 1 + x^2$$
$$x^2 - 2x - 3 = 0$$
$$(x - 3)(x + 1) = 0$$

$$x = 3 \quad \text{or} \quad x = -1$$

The shortest side is 3. The other two sides are 4 and 5.

65.
$$\frac{1}{x^3} - \frac{1}{3x^2} - \frac{1}{4x} + \frac{1}{12} = 0$$

$$12x^3\left(\frac{1}{x^3}\right) - 12x^3\left(\frac{1}{3x^2}\right) - 12x^3\left(\frac{1}{4x}\right) + 12x^3\left(\frac{1}{12}\right) = 0 \qquad \text{LCD} = 12x^3$$

$$12 - 4x - 3x^2 + x^3 = 0$$
$$4(3 - x) - x^2(3 - x) = 0$$
$$(4 - x^2)(3 - x) = 0$$
$$(2 + x)(2 - x)(3 - x) = 0$$

$$2 + x = 0 \quad \text{or} \quad 2 - x = 0 \quad \text{or} \quad 3 - x = 0$$
$$x = -2 \qquad\qquad 2 = x \qquad\qquad 3 = x$$

120

1. Let x = the smaller number. The larger number is 3x. Their reciprocals are $\frac{1}{x}$ and $\frac{1}{3x}$. The equation is

$$\frac{1}{x} + \frac{1}{3x} = \frac{20}{3}$$

Multiply both sides by the LCD 3x, we have

$$3x \cdot \frac{1}{x} + 3x \cdot \frac{1}{3x} = 3x \cdot \frac{20}{3}$$
$$3 + 1 = 20x$$
$$4 = 20x$$
$$\frac{4}{20} = x$$
$$x = \frac{1}{5}$$

The smaller number is $\frac{1}{5}$. The larger is $3(\frac{1}{5}) = \frac{3}{5}$. Adding their reciprocals, we have

$$\frac{5}{1} + \frac{5}{3} = \frac{15}{3} + \frac{5}{3} = \frac{20}{3}$$

The sum of the reciprocals of $\frac{1}{5}$ and $\frac{3}{5}$ is $\frac{20}{3}$.

5. Let x = the first integer. The second consecutive integer is x + 1. Their reciprocals are $\frac{1}{x}$ and $\frac{1}{x+1}$. The equation is

$$\frac{1}{x} + \frac{1}{x + 1} = \frac{7}{12}$$

Multiply both sides by the LCD 12x(x + 1), we have

$$12x(x + 1) \cdot \frac{1}{x} + 12x(x + 1) \cdot \frac{1}{x + 1} = 12x(x + 1) \cdot \frac{7}{12}$$
$$12(x + 1) + 12x = 7x(x + 1)$$
$$12x + 12 + 12x = 7x^2 + 7x$$
$$24x + 12 = x^2 + x$$
$$0 = 7x^2 - 17x - 12 \qquad \text{Standard form}$$
$$0 = (7x + 4)(x - 3)$$

$$7x + 4 = 0 \qquad \text{or} \qquad x - 3 = 0$$
$$7x = -4 \qquad\qquad\qquad x = 3$$
$$x = -\frac{4}{7}$$

The problem asks for integers and $-\frac{4}{7}$ is a fraction, therefore $-\frac{4}{7}$ is is an extraneous root. The solution x = 3 is an integer, therefore x = 3 and x + 1 = 4.

9. Let x = current

	d	r	$t = \dfrac{d}{r}$
Upstream	1.5	5 - x	$\dfrac{1.5}{5 - x}$
Downstream	3	5 + x	$\dfrac{3}{5 + x}$

Remember d = rt

$$\frac{d}{r} = t$$

Since the time moving upstream is equal to the time moving downstream, or

$$\frac{1.5}{5 - x} = \frac{3}{5 + x}$$

Multiplying both sides by the LCD (5 - x)(5 + x) gives

$$(5 + x)(1.5) = 3(5 - x)$$
$$7.5 + 1.5x = 15 - 3x$$
$$7.5 + 4.5x = 15$$
$$4.5x = 7.5$$
$$x = \frac{7.5}{4.5} = \frac{75}{45} = \frac{5}{3}$$

The speed of the current is $\frac{5}{3}$ miles/hour.

13. Let x = speed of train B.

	d	r	$t = \dfrac{d}{r}$
train A	150	x + 15	$\dfrac{150}{x + 15}$
train B	120	x	$\dfrac{120}{x}$

Since the times of both trains are the same, we have

$$\frac{150}{x + 15} = \frac{120}{x}$$

Multiplying both sides by the LCD x(x + 15) gives

$$\begin{aligned}
150x &= 120(x + 15) \\
150x &= 120x + 1800 \\
30x &= 1800 \\
x &= 60
\end{aligned}$$

The speed of train B is 60 miles/hour. Train A is 15 miles/hour greater which is 75 miles/hour.

17. Let x - one person's work. The other person works twice as fast = 2x. Since one worker wants to do the job alone, we must subtract the other worker's work.

In 1 hour

$$\begin{bmatrix} \text{Amount of work} \\ \text{by faster person} \end{bmatrix} - \begin{bmatrix} \text{Amount of work} \\ \text{by slower person} \end{bmatrix} = \begin{bmatrix} \text{Total amount} \\ \text{of work done} \end{bmatrix}$$

$$\frac{1}{2x} \quad + \quad \frac{1}{x} \quad = \quad \frac{1}{6}$$

Multiplying through by the LCD 12x, we have

$$12x \cdot \frac{1}{2x} + 12x \cdot \frac{1}{x} = 12x \cdot \frac{1}{6}$$

$$6 + 12 = 2x$$
$$18 = 2x$$
$$9 = x$$

The slower person can do it in 9 hours and the faster person can do it in $4\frac{1}{2}$ hours but the question asks how long for the faster person doing it alone. To answer this we must

$$\begin{bmatrix} \text{faster person's} \\ \text{work} \end{bmatrix} + \begin{bmatrix} \text{half of} \\ \text{slower person's work} \end{bmatrix} = [\text{total job}]$$

$$4\frac{1}{2} \text{ hrs} \quad + \quad 4\frac{1}{2} \text{ hrs} \quad = \quad 9 \text{ hrs}$$

It would take the faster person 9 hours alone to do the job.

21. Let x = time to fill a full pool

In 1 hour

$$\begin{bmatrix} \text{Amount full by} \\ \text{inlet pipe} \end{bmatrix} - \begin{bmatrix} \text{Amount empty by} \\ \text{outlet pipe} \end{bmatrix} = \begin{bmatrix} \text{Fraction of pool} \\ \text{filled by both} \end{bmatrix}$$

$$\frac{1}{10} \quad - \quad \frac{1}{15} \quad = \quad \frac{1}{x}$$

Multiplying by the LCD 30x, we have

$$30x \cdot \frac{1}{10} - 30x \cdot \frac{1}{15} = 30x \cdot \frac{1}{x}$$

$$3x - 2x = 30$$
$$x = 30$$

Since x = the time to fill a pool, then since the pool is half-full, we would take one-half of x.

If x = 30 hours

Then $\frac{x}{2} = \frac{30}{2}$ = 15 hours

It would take 15 hours to fill the pool.

25. $(x^2 - 9)\left(\frac{x + 2}{x + 3}\right)$

$$= \frac{x^2 - 9}{1} \cdot \frac{x + 2}{x + 3} \qquad \text{Write } x^2 - 9 \text{ with denominator 1}$$

$$= \frac{\cancel{(x + 3)}(x - 3)(x + 2)}{\cancel{(x + 3)}} \qquad \text{Factor}$$

$$= (x - 3)(x - 2) \qquad \text{Divide out common factors}$$

29. $\dfrac{\dfrac{1}{x} - \dfrac{1}{3}}{\dfrac{1}{x} + \dfrac{1}{3}}$ $\qquad$ LCD = 3x

$$= \frac{3x\left(\frac{1}{x} - \frac{1}{3}\right)}{3x\left(\frac{1}{x} + \frac{1}{3}\right)}$$

$$= \frac{(3x)\frac{1}{x} - (3x)\frac{1}{3}}{(3x)\frac{1}{x} + (3x)\frac{1}{3}} \qquad \text{Apply the distributive property to distribute 3x over both terms in the numerator and denominator}$$

$$= \frac{3 - x}{3 + x}$$

Chapter 4 Review

1. $\dfrac{\overset{25}{\cancel{125}}x^4 y \overset{1}{\cancel{z^3}}}{\underset{7}{\cancel{35}}x^2 y^4 \underset{1}{\cancel{z^3}}} = \dfrac{25x^2}{7y^3}$

5. $\dfrac{x^2 - 25}{x^2 + 10x + 25} = \dfrac{\overset{1}{\cancel{(x + 5)}}(x - 5)}{\underset{1}{\cancel{(x + 5)}}(x + 5)} = \dfrac{x - 5}{x + 5}$

9. $\dfrac{6 - x - x^2}{x^2 - 5x + 6} = \dfrac{-1(x^2 + x - 6)}{x^2 - 5x + 6}$ 　　Note: $6 - x - x^2 = -1(-6 + x + x^2)$
$= -1(x^2 + x - 6)$

$= \dfrac{-1(x + 3)\overset{1}{\cancel{(x - 2)}}}{(x - 3)\underset{1}{\cancel{(x - 2)}}}$

$= \dfrac{-1(x + 3)}{x - 3}$

$= -\dfrac{x + 3}{x - 3}$

13. $\dfrac{27a^2b^3 - 15a^3b^2 + 21a^4b^4}{-3a^2b^2}$

$= \dfrac{27a^2b^3}{-3a^2b^2} - \dfrac{15a^3b^2}{-3a^2b^2} + \dfrac{21a^4b^4}{-3a^2b^2}$

$= -9b + 5a - 7a^2b^2$

17. $\dfrac{x^2 - x - 6}{x - 3} = \dfrac{\overset{1}{\cancel{(x - 3)}}(x + 2)}{\underset{1}{\cancel{(x - 3)}}}$

$= x + 2$

21. $\dfrac{y^4 - 16}{y - 2} = \dfrac{(y^2 + 4)(y^2 - 4)}{y - 2}$

$= \dfrac{(y^2 + 4)(y + 2)\overset{1}{\cancel{(y - 2)}}}{\underset{1}{\cancel{y - 2}}}$

$= (y^2 + 4)(y + 2)$

$= y^3 + 2y^2 + 4y + 8$

25.

$$
\begin{array}{r}
y^2 - 3y \quad - 13 \\
2y - 3 \overline{)\ 2y^3 - 2y^2 - 17y + 39} \\
\end{array}
$$

$$
\begin{array}{c}
\quad - \quad + \\
\neq 2y^3 \neq 3y^2 \\
\hline
\quad - 6y^2 - 17y \\
\quad + \quad - \\
\neq 6y^2 \neq 9y \\
\hline
\quad - 26y + 39 \\
\quad + \quad - \\
\neq 26y \neq 39 \\
\hline
\quad 0
\end{array}
$$

Change signs

Change signs

Change signs

29.

$$
\begin{array}{r}
4x \quad + \quad 6 \\
x^2 - 5x + 6 \overline{)\ 4x^3 - 14x^2 - \ \ 6x + 36} \\
\end{array}
$$

$$
\begin{array}{c}
\quad - \quad + \quad - \\
\neq 4x^3 \neq 20x^2 \neq 24x \\
\hline
\quad 6x^2 - 30x + 36 \\
\quad - \quad + \quad - \\
\neq 6x^2 \neq 30x \neq 36 \\
\hline
\quad 0
\end{array}
$$

Change signs

Change signs

33.

$$\frac{3}{4} \cdot \frac{12}{15} \div \frac{1}{3} = \frac{3}{4} \cdot \frac{12}{15} \cdot \frac{3}{1}$$

Write division in terms of multiplication

$$= \frac{3(12)3}{4(15)1}$$

Multiply numerators and denominators

$$= \frac{\overset{1}{\cancel{3}} \cdot \overset{1}{\cancel{2}} \cdot \overset{1}{\cancel{2}} \cdot 3 \cdot 3}{\underset{1}{\cancel{2}} \cdot \underset{1}{\cancel{2}} \cdot \underset{1}{\cancel{3}} \cdot 5 \cdot 1}$$

Factor

$$= \frac{9}{5}$$

Divide out common factors

37.

$$\frac{x^3 - 1}{x^4 - 1} \quad \frac{x^2 - 1}{x^2 + x + 1}$$

$$= \frac{(x - 1)\cancel{(x^2 + x + 1)}\overset{1}{\cancel{(x^2 - 1)}}}{(x^2 + 1)\underset{1}{\cancel{(x^2 - 1)}}\underset{1}{\cancel{(x^2 + x + 1)}}}$$

Factor and multiply

$$= \frac{x - 1}{x^2 + 1}$$

41. $\dfrac{ax + bx + 2a + 2b}{ax - 3a + bx - 3b} \div \dfrac{ax - bx - 2a + 2b}{ax - bx - 3a + 3b}$

$\qquad = \dfrac{ax + bx + 2a + 2b}{ax - 3a + bx - 3b} \cdot \dfrac{ax - bx - 2a + 2b}{ax - bx - 3a + 3b}$ Change division and multiplication by the reciprocal

$\qquad = \dfrac{x(a + b) + 2(a + b)}{a(x - 3) + b(x - 3)} \cdot \dfrac{x(a - b) - 3(a - b)}{x(a - b) - 2(a - b)}$

$\qquad = \dfrac{(x + 2)\cancel{(a + b)}\cancel{(x - 3)}\cancel{(a - b)}}{\cancel{(a + b)}\cancel{(x - 3)}(x - 2)\cancel{(a - b)}}$ Factor by grouping

$\qquad = \dfrac{x + 2}{x - 2}$

45. $\dfrac{2x^3 + 3x^2 - 18x - 27}{10x^2 + 13x - 3} \cdot \dfrac{25x^2 - 1}{5x^2 - 14x - 3}$

$\qquad = \dfrac{x^2(2x + 3) - 9(2x + 3)}{(2x + 3)(5x - 1)} \cdot \dfrac{(5x + 1)(5x - 1)}{(5x + 1)(x - 3)}$

$\qquad = \dfrac{(x + 3)\cancel{(x - 3)}\cancel{(2x + 3)}\cancel{(5x + 1)}\cancel{(5x - 1)}}{\cancel{(2x + 3)}\cancel{(5x - 1)}\cancel{(5x + 1)}\cancel{(x - 3)}}$ Factor and multiply

$\qquad = x + 3$

49. $\dfrac{5}{x - 5} - \dfrac{x}{x - 5} = \dfrac{5 - x}{x - 5}$

$\qquad = \dfrac{-1\cancel{(x - 5)}}{\cancel{x - 5}}$ Note: $5 - x = -1(-5 + x) = -1(x\ 5)$

$\qquad = -1$

53. $\dfrac{x - 2}{x^2 + 5x + 4} - \dfrac{x - 4}{2x^2 + 12x + 16}$

$= \dfrac{x - 2}{(x + 1)(x + 4)} - \dfrac{x - 4}{2(x + 4)(x + 2)}$

$= \dfrac{x - 2}{(x + 1)(x + 4)} \cdot \dfrac{\mathbf{2(x + 2)}}{\mathbf{2(x + 2)}} - \dfrac{x - 4}{2(x + 4)(x + 2)} \cdot \dfrac{\mathbf{x + 1}}{\mathbf{x + 1}}$ LCD is $2(x + 1)(x + 2)(x + 4)$

$= \dfrac{2x^2 - 8}{2(x + 1)(x + 2)(x + 4)} - \dfrac{x^2 - 3x - 4}{2(x + 1)(x + 2)(x + 4)}$

$= \dfrac{2x^2 - 8 - x^2 + 3x + 4}{2(x + 1)(x + 2)(x + 4)}$

$= \dfrac{x^2 + 3x - 4}{2(x + 1)(x + 2)(x + 4)}$

$= \dfrac{(x + 4)(x - 1)}{2(x + 1)(x - 2)(x + 4)}$

$= \dfrac{x - 1}{2(x + 1)(x - 2)}$

57. $3 + \dfrac{4}{5x - 2}$

$= \dfrac{3}{1} + \dfrac{4}{5x - 2}$

$= \dfrac{3}{1} \cdot \dfrac{\mathbf{5x - 2}}{\mathbf{5x - 2}} + \dfrac{4}{5x - 2}$

$= \dfrac{15x - 6 + 4}{5x - 2}$

$= \dfrac{15x - 2}{5x - 2}$

61. $\dfrac{5}{3x^2 + x - 2} - \dfrac{1}{3x^2 + 5x + 2}$

$$= \dfrac{5}{(x + 1)(3x - 2)} - \dfrac{1}{(3x + 2)(x + 1)}$$

$$= \dfrac{3x + 2}{3x + 2} \cdot \dfrac{5}{(x + 1)(3x - 2)} - \dfrac{3x - 2}{3x - 2} \cdot \dfrac{1}{(3x + 2)(x + 1)}$$

$$= \dfrac{15x + 10}{(3x + 2)(x + 1)(3x - 2)} - \dfrac{3x - 2}{(3x - 2)(3x + 2)(x + 1)}$$

$$= \dfrac{12x + 12}{(3x + 2)(3x - 2)(x + 1)}$$

$$= \dfrac{12\overset{1}{\cancel{(x + 1)}}}{(3x + 2)(3x - 2)\underset{1}{\cancel{(x + 1)}}}$$

$$= \dfrac{12}{(3x + 2)(3x - 2)}$$

65. $\dfrac{1 + \frac{2}{3}}{1 - \frac{2}{3}} = \dfrac{3\left(1 + \frac{2}{3}\right)}{3\left(1 - \frac{2}{3}\right)}$ Multiply numerator and denominator by 3

$$= \dfrac{3 \cdot 1 + 3 \cdot \frac{2}{3}}{3 \cdot 1 - 3 \cdot \frac{2}{3}}$$ Distributive property

$$= \dfrac{3 + 2}{3 - 2}$$ Simplify

$$= \dfrac{5}{1}$$

$$= 5$$ Reduce

69. $1 + \dfrac{1}{x + \frac{1}{x}} = 1 + \dfrac{x \cdot 1}{x\left(x + \frac{1}{x}\right)}$

$$= 1 + \dfrac{x}{x^2 + 1}$$

$$= \dfrac{1}{1} \cdot \dfrac{x^2 + 1}{x^2 + 1} + \dfrac{x}{x^2 + 1}$$

$$= \dfrac{x^2 + x + 1}{x^2 + 1}$$

73.
$$\frac{3}{x - 1} = \frac{3}{5} \qquad \text{LCD is } 5(x - 1)$$

$$5(x - 1)\frac{3}{x - 1} = 5(x - 1)\frac{3}{5}$$

$$5 \cdot 3 = (x - 1)3$$
$$15 = 3x - 3$$
$$18 = 3x$$
$$x = 6$$

The solution set is {6}, which checks in the original equation.

77.
$$\frac{5}{y + 1} = \frac{4}{y + 2} \qquad \text{LCD is } (y + 1)(y + 2)$$

$$(y + 1)(y + 2)\frac{5}{y + 1} = (y + 1)(y + 2)\frac{4}{y + 2}$$

$$5(y + 2) = 4(y + 1)$$
$$5y + 10 = 4y + 4$$
$$y = -6$$

The solution set is {-6}, which checks in the original equation.

81.
$$\frac{4}{x^2 - x - 12} + \frac{1}{x^2 - 9} = \frac{2}{x^2 - 7x + 12}$$

$$\frac{4}{(x - 4)(x + 3)} + \frac{1}{(x + 3)(x - 3)} = \frac{2}{(x - 3)(x - 4)} \qquad \begin{array}{l}\text{LCD is} \\ (x - 4)(x + 3)(x - 3)\end{array}$$

$$(x - 4)(x + 3)(x - 3)\frac{4}{(x - 4)(x + 3)} + (x - 4)(x + 3)(x - 3)\frac{1}{(x + 3)(x - 3)}$$

$$= (x - 4)(x + 3)(x - 3)\frac{2}{(x - 3)(x - 4)}$$

$$4(x - 3) + 1(x - 4) = 2(x + 3)$$
$$4x - 12 + x - 4 = 2x + 6$$
$$5x - 16 = 2x + 6$$
$$3x = 22$$
$$x = \frac{22}{3}$$

The solution set is $\{\frac{22}{3}\}$ which checks in the original equation.

85.

$$\frac{3x}{x - 5} - \frac{2x}{x + 1} = \frac{-42}{x^2 - 4x - 5}$$

$$\frac{3x}{x - 5} - \frac{2x}{x + 1} = \frac{-42}{(x - 5)(x + 1)} \qquad \text{LCD is } (x - 5)(x + 1)$$

$$(x - 5)(x + 1)\frac{3x}{x - 5} - (x - 5)(x + 1)\frac{2x}{x + 1} = (x - 5)(x + 1)\frac{-42}{(x - 5)(x + 1)}$$

$$3x(x + 1) - 2x(x - 5) = -42$$
$$3x^2 + 3x - 2x^2 + 10x = -42$$
$$x^2 + 13x + 42 = 0$$
$$(x + 6)(x + 7) = 0$$

$$x + 6 = 0 \quad \text{or} \quad x + 7 = 0$$
$$x = -6 \qquad\qquad x = -7$$

The solution set is {-6,-7}, which checks in the original equation.

89. Let x = one number and 2x = another number. Their reciprocals are $\frac{1}{x}$ and $\frac{1}{2x}$, then

$$\frac{1}{x} + \frac{1}{2x} = \frac{1}{2}$$

$$2x\left(\frac{1}{x}\right) + 2x\left(\frac{1}{2x}\right) = 2x\left(\frac{1}{2}\right) \qquad \text{LCD is } 2x$$

$$2 + 1 = x$$
$$3 = x$$
$$6 = 2x$$

The numbers are 3 and 6.

93. Let x = the speed of the boat

	d	r	$t = \frac{d}{r}$
up river	26	x - 3	$\frac{26}{x - 3}$
down river	38	x + 3	$\frac{38}{x + 3}$

Remember d = rt

$$\frac{d}{r} = t$$

Since the time moving up river is equal to the time moving down river, or

$$\frac{26}{x - 3} = \frac{38}{x + 3}$$

Multiplying both sides by the LCD (x - 3)(x + 3) gives

$$26(x + 3) = 38(x - 3)$$
$$26x + 78 = 38x - 114$$
$$192 = 12x$$
$$16 = x$$

The speed of the boat is 16 mph.

97.   Let x = time to fill the tub

$$\left[\begin{array}{c}\text{Amount to fill}\\\text{by cold water}\end{array}\right] + \left[\begin{array}{c}\text{Amount to fill}\\\text{by hot water}\end{array}\right] = \left[\begin{array}{c}\text{Fraction of tub}\\\text{filled by both}\end{array}\right]$$

$$\frac{1}{10} + \frac{1}{12} = \frac{1}{x}$$

Multiplying by the LCD 60x, we have

$$60x \cdot \frac{1}{10} + 60x \cdot \frac{1}{12} = 60x \cdot \frac{1}{x}$$

$$6x + 5x = 60$$
$$11x = 60$$

$$x = \frac{60}{11}$$

It will take $\frac{60}{11}$ minutes to fill the tub.

101.   Let x = speed of the current

	d = rt	r	t
Upstream	6	9 - x	$\frac{6}{9-x}$
Downstream	6	9 + x	$\frac{6}{9+x}$

$$t = \frac{d}{r}$$

The total time for the trip up and back is 1.5 hours:

Time upstream + Time downstream = Total time

$$\frac{6}{9-x} + \frac{6}{9+x} = 1.5$$

Multiplying both sides by (9 - x)(9 + x), we have

$$6(9 + x) + 6(9 - x) = 1.5(81 - x^2)$$
$$54 + 6x + 54 - 6x = 121.5 - 1.5x^2$$
$$108 = 121.5 - 1.5x^2$$
$$1.5x^2 - 13.5 = 0$$
$$15x^2 - 135 = 0$$
$$15(x^2 - 9) = 0$$
$$15(x + 3)(x - 3) = 0$$

$$x = -3 \quad \text{or} \quad x = 3$$

The speed of the current is 3 mph.

Chapter 4 Test

1. $\dfrac{x^2 - y^2}{x - y} = \dfrac{(x + y)(x - y)}{x - y} = x + y$

5. $\dfrac{a^2 - 16}{5a - 15} \cdot \dfrac{10(a - 3)^2}{a^2 - 7a + 12}$

$\quad = \dfrac{(a + 4)(a - 4)}{5(a - 3)} \cdot \dfrac{\overset{2}{10}(a - 3)(a - 3)}{(a - 4)(a - 3)}$

$\quad = 2(a + 4)$

9. $\dfrac{3}{4} - \dfrac{1}{2} + \dfrac{5}{8} = \dfrac{3}{4} \cdot \dfrac{2}{2} - \dfrac{1}{2} \cdot \dfrac{4}{4} + \dfrac{5}{8}$ $\qquad$ LCD $= 8$

$\qquad = \dfrac{6}{8} - \dfrac{4}{8} + \dfrac{5}{8}$

$\qquad = \dfrac{7}{8}$

13. $\dfrac{2x + 8}{x^2 + 4x + 3} - \dfrac{x + 4}{x^2 + 5x + 6}$

$\quad = \dfrac{2x + 8}{(x + 1)(x + 3)} - \dfrac{x + 4}{(x + 2)(x + 3)}$

$\quad = \dfrac{2x + 8}{(x + 1)(x + 3)} \cdot \dfrac{x + 2}{x + 2} - \dfrac{x + 4}{(x + 2)(x + 3)} \cdot \dfrac{x + 1}{x + 1}$ $\quad$ LCD $=$ $(x + 1)(x + 2)(x + 3)$

$\quad = \dfrac{2x^2 + 12x + 16 - (x^2 + 5x + 4)}{(x + 1)(x + 2)(x + 3)}$

$\quad = \dfrac{2x^2 + 12x + 16 - x^2 - 5x - 4}{(x + 1)(x + 2)(x + 3)}$

$\quad = \dfrac{x^2 + 7x + 12}{(x + 1)(x + 2)(x + 3)}$

$\quad = \dfrac{(x + 3)(x + 4)}{(x + 1)(x + 2)(x + 3)}$

$\quad = \dfrac{x + 4}{(x + 1)(x + 2)}$

17.
$$\frac{x}{x-3} + 3 = \frac{3}{x-3}$$

$$(x-3)(\frac{x}{x-3}) + (x-3)(3) = (x-3)(\frac{3}{x-3}) \qquad LCD = x-3$$

$$x + 3x - 9 = 3$$
$$4x - 9 = 3$$
$$4x = 12$$
$$x = 3$$

Possible solution 3, which does not check $\emptyset$.

21. Let x = speed of the boat in still water

	d	r	t
upstream	8	x - 2	$\frac{8}{x-2}$
downstream	8	x + 2	$\frac{8}{x+2}$

$$t = \frac{d}{r}$$

The total time for the trip up and back is 3 hours:

Time upstream + Time downstream = Total time

$$\frac{8}{x-2} \qquad + \qquad \frac{8}{x+2} \qquad = \qquad 3$$

Multiplying both sides by (x - 2)(x + 3), we have

$$8(x + 2) + 8(x - 2) = 3(x^2 - 4)$$
$$8x + 16 + 8x - 16 = 3x^2 - 12$$
$$0 = 3x^2 - 16x - 12$$
$$0 = (3x + 2)(x - 6)$$

$$x = -\frac{2}{3} \quad or \quad x = 6$$

The speed of the motorboat in still water is 6 miles/hour.

Chapter 5

Section 5.1

1. To plot (1,2), we start at the origin and move 1 unit right, then 2 units up.

5. To plot (3,4), we start at the origin and move 3 units right, then 4 units up.

9. To plot (0,2), we start at the origin and move 2 down.

13. To plot ($\frac{1}{2}$,2), we start at the origin and move $\frac{1}{2}$ unit right, then 2 units down.

17. $(-\frac{5}{2},\frac{9}{2})$

21. $(-2,0)$

25. $(-2,-3)$

29. To find the y-intercept we let x = 0

    $$\begin{aligned} \text{When} \quad x &= 0 \\ \text{we have} \quad 2(0) - 3y &= 6 \\ -3y &= 6 \\ y &= -2 \end{aligned}$$

    The y-intercept is -2, and the graph crosses the y-axis at the point (0,-2).

    $$\begin{aligned} \text{When} \quad y &= 0 \\ \text{we have} \quad 2x - 3(0) &= 6 \\ 2x &= 6 \\ x &= 3 \end{aligned}$$

    The x-intercept is 3, so the graph crosses the x-axis at the point (3,0).

33. To find the y-intercept we let x = 0

    $$\begin{aligned} \text{When} \quad x &= 0 \\ \text{we have} \quad 4(0) - 5y &= 20 \\ -5y &= 20 \\ y &= -4 \end{aligned}$$

    The y-intercept is -4, and the graph crosses the y-axis at the point (0,-4).

    $$\begin{aligned} \text{When} \quad y &= 0 \\ \text{we have} \quad 4x - 5(0) &= 20 \\ 4x &= 20 \\ x &= 5 \end{aligned}$$

    The x-intercept is 5, so the graph crosses the x-axis at the point (5,0).

37.　To find the y-intercept we let x = 0

　　　When　x = 0
　we have　y = 2(0) + 3
　　　　　　y = 3

The y-intercept is 3, so the graph crosses the y-axis at the point (0,3).

　　　When　y = 0
　we have　　0 = 2x + 3
　　　　　　 -3 = 2x

　　　　　$-\frac{3}{2} = x$

The x-intercept is $-\frac{3}{2}$, so the graph crosses the x-axis at the point $(-\frac{3}{2},0)$.

41.　We need to find three ordered pairs that satisfy the equation.

　　　Let x = 3;　　　　　$y = \frac{1}{3}(3)$

　　　　　　　　　　　　$y = 1$

The ordered pair (3,1) is one solution.

　　　Let x = 0;　　　　　$y = \frac{1}{3}(0)$

　　　　　　　　　　　　$y = 0$

The ordered pair (0,0) is a second solution.

　　　Let x = -3;　　　　$y = \frac{1}{3}(-3)$

　　　　　　　　　　　　$y = -1$

The ordered pair (-3,-1) is a third solution.

Table

x	y
3	1
0	0
-3	-1

45.　　　Let y = 0;　　　-2x + 0 = -3
　　　　　　　　　　　　 -2x = -3

　　　　　　　　　　　　　$x = \frac{3}{2}$

The ordered pair $(\frac{3}{2},0)$ is one solution.

　　　Let x = 0;　　　-2(0) + y = -3
　　　　　　　　　　　　　　y = -3

The ordered pair (0,-3) is another solution.

Table

x	y
$\frac{3}{2}$	0
0	-3

49.   Let x = 0; $y = \frac{1}{2}(0) + 1$

                                 $y = 1$

The ordered pair (0,1) is one solution.

         Let y = 0; $0 = \frac{1}{2}x + 1$

                          $-1 = \frac{1}{2}x$

                          $-2 = x$

The ordered pair (-2,0) is another solution.

Table	
x	y
0	1
-2	0

53.   Let x = 0;   .02(0) + .03y = .06
                               .03y = .06
                                  y = 2

The ordered pair (0,2) is one solution.

         Let y = 0;   .02x + .03(0) = .06
                           .02x = .06
                              x = 3

The ordered pair (3,0) is another solution.

Table	
x	y
0	2
3	0

57.   See the graph on page A29 in the textbook.

61.   Some examples of points (-2,-2), (-1,-1), (0,0), (1,1), (2,2). See the graph on page A29 in the textbook.

65. $\dfrac{x^2 - 9}{x^4 - 81} = \dfrac{x^2 - 9}{(x^2 + 9)(x^2 - 9)} = \dfrac{1}{x^2 + 9}$

69.
$$
\begin{array}{r}
5x \;-\; 4 \\
2x + 3\;\overline{)\;10x^2 + 7x - 12} \\
\end{array}
$$

                 -       -   Change signs
            $\underline{10x^2 + 15x}$
                   - 8x - 12
                   +   +   Change signs
                 $\underline{- 8x - 12}$
                              0

73.   Let x = 0   0x + by = c
                        by = c

                         $y = \dfrac{c}{b}$ (x-intercept)

         Let y = 0   ax + 0y = c
                        ax = c

                         $x = \dfrac{c}{a}$ (y-intercept)

77.　Let $x = -5$;　　$y = |-5 + 2|$　　　Let $x = -4$;　　$y = |-4 + 2|$

　　　　　　　　　$y = |-3|$　　　　　　　　　　　　$y = |-2|$

　　　　　　　　　$y = 3$　　　　　　　　　　　　　$y = 2$

　　　Let $x = -3$;　　$y = |-3 + 2|$　　　Let $x = -2$;　　$y = |-2 + 2|$

　　　　　　　　　$y = |-1|$　　　　　　　　　　　　$y = |0|$

　　　　　　　　　$y = 1$　　　　　　　　　　　　　$y = 0$

　　　Let $x = -1$;　　$y = |-1 + 2|$　　　Let $x = 0$;　　$y = |0 + 2|$

　　　　　　　　　$y = |1|$　　　　　　　　　　　　$y = |2|$

　　　　　　　　　$y = 1$　　　　　　　　　　　　　$y = 2$

　　　Let $x = 1$;　　　$y = |1 + 2|$

　　　　　　　　　$y = |3|$

　　　　　　　　　$y = 3$

The completed ordered pairs are $(-5,3)$, $(-4,2)$, $(-3,1)$, $(-2,0)$, $(-1,1)$, $(0,2)$ and $(1,3)$.

Section 5.2

1.　Using the points on the graph $(0,2)$ and $(-3,0)$, we have

$$m = \frac{y_2 - y_1}{x_2 - x_1} = \frac{0 - 2}{-3 - 0} = \frac{-2}{-3} = \frac{2}{3}$$

5.　Using the point on the graph $(1,1)$ and $(4,3)$, we have

$$m = \frac{y_2 - y_1}{x_2 - x_1} = \frac{3 - 1}{4 - 1} = \frac{2}{3}$$

9.　Given $(1,4)$ and $(5,2)$, we have

$$m = \frac{y_2 - y_1}{x_2 - x_1} = \frac{2 - 4}{5 - 1} = \frac{-2}{4} = -\frac{1}{2}$$

13.　Given $(-3,-2)$ and $(1,3)$, we have

$$m = \frac{y_2 - y_1}{x_2 - x_1} = \frac{3 - (-2)}{1 - (-3)} = \frac{5}{4}$$

17.　Given $(2,-5)$ and $(3,-2)$, we have

$$m = \frac{y_2 - y_1}{x_2 - x_1} = \frac{-2 - (-5)}{3 - 2} = \frac{3}{1} = 3$$

21. Given $(2,6)$, $(3,y)$; $m = -7$, we have

$$m = \frac{y_2 - y_1}{x_2 - x_1}$$

$$-7 = \frac{y - 6}{3 - 2}$$

$$-7 = \frac{y - 6}{1}$$

$$-7 = y - 1$$
$$-1 = y$$

25. Given $(3,b^2)$ and $(4,3b)$ with slope 2 we write

$$\frac{b^2 - 3b}{3 - 4} = 2$$

$$\frac{b^2 - 3b}{-1} = 2$$

$$b^2 - 3b = -2$$
$$b^2 - 3b + 2 = 0$$
$$(b - 2)(b - 1) = 0$$

$$b - 2 = 0 \quad \text{or} \quad b - 1 = 0$$
$$b = 2 \quad \text{or} \quad b = 1$$

29. If the x-intercept is 4, then the ordered pair is $(4,0)$. If the y-intercept is 2, then the ordered pair is $(0,2)$.

$$m = \frac{y_2 - y_1}{x_2 - x_1} = \frac{2 - 0}{0 - 4} = \frac{2}{-4} = -\frac{1}{2}$$

33. Given $(5,-6)$ and $(5,2)$, we have

$$m = \frac{y_2 - y_1}{x_2 - x_1} = \frac{2 - (-6)}{5 - 5} = \frac{8}{0}$$

A line perpendicular to line 1 would have a slope of $-\frac{0}{8} = 0$.

37. Given $(2,y^2)$, $(1,y)$ and $m = -\frac{1}{6}$, we have

$$m = \frac{y_2 - y_1}{x_2 - x_1} = \frac{y - y^2}{1 - 2} = \frac{y - y^2}{-1} = -y + y^2$$

A line perpendicular to the given line has a slope of 6.

$$m = -y + y^2$$
$$6 = -y + y^2$$
$$0 = y^2 - y - 6 \qquad \text{Standard form}$$
$$0 = (y + 2)(y - 3) \qquad \text{Factor}$$

$$y + 2 = 0 \quad \text{or} \quad y - 3 = 0 \qquad \text{Factors set to 0}$$
$$y = -2 \qquad\qquad y = 3$$

140

41. $2x - 3y = 6$

$\qquad -3y = -2x + 6 \qquad$ Add $-2x$ to both sides

$\qquad y = \dfrac{-2x}{-3} + \dfrac{6}{-3} \qquad$ Divide by -3

$\qquad y = \dfrac{2}{3}x - 2$

45. $\dfrac{8xy^3}{9x^2y} \div \dfrac{16x^2y^2}{18xy^3} = \dfrac{8xy^3}{9x^2y} \cdot \dfrac{18xy^3}{16x^2y^2}$

$\qquad\qquad\qquad\qquad = \dfrac{144x^2y^6}{144x^4y^3}$

$\qquad\qquad\qquad\qquad = \dfrac{y^3}{x^2}$

49. $\dfrac{8x^3 + 27}{27x^3 + 1} \div \dfrac{6x^2 + 7x - 3}{9x^2 - 1}$

$\qquad = \dfrac{8x^3 + 27}{27x^3 + 1} \cdot \dfrac{9x^2 - 1}{6x^2 + 7x - 3}$

$\qquad = \dfrac{(2x)^3 + 3^3}{(3x)^3 + 1^3} \cdot \dfrac{(3x + 1)(3x - 1)}{(2x + 3)(3x - 1)}$

$\qquad = \dfrac{\cancel{(2x + 3)}(4x^2 - 6x + 9)}{\cancel{(3x + 1)}(9x^2 - 3x + 1)} \cdot \dfrac{\cancel{(3x + 1)}\cancel{(3x - 1)}}{\cancel{(2x + 3)}\cancel{(3x - 1)}}$

$\qquad = \dfrac{4x^2 - 6x + 9}{9x^2 - 3x + 1}$

53. $m = \dfrac{y_2 - y_1}{x_2 - x_1} = \dfrac{-.02 - .18}{.16 - .04} = \dfrac{-.20}{.12} = -\dfrac{5}{3}$

57. a. slope is 2 \qquad b. slope is $-\dfrac{1}{2}$

Section 5.3

1. $\qquad\qquad$ Let $\qquad m = 2, b = 3$

$\qquad$ the equation $\qquad y = mx + b$

$\qquad\qquad$ becomes $\qquad y = 2x + 3$

5. $\qquad\qquad$ Let $\qquad m = \dfrac{1}{2}, b = \dfrac{3}{2}$

$\qquad$ the equation $\qquad y = mx + b$

$\qquad\qquad$ becomes $\qquad y = \dfrac{1}{2}x + \dfrac{3}{2}$

9. The equation $y = 3x - 2$ has the form $y = mx + b$, therefore $m = 3$ and $b = -2$. The slope of any line perpendicular to the given line would be $-\frac{1}{3}$. See the graph on page A**32** in your textbook.

13. Solving the equation for y, we have

$$4x + 5y = 20$$
$$5y = -4x + 20$$
$$y = -\frac{4}{5}x + 4$$

The slope is $-\frac{4}{5}$, the y-intercept is 4 and the perpendicular slope is $\frac{5}{4}$.

17. Solving the equation for y, we have

$$3x + y = 3$$
$$y = -3x + 3$$

The slope is -3 and the y-intercept is 3.

21. Using $(x_1, y_1) = (-2, -5)$, $m = 2$

in $y - y_1 = m(x - x_1)$ Point-slope form

gives us $y + 5 = 2(x + 2)$ Note: $x - (-2) = x + 2$
$y - (-5) = y + 5$

$y + 5 = 2x + 4$ Multiply out right side
$y = 2x - 1$ Add -5 to each side

25. Using $(x_1, y_1) = (2, -3)$, $m = \frac{3}{2}$

in $y - y_1 = m(x - x_1)$ Point-slope form

gives us $y + 3 = \frac{3}{2}(x - 2)$ Note: $y - (-3) = y + 3$

$y + 3 = \frac{3}{2}x - 3$ Multiply out right side

$y = \frac{3}{2}x - 6$ Add -3 to each side

29. Using $(x_1, y_1) = (\frac{2}{5}, \frac{3}{2})$, $m = 1$

in $y - y_1 = m(x - x_1)$ Point-slope form

gives us $y - \frac{3}{2} = 1(x - \frac{2}{5})$

$y - \frac{3}{2} = x - \frac{2}{5}$

$y = x + \frac{11}{10}$ $-\frac{2}{5} + \frac{3}{2} = -\frac{4}{10} + \frac{15}{10} = \frac{11}{10}$

33. We begin by finding the slope of the line:

$$m = \frac{1 - (-5)}{2 - (-1)} = \frac{6}{3} = 2$$

Using $(x_1, y_1) = (-1, -5)$ and $m = 2$ in $y - y_1 = m(x - x_1)$ yields

$$y + 5 = 2(x + 1)$$
$$y + 5 = 2x + 2 \qquad \text{Multiply out right side}$$
$$y = 2x - 3 \qquad \text{Add -5 to each side}$$

We could use the point $(2,1)$ to obtain the same equation.

$$y - y_1 = m(x - x_1)$$
$$y - 1 = 2(x - 2)$$
$$y - 1 = 2x - 4 \qquad \text{Multiply out right side}$$
$$y = 2x - 3 \qquad \text{Add 1 to each side}$$

37. We begin by finding the slope of the line:

$$m = \frac{-1 - (-\frac{1}{5})}{\frac{1}{3} - (-\frac{1}{3})} = \frac{-\frac{4}{5}}{\frac{2}{3}} = -\frac{4}{5} \cdot \frac{3}{2} = -\frac{12}{10} = -\frac{6}{5}$$

Using $(x_1, y_1) = (-\frac{1}{3}, -\frac{1}{5})$ and $m = -\frac{6}{5}$ in

$$y - y_1 = m(x - x_1) \text{ yields}$$

$$y + \frac{1}{5} = -\frac{6}{5}(x + \frac{1}{3})$$

$$y + \frac{1}{5} = -\frac{6}{5}x - \frac{6}{15}$$

$$y = -\frac{6}{5}x - \frac{3}{5} \qquad -\frac{6}{15} - \frac{1}{5} = -\frac{6}{15} - \frac{3}{15} = -\frac{9}{15} = -\frac{3}{5}$$

We could use the point $(\frac{1}{3}, -1)$ to obtain the same equation.

$$y - y_1 = m(x - x_1)$$

$$y + 1 = -\frac{6}{5}(x - \frac{1}{3})$$

$$y + 1 = -\frac{6}{5}x + \frac{6}{15}$$

$$y = -\frac{6}{5}x - \frac{3}{5} \qquad \frac{6}{15} - 1 = \frac{6}{15} - \frac{15}{15} = -\frac{9}{15} = -\frac{3}{5}$$

To put in the equation form, we do the following

$$y = -\frac{6}{5}x - \frac{3}{5}$$

$$5y = -6x - 3 \qquad \text{Multiply by 5}$$
$$6x + 5y = -3$$

41. $y = -2$ in $y = mx + b$ form becomes $y = 0x - 2$. The slope is 0 and the y-intercept is -2.

45. To find the slope of $2x - 5y = 10$, we solve for y:

$$2x - 5y = 10$$
$$-5y = 2x + 10$$
$$y = \frac{2}{5}x - 2$$

The slope of this line is $\frac{2}{5}$. The line we are interested in is perpendicular to the line with slope $\frac{2}{5}$ and must, therefore have a slope of $-\frac{5}{2}$.

Using $(x_1, y_1) = (-4, -3)$ and $m = -\frac{5}{2}$, we have

$$y - y_1 = m(x - x_1)$$
$$y + 3 = -\frac{5}{2}(x + 4)$$
$$y + 3 = -\frac{5}{2}x - 10$$
$$y = -\frac{5}{2}x - 13$$

49. Given x-intercept 3 and y-intercept 2, the ordered pairs are (3,0) and (0,2). Finding the slope of the line, we have

$$m = \frac{2 - 0}{0 - 3} = -\frac{2}{3}$$

Using $(x_1, y_1) = (3, 0)$ and $m = -\frac{2}{3}$, we have

$$y - y_1 = m(x - x_1)$$
$$y - 0 = -\frac{2}{3}(x - 3)$$
$$y = -\frac{2}{3}x + 2$$

We could use the point (0,2) to obtain the same equation.

$$y - y_1 = m(x - x_1)$$
$$y - 2 = -\frac{2}{3}(x - 0)$$
$$y - 2 = -\frac{2}{3}x$$
$$y = -\frac{2}{3}x + 2$$

53. $\dfrac{2a - 4}{a + 2} - \dfrac{a - 6}{a + 2} = \dfrac{2a - 4 - a + 6}{a + 2}$

$$= \dfrac{a + 2}{a + 2}$$

$$= 1$$

57. $\dfrac{3}{2x - 5} - \dfrac{39}{8x^2 - 14x - 15}$

$$= \dfrac{3}{2x - 5} - \dfrac{39}{(2x - 5)(4x + 3)}$$

$$= \dfrac{3}{2x - 5} \cdot \dfrac{4x + 3}{4x + 3} - \dfrac{39}{(2x - 5)(4x + 3)}$$

$$= \dfrac{12x + 9}{(2x - 5)(4x + 3)} - \dfrac{39}{(2x - 5)(4x + 3)}$$

$$= \dfrac{12x - 30}{(2x - 5)(4x + 3)}$$

$$= \dfrac{6(2x - 5)}{(2x - 5)(4x + 3)}$$

$$= \dfrac{6}{4x + 3}$$

61. $2x + 5y = 100$
$5y = -2x + 100$

$$y = -\dfrac{2}{5}x + 20$$

Slope = $-\dfrac{2}{5}$, y-intercept is 20.

See the graph on page A33 in the textbook.

65. $\dfrac{x}{2} + \dfrac{y}{3} = 1$

$6\left(\dfrac{x}{2}\right) + 6\left(\dfrac{y}{3}\right) = 6(1)$ LCD is 6

$3x + 2y = 6$
$2y = -3x + 6$

$$y = -\dfrac{3}{2}x + 3$$

Slope = $-\dfrac{3}{2}$, y-intercept = 3, x-intercept = 2

69.
$$\frac{x}{a} + \frac{y}{b} = 1$$

$$ab\left(\frac{x}{a}\right) + ab\left(\frac{y}{b}\right) = ab(1) \qquad \text{LCD is } ab$$

$$bx + ay = ab$$
$$ay = -bx + ab$$

$$y = -\frac{b}{a}x + b$$

Slope $= -\dfrac{b}{a}$, y-intercept $= b$, x-intercept $= a$

Section 5.4

1. The boundary for the graph is the graph $x + y = 5$; the x- and y-intercepts are both 5. (Remember to let $y = 0$ for the x-intercept and $x = 0$ for the y-intercept.) The boundary is not included in the solution because the inequality symbol is <. Therefore, we use a broken line to represent the boundary.

A convenient test point is the origin:

$$\text{Using } (0,0)$$
$$\text{in } x + y \le 5$$
$$\text{we have } 0 + 0 < 5$$
$$0 < 5 \qquad \text{A true statement.}$$

Since our test point is true, the region below the boundary is shaded.

5. The boundary for the graph is the graph $2x + 3y = 6$; the x-intercept is 2 ($2(0) + 3y = 6, y = 2$) and the y-intercept is 3 ($2x + 3(0) = 6$, $x = 3$). The boundary is a broken line because the inequality symbol is <.

A convenient test point is the origin:

$$\text{Using } (0,0)$$
$$\text{in } 2x + 3y \le 6$$
$$\text{we have } 2(0) + 3(0) < 6$$
$$0 < 6 \qquad \text{A true statement.}$$

Since our test point is true, the region below the boundary is shaded.

9. The boundary for the graph is the graph $2x + y = 5$; the x-intercept is $\frac{5}{2}$ ($2x + 0 = 5, x = \frac{5}{2}$) and the y-intercept is 5 ($2(0) + y = 5, y = 5$).

   The boundary is a broken line because the inequality symbol is <.

   A convenient test point is the origin:

   ```
   Using  (0,0)
      in    2x + y < 5
   we have  2(0) + 0 < 5
              0 < 5      A true statement.
   ```

   Since our test point is true, the region below the boundary is shaded.

13. The boundary is the graph of $y = 2x - 1$: a line with slope 2 and y-intercept -1. The boundary is a broken line because the inequality symbol is <.

    A convenient test point is the origin:

    ```
    Using  (0,0)
       in   y < 2x - 1
    we have  0 < 2(0) - 1
             0 < -1         A false statement.
    ```

    Since our test point is false, the region below the boundary is shaded.

17. The boundary is $x = 3$, which is a vertical line. The boundary is a broken line because the inequality symbol is >. All points to the right have x-coordinates greater than 3.

21. The boundary for the graph is the graph $y = 2x$; the x-intercept is 0 ($0 = 2x, x = 0$) and the y-intercept is 0 ($y = 2(0), y = 0$). This only gives one set of coordinates, so substitute another point, if $x = 1$, $y = 1$ ($y = 2(1), y = 1$). The boundary is a broken line because the inequality symbol is <. We cannot use (0,0) because it is part of the line.

    ```
    Let's try  (1,-1)
       Using   (1,-1)
          in    y < 2x
    we have   -1 < 2(1)
              -1 < 1        A true statement
    ```

    Since our test point is true, the region below the boundary is shaded.

25. The boundary for the graph is the graph $y = \frac{3}{4}x - 2$; the x-intercept is $\frac{8}{3}$ $(0 = \frac{3}{4}x - 2, x = \frac{8}{3})$ and the y-intercept is -2 $(y = \frac{3}{4}(0) - 2, y = -2)$.

The boundary is a solid line because the inequality symbol is $\geq$.

A convenient test point is the origin:

Using $(0,0)$

in $y \geq \frac{3}{4}x - 2$

we have $0 \geq \frac{3}{4}(0) - 2$

$0 \geq -2$ A true statement

Since our test point is true, the region below the boundary is shaded.

29. The boundary for the graph is the graph $\frac{x}{3} + \frac{y}{2} = 1$; the x-intercept is 3 $(\frac{x}{3} + \frac{0}{2} > 1, x = 3)$ and the y-intercept is 2 $(\frac{0}{3} + \frac{y}{2} > 1, y = 2)$. The boundary is a broken line because the inequality symbol is $>$.

A convenient test point is the origin:

Using $(0,0)$

in $\frac{x}{3} + \frac{y}{2} > 1$

we have $\frac{0}{3} + \frac{0}{2} > 1$

$0 > 1$ A false statement

Since our test point is false, the region above the boundary is shaded.

33. $\dfrac{\frac{1}{4} - \frac{1}{3}}{\frac{1}{2} + \frac{1}{6}} = \dfrac{12(\frac{1}{4}) - 12(\frac{1}{3})}{12(\frac{1}{2}) + 12(\frac{1}{6})}$

$= \dfrac{3 - 4}{6 + 2}$

$= -\dfrac{1}{8}$

37. $\dfrac{4 + \frac{4}{x} + \frac{1}{x^2}}{4 - \frac{1}{x^2}} = \dfrac{x^2(4) + x^2(\frac{4}{x}) + x^2(\frac{1}{x^2})}{x^2(4) - x^2(\frac{1}{x^2})}$

$$= \frac{4x^2 + 4x + 1}{4x^2 - 1}$$

$$= \frac{(2x + 1)(2x + 1)}{(2x + 1)(2x - 1)}$$

$$= \frac{2x + 1}{2x - 1}$$

41. To graph $x + y \leq 3$:

The boundary for the graph is the graph $x + y = 3$; the x-intercept is $3(x + 0 = 3, x = 3)$ and the y-intercept is $3(0 + y = 3, y = 3)$. The boundary is a solid line because the inequality symbol is $\leq$.

A convenient test point is the origin:

Using (0,0)
in $x + y \leq 3$

we have $0 + 0 \leq 3$

$0 \leq 3$ A true statement

Since our test point is true, the region below the boundary is shaded.

To graph $x - y \leq 3$:

The boundary for the graph is the graph $x - y = 3$; the x-intercept is $3(x - 0 = 3, x = 3)$ and the y-intercept is $-3(0 - y = 3, y = -3)$. The boundary is a solid line because the inequality symbol is $\leq$.

A convenient test point is the origin:

Using (0,0)
in $x - y \leq 3$
we have $0 - 0 \leq 3$
$0 \leq 3$ A true statement

Since our test point is true, the region above the boundary is shaded.

45. Let x = 0 $y = |0 - 3|$
               $y = 3$

 Let x = 1 $y = |1 - 3|$
               $y = 2$

 Let x = 2 $y = |2 - 3|$
               $y = 1$

 Let x = 3 $y = |3 - 3|$
               $y = 0$

 Let x = 4 $y = |4 - 3|$
               $y = 1$

 Let x = 5 $y = |5 - 3|$
               $y = 2$

 Let x = 6 $y = |6 - 3|$
               $y = 3$

Table

x	y
0	3
1	2
2	1
3	0
4	1
5	2
6	3

The boundary is a broken line because the inequality is >.

A convenient test point is the origin:

Using (0,0)

in $y > |x - 3|$

we have $0 > |0 - 3|$

$0 > 3$ A false statement

Since our test point is false, the region above the boundary is shaded.

Section 5.5

1.   When $y = 10$
                and $x = 2$
 the equation $y = Kx$
     becomes $10 = K \cdot 2$
         or $K = 5$

The equation can now be written specifically as

$y = 5x$

Letting x = 6, we have

$y = 5 \cdot 6$
$y = 30$

5.

$$\text{When} \quad r = -3$$
$$\text{and} \quad s = 4$$

the equation $\quad r = \dfrac{K}{s}$

becomes $\quad -3 = \dfrac{K}{4}$

or $\quad K = -12$

The equation can now be written specifically as

$$r = \dfrac{-12}{s}$$

Letting s = 2, we have

$$r = \dfrac{-12}{2}$$

$$r = -6$$

9.

$$\text{When} \quad d = 10$$
$$\text{and} \quad r = 5$$

the equation $\quad d = Kr^2$

becomes $\quad 10 = K \cdot 25$

$$K = \dfrac{2}{5}$$

The equation can now be written specifically as

$$d = \dfrac{2}{5} r^2$$

Letting r = 10, we have

$$d = \dfrac{2}{5} \cdot 100$$

$$d = 40$$

13.

When $y = 45$
and $x = 3$

the equation $y = \dfrac{K}{x^2}$

becomes $45 = \dfrac{K}{3^2}$

$$45 = \dfrac{K}{9}$$

$$K = 405$$

The equation can now be written specifically as

$$y = \dfrac{405}{x^2}$$

Letting $x = 5$, we have

$$y = \dfrac{405}{5^2}$$

$$y = \dfrac{405}{25}$$

$$y = \dfrac{81}{5}$$

17. In the general equation

$$z = Kxy^2$$

substituting $z = 54$, $x = 3$ and $y = 3$, we have

$$54 = K(3)(3^2)$$
$$54 = 27K$$
$$K = 2$$

The specific equation is

$$z = 2xy^2$$

When $x = 2$ and $y = 4$, the last equation becomes

$$z = 2(2)(4^2)$$
$$z = 64$$

21.   Let F = force and s = spring
       When F = 5
       and s = 3
       the equation F = Ks
       becomes 5 = K · 3

$$\frac{5}{3} = K$$

The equation can now be written specifically as

$$F = \frac{5}{3} s$$

Letting s = 10, we have

$$F = \frac{5}{3} \cdot 10$$

$$F = \frac{50}{3} \, lb$$

25.   Let S = surface area, h = height, and r = radius. The equation is

S = Khr

When S = 94, h = 5, and r = 3, the equation becomes

94 = K(5)(3)
94 = 15K

or $K = \dfrac{94}{15}$

Using this value of S in our original equation, the result is

$$S = \frac{94}{15} hr$$

When h = 8 and r = 2, the equation becomes

$$S = \frac{94}{15}(8)(2)$$

$$S = \frac{1,504}{15} \, sq. \ in.$$

29. $y = -4x$ $y = 4x$

x	y
0	0
$\frac{1}{4}$	-1
$-\frac{1}{4}$	1
1	-4
-1	4
2	-8
-2	8

x	y
0	0
$\frac{1}{4}$	1
$-\frac{1}{4}$	1
1	4
-1	-4
2	8
-2	-8

See the graph on page A36 in the textbook.

33. $y = \frac{2}{3}x$ $y = -\frac{3}{2}x$

x	y
0	0
3	2
-3	-2
6	4
-6	-4

x	y
0	0
2	-3
-2	3
4	-6
-4	6

See the graph on page A36 in the textbook.

37. $y = \frac{-4}{x}$

x	y
1	-4
-1	4
2	-2
-2	2
4	-1
-4	1

See the graph on page A36 in the textbook.

41. $y = \frac{3}{x}$ $x + y = 4$

x	y
0	undefined
undefined	0
1	3
3	1
-1	-3
-3	-1

x	y
0	4
4	0
1	3
3	1

The two graphs intersect at (1,3) and (3,1). See the graph on page A36 in the textbook.

45.
$$\frac{x}{x + 1} - \frac{1}{x + 1} = \frac{1}{2}$$

$$2(x + 1)(\frac{x}{x + 1}) - 2(x + 1)(\frac{1}{x + 1}) = 2(x + 1)(\frac{1}{2}) \qquad \text{LCD is } 2(x + 1)$$

$$2x - 2(1) = (x + 1)1$$
$$2x - 2 = x + 1$$
$$x - 2 = 1$$
$$x = 3$$

49.   Let x = a number, the reciprocal is $\frac{1}{x}$.

$$x + \frac{1}{x} = \frac{41}{20}$$

$$20x(x) + 20x(\frac{1}{x}) = 20x(\frac{41}{20}) \qquad \text{LCD is } 20x$$

$$20x^2 + 20 = 41x$$
$$20x^2 - 41x + 20 = 0$$
$$(5x - 4)(4x - 5) = 0$$

$$5x - 4 = 0 \quad \text{or} \quad 4x - 5 = 0$$
$$5x = 4 \qquad\qquad 4x = 5$$
$$x = \frac{4}{5} \qquad\qquad x = \frac{5}{4}$$

1.
$$3x + 2y = 6$$

When $x = 0$
the equation $3x + 2y = 6$
becomes $3(0) + 2y = 6$
$$2y = 6$$
$$y = 3$$

The ordered pair $(0,3)$ is a solution.

When $y = 0$
the equation $3x + 2y = 6$
becomes $3x + 2(0) = 6$
$$3x = 6$$
$$x = 2$$

The ordered pair $(2,0)$ is a solution.

See the graph on page A37 in the textbook.

5.
$$y = -\frac{3}{2}x + 1$$

When $x = 2$
the equation $y = -\frac{3}{2}x + 1$

becomes $y = -\frac{3}{2}(2) + 1$

$$y = -3 + 1$$
$$y = -2$$

The ordered pair $(2,-2)$ is a solution.

When $y = 1$
the equation $y = -\frac{3}{2}x + 1$

becomes $1 = -\frac{3}{2}x + 1$

$0 = -\frac{3}{2}x$ Add -1 to both sides

$0 = x$ Multiply by $-\frac{2}{3}$

The ordered pair $(0,1)$ is a solution.

See the graph on page A37 in the textbook.

9. $(-4, 2), (3, 2)$

$$m = \frac{y_2 - y_1}{x_2 - x_1} = \frac{2 - 2}{3 - (-4)} = \frac{0}{7} = 0$$

The slope is 0.

156

13. $(-4, 7)$, $(2, x)$ $m = -\dfrac{1}{3}$

$$m = \frac{y_2 - y_1}{x_2 - x_1}$$

$$-\frac{1}{3} = \frac{x - 7}{2 - (-4)}$$

$$-\frac{1}{3} = \frac{x - 7}{6}$$

$\quad -2 = x - 7 \qquad$ Multiply both sides by 6

$\quad\ \ 5 = x \qquad\quad$ Add 7 to each side

17. The slope of the line through points $(5, y^2)$ and $(2, y)$ is

$$m = \frac{y - y^2}{2 - 5} = \frac{y - y^2}{-3}$$

Since the line through the two points is parallel, the slopes are the same value of 4.

$$4 = \frac{y - y^2}{-3}$$

$\quad -12 = y - y^2 \qquad$ Multiply by -3

$y^2 - y - 12 = 0 \qquad$ Standard form

$(y - 4)(y + 3) = 0 \qquad$ Factor

$y - 4 = 0 \ $ or $ \ y + 3 = 0$

$\quad\ y = 4 \qquad\qquad y = 3$

The values for y are -3 and 4.

21. $m = -2$, $b = 0$

$y = mx + b$

$y = -2x + 0$

$y = -2x$

25. $2x - 3y = 9$

$\quad\ -3y = -2x + 9 \qquad$ Add -2x to both sides

$\quad\quad\ y = \dfrac{2}{3}x - 3 \qquad$ Divide by -3

The slope is $\dfrac{2}{3}$ and the y-intercept is -3.

29.   Using $(x_1, y_1) = (-3, 1)$ and $m = -\frac{1}{3}$

in $y - y_1 = m(x - x_1)$ Point-slope form

gives us $y - 1 = -\frac{1}{3}(x + 3)$ Note: $x - (-3) = x + 3$

$y - 1 = -\frac{1}{3}x - 1$ Multiply by $-\frac{1}{3}$

$y = -\frac{1}{3}x + 0$ Add $+1$ to each side

$x = -\frac{1}{3}x$

33. $(-3, 7), (4, 7)$

$m = \frac{7 - 7}{4 - (-3)} = \frac{0}{7} = 0$

Using $(x_1, y_1) = (-3, 7)$ and $m = 0$ in

$y - y_1 = m(x - x_1)$ yields

$y - 7 = 0(x + 3)$

$y - 7 = 0$ Multiply out right side

$y = 7$ Add $+7$ to each side

37.   To find the slope of $2x - y = 4$, we solve for y:

$2x - y = 4$

$y = 2x - 4$

The slope is the same since the lines are parallel.

Using $(x_1, y_1) = (2, -3)$ and $m = 2$, we have

$y - y_1 = m(x - x_1)$

$y + 3 = 2(x - 2)$

$y + 3 = 2x - 4$ Multiply out each side

$y = 2x - 4 - 3$ Add -3 to each side

$y = 2x - 7$

41. $y \leq 2x - 3$

> When $x = 0$
> the equation $y = 2x - 3$
> becomes $y = 2(0) - 3$
> $y = -3$

The y-intercept is at -3.

> When $y = 0$
> the equation $y = 2x - 3$
> becomes $0 = 2x - 3$
> $3 = 2x$
> $\dfrac{3}{2} = x$

The x-intercept is at $\dfrac{3}{2}$.

The boundary is included in the solution set because the inequality symbol is $\leq$.

> Substituting $(0,0)$
> into $y \leq 2x - 3$
> gives us $0 \leq 2(0) - 3$
> $0 \leq -3$ \qquad A false statement

Since our test gives us a false statement and it lies above the boundary, the solution set must lie on the other side of the boundary. See the graph on page A37 in the textbook.

45.

> $y = Kx$
> When $y = 6$
> and $x = 2$
> the equation $y = Kx$
> becomes $y = K \cdot 2$
> or $K = 3$

The equation can now be written specifically as

> $y = 3x$

Letting $x = 8$, we have

> $y = 3 \cdot 8$
> $y = 24$

49.

$$z = Kxy^3$$

When $z = 6$

and $x = 2$

and $y = -1$

the equation $x = Kxy^3$

becomes $6 = K(2)(-1)^3$

or $6 = -2K$

and $K = -3$

Now we write the equation again as

$$z = -3xy^3$$

We finish by substituting $x = 3$ and $y = 2$ into the last equation.

$$z = -3(3)(2)^3$$
$$z = -72$$

53. We will let t = tension and d = distance. Since the tension varies directly with the distance, we have

$$t = Kd$$

When $t = 42$

and $d = 2$

the equation $t = Kd$

becomes $42 = K(2)$

and $K = 21$

Specifically, then the relationship between t and d is

$$t = 21d$$

Finally, we find t when $d = 4$:

$$t = 21(4)$$
$$t = 84$$

The tension is 84 pounds.

57. $y = 3x$

x	y	
1	3	Let $x = 1$, then $y = 3(1) = 3$
0	0	Let $x = 0$, then $y = 3(0) = 0$
-1	-3	Let $x = -1$, then $y = 3(-1) = -3$

See the graph on page A38 in the textbook.

Chapter 5 Test

1. The x-intercept is 3($2x + 0 = 6, x = 3$) and the y-intercept is 6($2 \cdot 0 + y = 6, y = 6$). To find the slope put the equation in the slope-intercept form.

$$2x + y = 6$$
$$y = -2 + 6$$

The slope is -2.

See the graph on page A39 in the textbook.

5. Using $(x_1, y_1) = (-1, 3)$, $m = 2$

in $\quad y - y_1 = m(x - x_1) \qquad$ Point-slope form

gives us $\quad y - 3 = 2(x + 1)$
$$y - 3 = 2x + 2$$
$$y = 2x + 5$$

9. The vertical line is when x is a constant, so $x = 4$.

13. $$z = Kxy^3$$

When $\quad z = 15$
and $\quad x = 5$
and $\quad y = 2$
the equation $\quad z = Kxy^3$
becomes $\quad 15 = K(5)(2)^3$
or $\quad 15 = 40K$

and $\quad K = \dfrac{15}{40} = \dfrac{3}{8}$

Now we write the equation as

$$z = \frac{3}{8} xy^3$$

We finish by substituting $x = 2$ and $y = 3$ into the last equation.

$$z = \frac{3}{8}(2)(3)^3$$

$$z = \frac{81}{4}$$

Chapter 6

Section 6.1

1. $\sqrt{144} = 12$ because $12^2 = 144$

5. $-\sqrt{49} = -7$ This is the negative square root of 49.

9. $\sqrt[4]{16} = 2$ because $2^4 = 16$

13. $\sqrt{.04} = .2$ because $(.2)^2 = .04$

17. $\sqrt{36a^8} = 6a^4$ because $(6a^4)^2 = 36a^8$

21. $\sqrt[3]{x^3y^6} = xy^2$ because $(xy^2)^3 = x^3y^6$

25. $\sqrt[4]{16a^{12}b^{20}} = 2x^3b^5$ because $(2x^3b^5)^4 = 16a^{12}b^{20}$

29. $-9^{1/2} = -\sqrt{9} = -3$

33. $(-8)^{1/3} = \sqrt[3]{-8} = -2$

37. $\left(\dfrac{81}{25}\right)^{1/2} = \sqrt{\dfrac{81}{25}} = \dfrac{9}{5}$

41. $\begin{aligned} 27^{2/3} &= (27^{1/3})^2 && \text{Theorem 6.1} \\ &= 3^2 && \text{Definition of fractional exponents} \\ &= 9 && \text{The square of 3 is 9} \end{aligned}$

45. $\begin{aligned} 16^{3/4} &= (16^{1/4})^3 && \text{Theorem 5.1} \\ &= 2^3 && \text{Definition of fractional exponents} \\ &= 8 \end{aligned}$

49. $\begin{aligned} 81^{-3/4} &= (81^{1/4})^{-3} && \text{Theorem 6.1} \\ &= 3^{-3} && \text{Definition of fractional exponents} \\ &= \dfrac{1}{3^3} && \text{Property 4} \\ &= \dfrac{1}{27} && \text{The cube of 3 is 27} \end{aligned}$

53. $\begin{aligned} \left(\dfrac{81}{16}\right)^{-3/4} &= \left[\left(\dfrac{81}{16}\right)^{1/4}\right]^{-3} && \text{Theorem 6.1} \\ &= \left(\dfrac{3}{2}\right)^{-3} && \text{Definition of fractional exponents} \\ &= \left(\dfrac{2}{3}\right)^3 && \text{Property 4} \\ &= \dfrac{8}{27} && \left(\dfrac{2}{3}\right)^3 = \dfrac{2^3}{3^3} = \dfrac{8}{27} \end{aligned}$

162

57. $8^{-2/3} + 4^{-1/2} = (8^{1/3})^{-2} + (4^{1/2})^{-1}$ Theorem 6.1

$\qquad\qquad\qquad = 2^{-2} + 2^{-1}$ Definition of fractional exponents

$\qquad\qquad\qquad = \dfrac{1}{2^2} + \dfrac{1}{2}$ Property 4

$\qquad\qquad\qquad = \dfrac{1}{4} + \dfrac{1}{2}$ The square of 2 is 4

$\qquad\qquad\qquad = \dfrac{3}{4}$

61. $(a^{3/4})^{4/3} = a^{(3/4)(4/3)}$ Property 2

$\qquad\qquad\quad = a$ Multiply fractions: $\dfrac{3}{4} \cdot \dfrac{4}{3} = \dfrac{12}{12} = 1$

65. $\dfrac{x^{5/6}}{x^{2/3}} = x^{5/6 - 2/3}$ Property 6

$\qquad\quad = x^{5/6 - 4/6}$ LCD is 6

$\qquad\quad = x^{1/6}$

69. $\dfrac{a^{3/4}b^2}{a^{7/8}b^{1/4}} = a^{3/4 - 7/8} \cdot b^{2 - 1/4}$ Property 6

$\qquad\qquad = a^{6/8 - 7/8} \cdot b^{8/4 - 1/4}$ LCD are 8 and 4

$\qquad\qquad = a^{-1/8} \cdot b^{7/8}$

$\qquad\qquad = \dfrac{b^{7/8}}{a^{1/8}}$ Property 4

73. $\left(\dfrac{a^{-1/4}}{b^{1/2}}\right)^8 = \dfrac{(a^{-1/4})^8}{(b^{1/2})^8}$ Property 5

$\qquad\qquad = \dfrac{a^2}{b^4}$ Property 2

$\qquad\qquad = \dfrac{1}{a^2 b^4}$ Property 4

77. $\dfrac{(25a^6b^4)^{1/2}}{(8a^{-9}b^3)^{-1/3}} = \dfrac{(25)^{1/2}(a^6)^{1/2}(b^4)^{1/2}}{(8)^{-1/3}(a^{-9})^{-1/3}(b^3)^{-1/3}}$ Property 3

$\qquad\qquad\qquad = \dfrac{5a^3b^2}{2^{-1}a^3b^{-1}}$ Definition of fractional exponents

$\qquad\qquad\qquad = \dfrac{5 \cdot 2a^3b^2b}{a^3}$ Property 4

$\qquad\qquad\qquad = 10b^3$ Properties 1 and 6

81. $(a^{1/r})^{r^2} = a^r$ Property 2

85. $\dfrac{(r^{1/n}s^n)^n}{r^n s^{n^2}} = \dfrac{(r^{1/n})^n (s^n)^n}{r^n s^{n^2}}$ Property 3

$\qquad = \dfrac{rs^{n^2}}{r^n s^{n^2}}$ Property 2

$\qquad = r^{1-n}$ Property 6

89.

If $a = 9$ and $b = 4$

Then $(a^{1/2} + b^{1/2})^2$

Becomes $(9^{1/2} + 4^{1/2})^2$

$\qquad = (3 + 2)^2$

$\qquad = 5^2$

$\qquad = 25$

If $a = 9$ and $b = 4$

Then $a + b$

Becomes $9 + 4 = 13$

Since $25 \neq 13$ then $(a^{1/2} + b^{1/2})^2 \neq a + b$.

93. $\quad \sqrt{\sqrt{a}} = \sqrt[4]{a}$ $(a \geq 0)$

$(a^{1/2})^{1/2} = a^{1/4}$ Definition of fractional exponents

$a^{1/4} = a^{1/4}$ Property 2

97. $x^2(x^4 - x) = x^2(x^4) - x^2(x)$

$\qquad = x^6 - x^3$

101. $(x^2 - 5)^2 = (x^2 - 5)(x^2 - 5)$

$\qquad = x^4 - (2)5x^2 + 5^2$ $\qquad (a - b)^2 = a^2 - 2ab + b^2$

$\qquad = x^4 - 10x^2 + 25$

105. Let $x = 0$,

$5x - 4y = 10$

$5(0) - 4y = 10$

$-4y = 10$

$y = -\dfrac{10}{4}$

$y = -\dfrac{5}{2}$

y-intercept $-\dfrac{5}{2}$

Let $y = 0$

$5x - 4y = 10$

$5x - 4(0) = 10$

$5x = 10$

$x = 2$

x-intercept 2

109. a. Let A = 3 and t = 5,000

$$A \cdot 2^{-t/5,600} = 3 \cdot 2^{-5,000/5,600} = 1.62$$

b. Let A = 3 and t = 10,000

$$A \cdot 2^{-t/5,600} = 3 \cdot 2^{-10,000/5,600} = .87$$

c. Let A = 3 and t = 56,000

$$A \cdot 2^{-t/5,600} = 3 \cdot 2^{-56,000/5,600} = .00293$$

d. Let A = 3 and t = 112,000

$$A \cdot 2^{-t/5,600} = 3 \cdot 2^{-112,000/5,600} = 3 \cdot 2^{-20} = 1.62 \text{ micrograms}$$

Section 6.2

1. $x^{2/3}(x^{1/3} + x^{4/3}) = x^{2/3}x^{1/3} + x^{2/3}x^{4/3}$ Distributive property

$$= x^{3/3} + x^{6/3}$$ Add exponents

$$= x + x^2$$ Simplify

5. $2x^{1/3}(3x^{8/3} - 4x^{5/3} + 5x^{2/3})$

$$= 2x^{1/3}3x^{8/3} - 2x^{1/3}4x^{5/3} + 2x^{1/3}5x^{2/3}$$

$$= 6x^{9/3} - 8x^{2/6+10/6} + 10x^{3/3}$$

$$= 6x^3 - 8x^2 + 10x$$

9. $(x^{2/3} - 4)(x^{2/3} + 2)$

$$= x^{2/3}x^{2/3} + 2x^{2/3} - 4x^{2/3} - 8$$ FOIL Method

$$= x^{4/3} - 2x^{2/3} - 8$$

13. $(4y^{1/3} - 3)(5y^{1/3} + 2)$

$$= 4y^{1/3}5y^{1/3} + 4y^{1/3} \cdot 2 - 3 \cdot 5y^{1/3} - 3 \cdot 2$$

$$= 20y^{2/3} + 8y^{1/3} - 15y^{1/3} - 6$$

$$= 20y^{2/3} - 7y^{1/3} - 6$$

17. $(t^{1/2} + 5)^2 = (t^{1/2} + 5)(t^{1/2} + 5)$ Expanded form

$$= t^{1/2}t^{1/2} + 5t^{1/2} + 5t^{1/2} + 25$$ FOIL Method

$$= t + 10t^{1/2} + 25$$

We can obtain the same result by using the formula for the square of a binomial, $(a + b)^2 = a^2 + 2ab + b^2$:

$$(t^{1/2} + 5)^2 = (t^{1/2})^2 + 2t^{1/2} \cdot 5 + 5^2$$

$$= t + 10t^{1/2} + 25$$

21. $(a^{1/2} - b^{1/2})^2$

$= (a^{1/2} - b^{1/2})(a^{1/2} - b^{1/2})$ Expanded form

$= a^{1/2}a^{1/2} - a^{1/2}b^{1/2} - a^{1/2}b^{1/2} + b^{1/2}b^{1/2}$ FOIL Method

$= a - 2a^{1/2}b^{1/2} + b$

We can obtain the same result by using the formula for the square of a binomial, $(a - b)^2 = a^2 - 2ab + b^2$:

$$(a^{1/2} - b^{1/2})^2 = (a^{1/2})^2 - 2a^{1/2}b^{1/2} + (b^{1/2})^2$$
$$= a - 2a^{1/2}b^{1/2} + b$$

25. This product has the form $(a - b)(a + b)$, which will result in the difference of two squares, $a^2 - b^2$:

$$(a^{1/2} - 3^{1/2})(a^{1/2} + 3^{1/2}) = (a^{1/2})^2 - (3^{1/2})^2 = a - 3$$

29. This product has the form $(a - b)(a + b)$, which will result in the difference of two squares, $a^2 - b^2$:

$$(t^{1/2} - 2^{3/2})(t^{1/2} + 2^{3/2}) = (t^{1/2})^2 - (2^{3/2})^2$$
$$= t - 2^3$$
$$= t - 8$$

33.
$$\begin{array}{r} x^{2/3} - x^{1/3}y^{1/3} + y^{2/3} \\ x^{1/3} + y^{1/3} \\ \hline x^{2/3}y^{1/3} - x^{1/3}y^{2/3} + y \\ x - x^{2/3}y^{1/3} + x^{1/3}y^{1/3} \\ \hline x \qquad\qquad\qquad + y \end{array}$$

The product is $x + y$.

37.
$$\begin{array}{r} 4x^{2/3} - 2x^{1/3} + 1 \\ 2x^{1/3} + 1 \\ \hline 4x^{2/3} - 2x^{1/3} + 1 \\ 8x - 4x^{2/3} \quad 2x^{1/3} \\ \hline 8x \qquad\qquad + 1 \end{array}$$

The product is $8x + 1$.

41. $\dfrac{18x^{3/4} + 27x^{1/4}}{9x^{1/4}} = \dfrac{18x^{3/4}}{9x^{1/4}} + \dfrac{27x^{1/4}}{9x^{1/4}}$

$$= 2x^{1/2} + 3$$

45. $\dfrac{21a^{7/5}b^{3/5} - 14a^{2/5}b^{8/5}}{7a^{2/5}b^{3/5}} = \dfrac{21a^{7/5}b^{3/5}}{7a^{2/5}b^{3/5}} - \dfrac{14a^{2/5}\,b^{8/5}}{7a^{2/5}b^{3/5}}$

$$= 3a - 2b$$

49. This solution is similar to factoring out the greatest common factor:

$5(x - 3)^{12/5} - 15(x - 3)^{7/5}$

$\quad = 5(x - 3)^{2/5}[(x - 3)^2 - 3(x - 3)]$

$\quad = 5(x - 3)^{2/5}(x^2 - 6x + 9 - 3x + 9)$

$\quad = 5(x - 3)^{2/5}(x^2 - 9x + 18)$

$\quad = 5(x - 3)^{2/5}(x - 3)(x - 6)$

53. We can think of the expression in question as a trinomial in $x^{1/3}$:

$$x^{2/3} - 5x^{1/3} + 6 = (x^{1/3} - 2)(x^{1/3} - 3)$$

57. We can think of the expression in question as a trinomial in $y^{1/3}$:

$$2y^{2/3} - 5y^{1/3} - 3 = (2y^{1/3} + 1)(y^{1/3} - 3)$$

61. We can think of the expression in question as a trinomial in $x^{1/7}$:

$4x^{2/7} + 20x^{1/7} + 25 = (2x^{1/7} + 5)(2x^{1/7} + 5)$

$\qquad\qquad\qquad\quad = (2x^{1/7} + 5)^2$

65. To combine these two expressions, we need to find a least common denominator, change to equivalent fractions, and subtract numerators. The least common denominator is $x^{1/3}$.

$x^{2/3} + \dfrac{5}{x^{1/3}}$

$\quad = \dfrac{x^{2/3}}{1} \cdot \dfrac{x^{1/3}}{x^{1/3}} + \dfrac{5}{x^{1/3}}$

$\quad = \dfrac{x + 5}{x^{1/3}}$

69. To combine these two expressions, we need to find a least common denominator, change to equivalent fractions, and subtract numerators. The least common demonimator is $(x^2 + 4)^{1/2}$.

$$\frac{x^2}{(x^2 + 4)^{1/2}} - (x^2 + 4)^{1/2}$$

$$= \frac{x^2}{(x^2 + 4)^{1/2}} - \frac{(x^2 + 4)^{1/2}}{1} \cdot \frac{(x^2 + 4)^{1/2}}{(x^2 + 4)^{1/2}}$$

$$= \frac{x^2 - (x^2 + 4)}{(x^2 + 4)^{1/2}}$$

$$= \frac{-4}{(x^2 + 4)^{1/2}}$$

73. Press the keys in the following sequence:

$$1.5 \boxed{y^x} \, 9 \, \boxed{=}$$

$$9^{1.5} = 27$$

77. Using A = $900, P = $500 and t = 4 in the following formula:

$$r = \left(\frac{A}{P}\right)^{1/t} - 1$$

we have

$$r = \left(\frac{900}{500}\right)^{1/4} - 1$$

$$= (1.8)^{.25} - 1$$
$$= 1.158 - 1$$
$$= .158$$
$$= 15.8\%$$

The annual rate of return was 15.8%.

81. $m = \dfrac{y_2 - y_1}{x_2 - x_1} = \dfrac{5 - (-1)}{-2 - (-4)} = \dfrac{6}{2} = 3$

85. Slope = $\dfrac{2}{3}$ (Parallel lines have the same slope.)

1. $\sqrt{8} = \sqrt{4 \cdot 2}$

    $= \sqrt{4} \ \sqrt{2}$

    $= 2\sqrt{2} \qquad \sqrt{4} = 2$

5. $\sqrt{288} = \sqrt{144 \cdot 2}$

    $= \sqrt{144} \cdot \sqrt{2}$

    $= 12\sqrt{2}$

9. $\sqrt{48} = \sqrt{16 \cdot 3}$

    $= \sqrt{16} \cdot \sqrt{3}$

    $= 4\sqrt{3}$

13. $\sqrt[3]{54} = \sqrt[3]{27 \cdot 2}$

    $= \sqrt[3]{27} \cdot \sqrt[3]{2}$

    $= 3 \ \sqrt[3]{2}$

17. $\sqrt[3]{432} = \sqrt[3]{216 \cdot 2}$

    $= \sqrt[3]{216} \cdot \sqrt[3]{2}$

    $= 6 \ \sqrt[3]{2}$

21. $\sqrt{18x^3} = \sqrt{9x^2 \cdot 2x}$

    $= \sqrt{9x^2} \ \sqrt{2x}$

    $= 3x\sqrt{2x} \qquad \sqrt{9x^2} = 3x$

25. $\sqrt[3]{40x^4y^7} = \sqrt[3]{8x^3y^6 \cdot 5xy}$

    $= \sqrt[3]{8x^3y^6} \ \sqrt[3]{5xy}$

    $= 2xy^2 \ \sqrt[3]{5xy} \qquad \sqrt[3]{8x^3y^6} = 2xy^2$

29. $\sqrt[3]{48a^2b^3c^4} = \sqrt[3]{8b^3c^3 \cdot 6a^2c}$

    $= \sqrt[3]{8b^3c^3} \ \sqrt[3]{6a^2c}$

    $= 2bc \ \sqrt[3]{6a^2c}$

33. $\sqrt[5]{243x^7y^{10}z^5} = \sqrt[5]{243x^5y^{10}z^5 \cdot x^2}$

    $= \sqrt[5]{243x^5y^{10}z^5} \cdot \sqrt[5]{x^2}$

    $= 3xy^2z \ \sqrt[5]{x^2}$

37.   If $a = 1$, $b = 2$ and $c = 6$

    Then $\sqrt{b^2 - 4ac}$

    Becomes $\sqrt{2^2 - 4(1)(6)}$

    $= \sqrt{4 - 24}$

    $= -20$, which is not a real number

169

41. $\dfrac{2}{\sqrt{3}} = \dfrac{2}{\sqrt{3}} \cdot \dfrac{\sqrt{3}}{\sqrt{3}} = \dfrac{2\sqrt{3}}{\sqrt{3^2}} = \dfrac{2\sqrt{3}}{3}$

45. $\sqrt{\dfrac{1}{2}} = \dfrac{\sqrt{1}}{\sqrt{2}} = \dfrac{1}{\sqrt{2}} \cdot \dfrac{\sqrt{2}}{\sqrt{2}} = \dfrac{\sqrt{2}}{\sqrt{2^2}} = \dfrac{\sqrt{2}}{2}$

49. $\dfrac{4}{\sqrt[3]{2}} = \dfrac{4}{\sqrt[3]{2}} \cdot \dfrac{\sqrt[3]{2^2}}{\sqrt[3]{2^2}}$

$\qquad = \dfrac{4\sqrt[3]{2^2}}{\sqrt[3]{2^3}}$

$\qquad = \dfrac{4\sqrt[3]{4}}{2}$

$\qquad = 2\sqrt[3]{4}$

53. $\sqrt[4]{\dfrac{3}{2x^2}} = \dfrac{\sqrt[4]{3}}{\sqrt[4]{2x^2}} \cdot \dfrac{\sqrt[4]{2^3x^2}}{\sqrt[4]{2^3x^2}}$

$\qquad = \dfrac{\sqrt[4]{3 \cdot 2^3x^2}}{\sqrt[4]{2^4x^4}}$

$\qquad = \dfrac{\sqrt[4]{24x^2}}{2x}$

57. $\sqrt[3]{\dfrac{4x}{3y}} = \dfrac{\sqrt[3]{4x}}{\sqrt[3]{3y}} \cdot \dfrac{\sqrt[3]{3^2y^2}}{\sqrt[3]{3^2y^2}}$

$\qquad = \dfrac{\sqrt[3]{4 \cdot 3^2xy^2}}{\sqrt[3]{3^3y^3}}$

$\qquad = \dfrac{\sqrt[3]{36xy^2}}{3y}$

61. $\sqrt[4]{\dfrac{1}{8x^3}} = \dfrac{\sqrt[4]{1}}{\sqrt[4]{2^3x^3}} \cdot \dfrac{\sqrt[4]{2x}}{\sqrt[4]{2x}}$

$\qquad = \dfrac{\sqrt[4]{2x}}{\sqrt[4]{2^4x^4}}$

$\qquad = \dfrac{\sqrt[4]{2x}}{2x}$

65. $\sqrt{\dfrac{75x^3y^2}{2z}} = \dfrac{\sqrt{75x^3y^2}}{\sqrt{2z}}$

$= \dfrac{\sqrt{25x^2y^2}\,\sqrt{3x}}{\sqrt{2z}}$ Simplify the numerator first

$= \dfrac{5xy\sqrt{3x}}{\sqrt{2z}} \cdot \dfrac{\sqrt{2z}}{\sqrt{2z}}$

$= \dfrac{5xy\sqrt{6xz}}{\sqrt{2^2z^2}}$

$= \dfrac{5xy\sqrt{6xz}}{2z}$

69. $\sqrt[3]{\dfrac{8x^3y^6}{9z}} = \dfrac{\sqrt[3]{8x^3y^6}}{\sqrt[3]{9z}}$

$= \dfrac{2xy^2}{\sqrt[3]{9z}}$ Simplify the numerator first

$= \dfrac{2xy^2}{\sqrt[3]{3^2z}} \cdot \dfrac{\sqrt[3]{3z^2}}{\sqrt[3]{3z^2}}$

$= \dfrac{2xy^2\,\sqrt[3]{3z^2}}{\sqrt[3]{3^3z^3}}$

$= \dfrac{2xy^2\,\sqrt[3]{3z^2}}{3z}$

73. $\sqrt{27x^3y^2} = \sqrt{9x^2y^2 \cdot 3x}$

$= \sqrt{9x^2y^2} \cdot \sqrt{3x}$

$= 3|xy|\sqrt{3x}$

77. $\sqrt{4x^2 + 12x + 9} = \sqrt{(2x+3)^2}$

$= |2x + 3|$

81. $\sqrt{4x^3 - 8x^2} = \sqrt{4x^2(x-2)}$

$= \sqrt{4x^2} \cdot \sqrt{x-2}$

$= 2|x|\sqrt{x-2}$

85. Let $l = 10$ and $w = 15$

$d = \sqrt{l^2 + w^2}$

$d = \sqrt{10^2 + 15^2}$

$d = \sqrt{100 + 225}$

$d = \sqrt{325}$

$d = \sqrt{25 \cdot 13} = \sqrt{25} \cdot \sqrt{13} = 5\sqrt{13}$ feet

89. The boundary for the graph is the graph $y = -3x - 4$, the x-intercept is $-\frac{4}{3}$ $(0 = -3x - 4, x = -\frac{4}{3})$ and the y-intercept is -4 $(y = -3(0) = 4, y = -4)$. The boundary is a solid line because the inequality symbol is $\geq$. A convenient test point is the origin:

$$\text{Using } (0,0)$$
$$\text{in } y \geq -3x - 4$$
$$\text{we have } 0 \geq -3(0) - 4$$
$$0 \geq -4 \qquad \text{A true statement}$$

Since our test point is true, the region above the boundary is shaded.

93. $\sqrt[3]{8640} = \sqrt[3]{2^6 \cdot 3^3 \cdot 5}$

$\qquad = \sqrt[3]{2^6 \cdot 3^3} \cdot \sqrt[3]{5}$

$\qquad = 2^2 \cdot 3 \sqrt[3]{5}$

$\qquad = 12 \sqrt[3]{5}$

97. $\dfrac{1}{\sqrt[10]{a^3}} = \dfrac{1}{\sqrt[10]{a^3}} \cdot \dfrac{\sqrt[10]{a^7}}{\sqrt[10]{a^7}}$

$\qquad = \dfrac{\sqrt[10]{a^7}}{\sqrt[10]{a^{10}}}$

$\qquad = \dfrac{\sqrt[10]{a^7}}{a}$

101. $\dfrac{1}{\sqrt[n]{a^m}} = \dfrac{1}{\sqrt[n]{a^m}} \cdot \dfrac{\sqrt[n]{a^{n-m}}}{\sqrt[n]{a^{n-m}}}$

$\qquad = \dfrac{\sqrt[n]{a^{n-m}}}{\sqrt[n]{a^n}}$

$\qquad = \dfrac{\sqrt[n]{a^{n-m}}}{a}$

Section 6.4

1. $3\sqrt{5} + 4\sqrt{5} = (3 + 4)\sqrt{5} = 7\sqrt{5}$

5. $5\sqrt[3]{10} - 4\sqrt[3]{10} = (5 - 4)\sqrt[3]{10} = \sqrt[3]{10}$

9. $3x\sqrt{2} - 4x\sqrt{2} + x\sqrt{2} = (3x - 4x + x)\sqrt{2} = 0\sqrt{2} = 0$

13. $4\sqrt{8} - 2\sqrt{50} - 5\sqrt{72}$

$= 4\sqrt{4}\sqrt{2} - 2\sqrt{25}\sqrt{2} - 5\sqrt{36}\sqrt{2}$

$= 8\sqrt{2} - 10\sqrt{2} - 30\sqrt{2}$

$= (8 - 10 - 30)\sqrt{2}$

$= -32\sqrt{2}$

17. $5\sqrt[3]{16} - 4\sqrt[3]{54}$

$= 5\sqrt[3]{8}\sqrt[3]{2} - 4\sqrt[3]{27}\sqrt[3]{2}$

$= 5 \cdot 2\sqrt[3]{2} - 4 \cdot 3\sqrt[3]{2}$

$= 10\sqrt[3]{2} - 12\sqrt[3]{2}$

$= -2\sqrt[3]{2}$

21. $5a^2\sqrt{27ab^3} - 6b\sqrt{12a^5b}$

$= 5a^2\sqrt{9b^2}\sqrt{3ab} - 6b\sqrt{4a^4}\sqrt{3ab}$

$= 5a^2 \cdot 3b\sqrt{3ab} - 6b \cdot 2a^2\sqrt{3ab}$

$= 15a^2b\sqrt{3ab} - 12a^2b\sqrt{3ab}$

$= 3a^2b\sqrt{3ab}$

25. $5x\sqrt[4]{3y^5} + y\sqrt[4]{243x^4y} + \sqrt[4]{48x^4y^5}$

$= 5x\sqrt[4]{y^4}\sqrt[4]{3y} + y\sqrt[4]{81x^4}\sqrt[4]{3y} + \sqrt[4]{16x^4y^4}\sqrt[4]{3y}$

$= 5x \cdot y\sqrt[4]{3y} + y \cdot 3x\sqrt[4]{3y} + 2xy\sqrt[4]{3y}$

$= 5xy\sqrt[4]{3y} + 3xy\sqrt[4]{3y} + 2xy\sqrt[4]{3y}$

$= 10xy\sqrt[4]{3y}$

29. $\dfrac{\sqrt{5}}{3} + \dfrac{1}{\sqrt{5}} = \dfrac{\sqrt{5}}{3} + \dfrac{1}{\sqrt{5}} \cdot \dfrac{\sqrt{5}}{\sqrt{5}}$

$= \dfrac{\sqrt{5}}{3} + \dfrac{\sqrt{5}}{5}$

$= \dfrac{1}{3}\sqrt{5} + \dfrac{1}{5}\sqrt{5}$

$= \left(\dfrac{1}{3} + \dfrac{1}{5}\right)\sqrt{5}$

$= \dfrac{8}{15}\sqrt{5}$ $\dfrac{1}{3} + \dfrac{1}{5} = \dfrac{5}{15} + \dfrac{3}{15} = \dfrac{8}{15}$

33. $\dfrac{\sqrt{18}}{6} + \sqrt{\dfrac{1}{2}} + \dfrac{\sqrt{2}}{2} = \dfrac{\sqrt{18}}{6} + \dfrac{1}{\sqrt{2}} \cdot \dfrac{\sqrt{2}}{\sqrt{2}} + \dfrac{\sqrt{2}}{2}$

$$= \dfrac{3\sqrt{2}}{6} + \dfrac{\sqrt{2}}{2} + \dfrac{\sqrt{2}}{2}$$

$$= \dfrac{\sqrt{2}}{2} + \dfrac{\sqrt{2}}{2} + \dfrac{\sqrt{2}}{2}$$

$$= \dfrac{3\sqrt{2}}{2}$$

37. $\sqrt[3]{25} + \dfrac{3}{\sqrt[3]{5}} = \sqrt[3]{25} + \dfrac{3}{\sqrt[3]{5}} \cdot \dfrac{\sqrt[3]{25}}{\sqrt[3]{25}}$

$$= \sqrt[3]{25} + \dfrac{3\,\sqrt[3]{25}}{5}$$

$$= \left(1 + \dfrac{3}{5}\right) \sqrt[3]{25}$$

$$= \dfrac{8\,\sqrt[3]{25}}{5}$$

41. $\sqrt{8} + \sqrt{18} = 2.828 + 4.243 = 7.071$

$\quad\quad \sqrt{50} = 7.071$

$\quad\quad \sqrt{26} = 5.099$

$\sqrt{8} + \sqrt{18}$ is equal to the decimal approximation for $\sqrt{50}$.

45. $\sqrt{9 + 16} = \sqrt{25} = 5$ The right side corrected

49. Using the point slope-form, we have

$$y - y_1 = m(x - x_1)$$

$$y - 2 = \dfrac{2}{3}(x + 6)$$

$$y - 2 = \dfrac{2}{3}x + 4$$

$$y = \dfrac{2}{3}x + 6$$

53. If the x-intercept is -4, the coordinate is (-4,0). Using the point-slope form, we have

$$y - y_1 = m(x - x_1)$$

$$y - 0 = \dfrac{3}{4}(x + 4)$$

$$y = \dfrac{3}{4}x + 3$$

57. $5\sqrt{x^3 + 4x^2} - x\sqrt{25x + 100}$

$= 5\sqrt{x^2(x + 4)} - x\sqrt{25(x + 4)}$

$= 5\sqrt{x^2} \cdot \sqrt{x + 4} - x\sqrt{25} \cdot \sqrt{x + 4}$

$= 5x\sqrt{x + 4} - 5x\sqrt{x - 4}$

$= 0\sqrt{x + 4}$

$= 0$

Section 6.5

1. $\sqrt{6}\,\sqrt{3} = \sqrt{6 \cdot 3} = \sqrt{18} = \sqrt{9 \cdot 2} = 3\sqrt{2}$

5. $(4\sqrt{6})(2\sqrt{15})3\sqrt{10} = (4 \cdot 2 \cdot 3)(\sqrt{6} \cdot \sqrt{15} \cdot \sqrt{10})$

$= 24\sqrt{900}$

$= 24 \cdot 30$

$= 720$

9. $\sqrt{3}(\sqrt{2} - 3\sqrt{3}) = \sqrt{3} \cdot \sqrt{2} - \sqrt{3} \cdot 3\sqrt{3}$

$= \sqrt{6} - 3\sqrt{9}$

$= \sqrt{6} - 3 \cdot 3$

$= \sqrt{6} - 9$

13. $(\sqrt{3} + \sqrt{2})(3\sqrt{3} - \sqrt{2})$

$\qquad\qquad$ F $\qquad$ O $\qquad$ I $\qquad$ L

$= \sqrt{3} \cdot 3\sqrt{3} - \sqrt{3}\,\sqrt{2} + \sqrt{2} \cdot 3\sqrt{3} - \sqrt{2}\,\sqrt{2}$

$= 3 \cdot 3 - \sqrt{6} + 3\sqrt{6} - 2$

$= 9 + 2\sqrt{6} - 2$

$= 2\sqrt{6} + 7$

17. $(3\sqrt{6} + 4\sqrt{2})(\sqrt{6} + 2\sqrt{2})$

$\qquad\qquad$ F $\qquad\qquad$ O $\qquad\qquad$ I $\qquad\qquad$ L

$= 3\sqrt{6} \cdot \sqrt{6} + 3\sqrt{6} \cdot 2\sqrt{2} + 4\sqrt{2} \cdot \sqrt{6} + 4\sqrt{2} \cdot 2\sqrt{2}$

$= 3 \cdot 6 + 6\sqrt{12} + 4\sqrt{12} + 8 \cdot 2$

$= 18 + 10\sqrt{12} + 16$

$= 10\sqrt{12} + 34$

$= 10\sqrt{4 \cdot 3} + 34$

$= 20\sqrt{3} + 34$

21. Use the formula for the square of a difference,
$(a - b)^2 = a^2 - 2ab + b^2$:

$(\sqrt{x} - 3)^2 - (\sqrt{x})^2 - 2(\sqrt{x})(3) + (3)^2 = x^2 - 6\sqrt{x} + 9$

25. Use the formula for the square of a sum,
$(a + b)^2 = a^2 + 2ab + b^2$:

$$(\sqrt{x - 4} + 2)^2 = (\sqrt{x - 4})^2 + 2(\sqrt{x - 4})(2) + 2^2$$
$$= x - 4 + 4\sqrt{x - 4} + 4$$
$$= x + 4\sqrt{x - 4}$$

29. We notice the product is of the form $(a + b)(a - b)$, which always gives the difference of two squares, $a^2 - b^2$:

$$(\sqrt{3} - \sqrt{2})(\sqrt{3} + \sqrt{2}) = (\sqrt{3})^2 - (\sqrt{2})^2 = 3 - 2 = 1$$

33. $(5 - \sqrt{x})(5 + \sqrt{x}) = 5^2 - (\sqrt{x})^2$
$$= 25 - x$$

37. $(\sqrt{3} + 1)^3 = (\sqrt{3} + 1)(\sqrt{3} + 1)^2$
$$= (\sqrt{3} + 1)(3 + 2\sqrt{3} + 1)$$
$$= (\sqrt{3} + 1)(2\sqrt{3} + 4)$$
$$= 2 \cdot 3 + 4\sqrt{3} + 2\sqrt{3} + 4$$
$$= 10 + 6\sqrt{3}$$

41. $\dfrac{\sqrt{5}}{\sqrt{5} + 1} = \dfrac{\sqrt{5}}{\sqrt{5} + 1} \cdot \dfrac{\sqrt{5} - 1}{\sqrt{5} - 1}$

$$= \dfrac{\sqrt{5}\sqrt{5} - \sqrt{5} \cdot 1}{(\sqrt{5})^2 - \sqrt{1}^2}$$

$$= \dfrac{5 - \sqrt{5}}{5 - 1}$$

$$= \dfrac{5 - \sqrt{5}}{4}$$

45. $\dfrac{\sqrt{5}}{2\sqrt{5} - 3} = \dfrac{\sqrt{5}}{2\sqrt{5} - 3} \cdot \dfrac{2\sqrt{5} + 3}{2\sqrt{5} + 3}$

$$= \dfrac{\sqrt{5} \cdot 2\sqrt{5} + \sqrt{5} \cdot 3}{(2\sqrt{5})^2 - 3^2}$$

$$= \dfrac{10 + 3\sqrt{5}}{20 - 9}$$

$$= \dfrac{10 + 3\sqrt{5}}{11}$$

49. $\dfrac{\sqrt{6} + \sqrt{2}}{\sqrt{6} - \sqrt{2}} = \dfrac{\sqrt{6} + \sqrt{2}}{\sqrt{6} - \sqrt{2}} \cdot \dfrac{\sqrt{6} + \sqrt{2}}{\sqrt{6} + \sqrt{2}}$

$$= \dfrac{(\sqrt{6})^2 + 2\sqrt{2}\sqrt{6} + (\sqrt{2})^2}{(\sqrt{6})^2 - (\sqrt{2})^2}$$

$$= \dfrac{6 + 2\sqrt{12} + 2}{6 - 2}$$

$$= \dfrac{2\sqrt{12} + 8}{4}$$

$$= \dfrac{2\sqrt{4 \cdot 3} + 8}{4}$$

$$= \dfrac{4\sqrt{3} + 8}{4}$$

$$= \dfrac{4(\sqrt{3} + 2)}{4}$$

$$= \sqrt{3} + 2$$

53. $\dfrac{\sqrt{a} + \sqrt{b}}{\sqrt{a} - \sqrt{b}} = \dfrac{\sqrt{a} + \sqrt{b}}{\sqrt{a} - \sqrt{b}} \cdot \dfrac{\sqrt{a} + \sqrt{b}}{\sqrt{a} + \sqrt{b}}$

$$= \dfrac{(\sqrt{a})^2 + 2\sqrt{a}\sqrt{b} + (\sqrt{b})^2}{(\sqrt{a})^2 - (\sqrt{b})^2}$$

$$= \dfrac{a + 2\sqrt{ab} + b}{a - b}$$

57. $\dfrac{2\sqrt{3} - \sqrt{7}}{3\sqrt{3} + \sqrt{7}} = \dfrac{2\sqrt{3} - \sqrt{7}}{3\sqrt{3} + \sqrt{7}} \cdot \dfrac{3\sqrt{3} - \sqrt{7}}{3\sqrt{3} - \sqrt{7}}$

$$= \dfrac{2\sqrt{3} \cdot 3\sqrt{3} - 2\sqrt{3} \cdot \sqrt{7} - \sqrt{7} \cdot 3\sqrt{3} + (\sqrt{7})^2}{(3\sqrt{3})^2 - (\sqrt{7})^2}$$

$$= \dfrac{6 \cdot 3 - 2\sqrt{21} - 3\sqrt{21} + 7}{9 \cdot 3 - 7}$$

$$= \dfrac{18 - 5\sqrt{21} + 7}{27 - 7}$$

$$= \dfrac{25 - 5\sqrt{21}}{20}$$

$$= \dfrac{5(5 - \sqrt{21})}{20}$$

$$= \dfrac{5 - \sqrt{21}}{4}$$

61.

$$\sqrt[3]{4} \ - \ \sqrt[3]{6} \ + \ \sqrt[3]{9}$$
$$\sqrt[3]{2} \ + \ \sqrt[3]{3}$$
$$\overline{\sqrt[3]{12} \ - \ \sqrt[3]{18} \ + \ 3}$$
$$\frac{2 \ - \ \sqrt[3]{12} \ + \ \sqrt[3]{18}}{2 \qquad\qquad\qquad + \quad 3}$$

The product is 5.

65. $(\sqrt{x} + 3)^2 = (\sqrt{x} + 3)(\sqrt{x} + 3)$

$$= (\sqrt{x})^2 + 2\sqrt{x} \cdot 3 + 3^2$$

$$= x + 6\sqrt{x} + 9 \qquad\qquad \text{The right side corrected}$$

69. If $h = 50$

Then $t = \dfrac{\sqrt{100 - h}}{4}$

Becomes $t = \dfrac{\sqrt{100 - 50}}{4}$

$$= \frac{\sqrt{50}}{4}$$

$$= \frac{\sqrt{25 \cdot 2}}{4}$$

$$= \frac{5\sqrt{2}}{4}$$

It takes $\dfrac{5\sqrt{2}}{2}$ seconds for the object to be 50 feet from the ground.

If $h = 0$

Then $t = \dfrac{\sqrt{100 - h}}{4}$

Becomes $t = \dfrac{\sqrt{100 - 0}}{4}$

$$= \frac{\sqrt{100}}{4}$$

$$= \frac{10}{4}$$

$$= \frac{5}{2}$$

It takes the object $\dfrac{5}{2}$ seconds to reach the ground.

73. In $y = -\frac{2}{3}x + 1$, the slope is $-\frac{2}{3}$ and y-intercept is 1. See the graph on page A42 in the textbook.

77. $\dfrac{\sqrt{x-4}}{\sqrt{x-4}+2} = \dfrac{\sqrt{x-4}}{\sqrt{x-4}+2} \cdot \dfrac{\sqrt{x-4}-2}{\sqrt{x-4}-2}$

$$= \frac{x-4-2\sqrt{x-4}}{x-4-4}$$

$$= \frac{x-4-2\sqrt{x-4}}{x-8}$$

81. $\dfrac{1}{\sqrt[3]{x}+2} = \dfrac{1}{\sqrt[3]{x}+2} \cdot \dfrac{\sqrt[3]{x^2}-2\sqrt[3]{x}+4}{\sqrt[3]{x^2}-2\sqrt[3]{x}+4}$

$$= \frac{\sqrt[3]{x^2}-2\sqrt[3]{x}+4}{x+8}$$

Remember $a^3 + b^3 = (a+b)(a^2 - ab + b^2)$

Section 6.6

1. $\sqrt{2x+1} = 3$

$(\sqrt{2x+1})^2 = 3^2$

$2x + 1 = 9$

$2x = 8$

$x = 4$

Checking $x = 4$ in the original equation gives

$\sqrt{2(4)+1} = 3$

$\sqrt{8+1} = 3$

$\sqrt{9} = 3$

$3 = 3$

The solution $x = 4$ satisfies the original equation.

5. $\sqrt{2y-1} = 3$

$(\sqrt{2y-1})^2 = 3^2$

$2y - 1 = 9$

$2y = 10$

$y = 5$

Checking $y = 5$ in the original equation gives

$\sqrt{2(5)-1} = 3$

$\sqrt{10-1} = 3$

$\sqrt{9} = 3$

$3 = 3$

The solution $y = 5$ satisfies the original equation.

9. $\sqrt{2x - 3} - 2 = 4$

$\sqrt{2x - 3} = 6$

$(\sqrt{2x - 3})^2 = 6^2$

$2x - 3 = 36$

$2x = 39$

$x = \dfrac{39}{2}$

Checking $x = \dfrac{39}{2}$ in the original equation, we have

$$\sqrt{2\left(\dfrac{39}{2}\right) - 3} - 2 = 4$$

$$\sqrt{39 - 3} - 2 = 4$$

$$\sqrt{36} - 2 = 4$$

$$6 - 2 = 4$$

$$4 = 4$$

The solution $x = \dfrac{39}{2}$ satisfies the original equation.

13. $\sqrt[4]{3x + 1} = 2$

$(\sqrt[4]{3x + 1})^4 = 2^4$

$3x + 1 = 16$

$3x = 15$

$x = 5$

Checking $x = 5$ in the original equation, we have

$$\sqrt[4]{3(5) + 1} = 2$$

$$\sqrt[4]{15 + 1} = 2$$

$$\sqrt[4]{16} = 2$$

$$2 = 2$$

The solution $x = 5$ satisfies the original equation.

17. $\sqrt[3]{3a + 5} = -3$

$(\sqrt[3]{3a + 5})^3 = (-3)^3$

$3a + 5 = -27$

$3a = -32$

$a = -\dfrac{32}{3}$

Raising both sides of an equation to an odd power will not produce extraneous solutions, therefore the solution is $a = -\dfrac{32}{3}$.

21.
$$\sqrt{a + 2} = a + 2$$
$$(\sqrt{a + 2})^2 = (a + 2)^2 \qquad \text{Square both sides}$$
$$a + 2 = a^2 + 4a + 4$$
$$0 = a^2 + 3a + 2 \qquad \text{Standard form}$$
$$0 = (a + 1)(a + 2) \qquad \text{Factor the right side}$$

$$a + 1 = 0 \quad \text{or} \quad a + 2 = 0 \qquad \text{Set factors equal to 0}$$
$$a = -1 \qquad\qquad a = -2$$

We must check each solution in the original equation:

Check $a = -1$
$$\sqrt{-1 + 2} \overset{?}{=} -1 + 2$$
$$\sqrt{1} = 1$$
$$1 = 1$$
A true statement

Check $a = -2$
$$\sqrt{-2 + 2} = -2 + 2$$
$$\sqrt{0} = 0$$
$$0 = 0$$
A true statement

The solutions are -1 and -2.

25.
$$\sqrt{4a + 7} = -\sqrt{a + 2}$$
$$(\sqrt{4a + 7})^2 = (-\sqrt{a + 2})^2 \qquad \text{Square both sides}$$
$$4a + 7 = a + 2$$
$$3a + 7 = 2$$
$$3a = -5$$
$$a = -\frac{5}{3}$$

Checking $a = -\frac{5}{3}$ in our original equation, we have

$$\sqrt{4(-\frac{5}{3}) + 7} = -\sqrt{-\frac{5}{3} + 2}$$

$$\sqrt{-\frac{20}{3} + \frac{21}{3}} = -\sqrt{-\frac{5}{3} + \frac{6}{3}}$$

$$\sqrt{\frac{1}{3}} = -\sqrt{\frac{1}{3}}$$

A false statement

Since $a = -\frac{5}{3}$ was our only possible solution, there is no solution to our equation.

29.
$$x + 1 = \sqrt{5x + 1}$$

$$(x + 1)^2 = (\sqrt{5x + 1})^2 \qquad \text{Square both sides}$$

$$x^2 + 2x + 1 = 5x + 1$$

$$x^2 - 3x = 0 \qquad \text{Standard form}$$

$$x(x - 3) = 0 \qquad \text{Factor the left side}$$

$$x = 0 \quad \text{or} \quad x - 3 = 0 \qquad \text{Set factors equal to 0}$$

$$x = 3$$

We must check each solution in the original equation.

Check $x = 0$ Check $x = 3$

$$0 + 1 = \sqrt{5(0) + 1} \qquad\qquad 3 + 1 = \sqrt{5(3) + 1}$$

$$1 = \sqrt{1} \qquad\qquad\qquad\quad 4 = \sqrt{15 + 1}$$

$$1 = 1 \qquad\qquad\qquad\qquad 4 = \sqrt{16}$$

$$4 = 4$$

A true statement   A true statement

The solutions are 0 and 3.

33.
$$\sqrt{y - 8} = \sqrt{8 - y}$$

$$(\sqrt{y - 8})^2 = (\sqrt{8 - y})^2$$

$$y - 8 = 8 - y$$

$$2y - 8 = 8$$

$$2y = 16$$

$$y = 8$$

Checking $y = 8$ in the original equation, we have

$$\sqrt{8 - 8} = \sqrt{8 - 8}$$

$$\sqrt{0} = \sqrt{0}$$

$$0 = 0$$

The solution $y = 8$ satisfies the original equation.

37. $\sqrt{x - 8} = \sqrt{x} - 2$

$(\sqrt{x - 8})^2 = (\sqrt{x} - 2)^2$ Square both sides

$x - 8 = (\sqrt{x} - 2)(\sqrt{x} - 2)$

$x - 8 = x - 4\sqrt{x} + 4$

$-8 = -4\sqrt{x} + 4$ Add -x to both sides

$-12 = -4\sqrt{x}$ Add -4 to both sides

$3 = \sqrt{x}$ Divide each side by -4

$9 = x$ Square each side

Checking x = 9 in the original equation, we have

$\sqrt{9 - 8} = \sqrt{9} - 2$

$\sqrt{1} = 3 - 2$

$1 = 1$

Our solution checks.

41. $\sqrt{x + 8} = \sqrt{x - 4} + 2$

$(\sqrt{x + 8})^2 = (\sqrt{x - 4} + 2)^2$ Square both sides

$x + 8 = (\sqrt{x - 4} + 2)(\sqrt{x - 4} + 2)$

$x + 8 = x - 4 + 4\sqrt{x - 4} + 4$

$x + 8 = x + 4\sqrt{x - 4}$ Simplify the right side

$8 = 4\sqrt{x - 4}$ Add -x to both sides

$2 = \sqrt{x - 4}$ Divide each side by 4

$2^2 = (\sqrt{x - 4})^2$ Square both sides

$4 = x - 4$

$8 = x$

Checking x = 8 in the original equation, we have

$\sqrt{8 + 8} = \sqrt{8 - 4} + 2$

$\sqrt{16} = \sqrt{4} + 2$

$4 = 2 + 2$

$4 = 4$

Our solution checks.

45.
$$\sqrt{x + 4} = 2 - \sqrt{2x}$$
$$(\sqrt{x + 4})^2 = (2 - \sqrt{2x})^2 \qquad \text{Square both sides}$$
$$x + 4 = 4 - 4\sqrt{2x} + 2x$$
$$-x + 4 = 4 - 4\sqrt{2x} \qquad \text{Add } -2x \text{ to each side}$$
$$-x = -4\sqrt{2x} \qquad \text{Add } -4 \text{ to each side}$$
$$\frac{x}{4} = \sqrt{2x} \qquad \text{Divide each side by } -4$$

$$\left(\frac{x}{4}\right)^2 = (\sqrt{2x})^2 \qquad \text{Square both sides}$$

$$\frac{x^2}{16} = 2x$$

$$x^2 = 32x \qquad \text{Multiply each side by 16}$$
$$x^2 - 32x = 0 \qquad \text{Standard form}$$
$$x(x - 32) = 0 \qquad \text{Factor the left side}$$

$$x = 0 \quad \text{or} \quad x - 32 = 0 \qquad \text{Set factors equal to 0}$$
$$x = 32$$

We must check each solution in the original equation.

Check $x = 0$

$$\sqrt{0 + 4} = 2 - \sqrt{2(0)}$$
$$\sqrt{4} = 2 - \sqrt{0}$$
$$2 = 2$$

A true statement

Check $x = 32$

$$\sqrt{32 + 4} = 2 - \sqrt{2(32)}$$
$$\sqrt{36} = 2 - \sqrt{64}$$
$$6 = 2 - 8$$
$$6 = -6$$

A false statement

Since $x = 32$ does not check, our only solution is $x = 0$.

49. Solving the following formula for h:

$$t = \frac{\sqrt{100 - h}}{4}$$

$$4t = \sqrt{100 - h} \qquad \text{Multiply both sides by 4}$$
$$(4t)^2 = (\sqrt{100 - h})^2 \qquad \text{Square both sides}$$
$$16t^2 = 100 - h \qquad \text{Add } -100 \text{ to each side}$$
$$16t^2 - 100 = -h \qquad \text{Divide both sides by } -1$$
$$-16t^2 + 100 = h$$
$$h = 100 - 16t^2$$

53. Let $x = 0$,

    $y = 2\sqrt{x}$

    $y = 2\sqrt{0}$

    $y = 0$

    Let $x = 1$,

    $y = 2\sqrt{x}$

    $y = 2\sqrt{1}$

    $y = 2$

    Let $x = 4$

    $y = 2\sqrt{x}$

    $y = 2\sqrt{4}$

    $y = 4$

    Table

x	y
0	0
1	2
4	4

    See the graph on page A42 in the textbook.

57. Let $x = 2$

    $y = \sqrt{x - 2}$

    $y = \sqrt{2 - 2}$

    $y = 0$

    Let $x = 3$

    $y = \sqrt{x - 2}$

    $y = \sqrt{3 - 2}$

    $y = 1$

    Let $x = 6$

    $y = \sqrt{x - 2}$

    $y = \sqrt{6 - 2}$

    $y = 2$

    Table

x	y
2	0
3	1
6	2

    See the graph on page A42 in the textbook.

61. Let $x = -1$

$y = 3\sqrt[3]{x}$

$y = 3\sqrt[3]{-1}$

$y = -1$

Let $x = 0$

$y = 3\sqrt[3]{x}$

$y = 3\sqrt[3]{0}$

$y = 0$

Let $x = 1$

$y = 3\sqrt[3]{x}$

$y = 3\sqrt[3]{1}$

$y = 3$

Table

x	y
-1	-3
0	0
1	3

See the graph on page A43 in the textbook.

65. $\sqrt{2}(\sqrt{3} - \sqrt{2}) = \sqrt{2}(\sqrt{3}) - \sqrt{2}(\sqrt{2}) = \sqrt{6} - 2$

69. $\dfrac{\sqrt{x}}{\sqrt{x} + 3} = \dfrac{\sqrt{x}}{\sqrt{x} + 3} \cdot \dfrac{\sqrt{x} - 3}{\sqrt{x} - 3}$

$= \dfrac{x - 3\sqrt{x}}{x - 9}$

73.

$$x + 1 = \sqrt[3]{4x + 4}$$

$$(x + 1)^3 = (\sqrt[3]{4x + 4})^3$$

$$x^3 + 3x^2 + 3x + 1 = 4x + 4$$

$$x^3 + 3x^2 - x - 3 = 0$$

$$x^2(x + 3) - 1(x + 3) = 0$$

$$(x^2 - 1)(x + 3) = 0$$

$$(x + 1)(x - 1)(x + 3) = 0$$

$x + 1 = 0$ or $x - 1 = 0$ or $x + 3 = 0$

$x = -1$ $x = 1$ $x = -3$

Remember $(x + 1)^3 = (x + 1)^2(x + 1)$

$= (x^2 + 2x + 1)(x + 1)$

$= x^3 + 3x^2 + 3x + 1$

Section 6.7

1. $\sqrt{-36} = \sqrt{36(-1)} = \sqrt{36}\sqrt{-1} = 6i$

5. $\sqrt{-72} = \sqrt{72(-1)} = \sqrt{72}\sqrt{-1} = 6\sqrt{2}\,i = 6i\sqrt{2}$

9. $i^{28} = (i^2)^{14} = (-1)^{14} = 1$

13. $i^{75} = (i^2)^{37} \cdot i = (-1)^{37} \cdot i = -1 \cdot i = -i$

17. Since the two complex numbers are equal, their real parts are equal and their imaginary parts are equal:

Given $2 - 5i = -x + 10yi$
Then $2 = -x$ and $-5i = 10yi$

$$-2 = x \qquad -\frac{1}{2} = y$$

21. Because of the statement given in problem 17, we have:

Given $(2x - 4) - 3i = 10 - 6yi$
Then $2x - 4 = 10$ and $-3i = -6yi$

$$2x = 14 \qquad \frac{1}{2} = y$$
$$x = 7$$

25. $(2 + 3i) + (3 + 6i) = (2 + 3) + (3 + 6)i = 5 + 9i$

29. $(5 + 2i) - (3 + 6i) = (5 - 3) + (2 - 6)i = 2 - 4i$

33. $[(3 + 2i) - (6 + i)] + (5 + i)$
$= [(3 - 6) + (2 - 1)i] + (5 + i)$
$= (-3 + i) + (5 + i)$
$= (-3 + 5) + (i + i)$
$= 2 + 2i$

37. $(3 + 2i) - [(3 - 4i) - (6 + 2i)]$
$= (3 + 2i) - [(3 - 6) + (-4 - 2)i]$
$= (3 + 2i) - (-3 - 6i)$
$= (3 + 3) + (2 + 6)i$
$= 9 + 8i$

41. $3i(4 + 5i) = 3i \cdot 4 + 3i(5i)$
$= 12i + 15i^2$
$= -15 + 12i$

45. $(3 + 2i)(4 + i)$

$$\qquad\quad \text{F} \qquad \text{O} \qquad \text{I} \qquad \text{L}$$
$= 3 \cdot 4 + 3 \cdot i + 2i \cdot 4 + 2i \cdot i$
$= 12 + 3i + 8i + 2i^2$

Combining similar terms and using the fact that $i^2 = -1$, we can simplify as follows:

$12 + 3i + 8i + 2i^2 = 12 + 11i + 2(-1)$
$= 12 + 11i - 2$
$= 10 + 11i$

49.
$$
\begin{aligned}
(1 + i)^3 &= (1 + i)^2(1 + i) \\
&= (1 + 2i + i^2)(1 + i) \\
&= (1 + 2i - 1)(1 + i) \\
&= (2i)(1 + i) \\
&= 2i + 2i^2 \\
&= 2i + 2(-1) \\
&= -2 + 2i
\end{aligned}
$$

53. Remember: $(a + b)^2 = a^2 + 2ab + b^2$

$$
\begin{aligned}
(2 + 5i)^2 &= 2^2 + 2(2)(5i) + (5i)^2 \\
&= 4 + 20i + 25i^2 \\
&= 4 + 20i + 25(-1) \\
&= 4 + 20i - 25 \\
&= -21 + 20i
\end{aligned}
$$

57. Remember: $(a - b)^2 = a^2 - 2ab + b^2$

$$
\begin{aligned}
(3 - 4i)^2 &= 3^2 - 2(3)(4i) + (4i)^2 \\
&= 9 - 24i + 16i^2 \\
&= 9 - 24i + 16(-1) \\
&= 9 - 24i - 16 \\
&= -7 - 24i
\end{aligned}
$$

61. Remember: $(a + b)(a - b) = a^2 - b^2$

$$
\begin{aligned}
(6 - 2i)(6 + 2i) &= 6^2 - (2i)^2 \\
&= 36 - 4i^2 \\
&= 36 - 4(-1) \\
&= 36 + 4 \\
&= 40
\end{aligned}
$$

65. Remember: $(a + b)(a - b) = a^2 - b^2$

$$
\begin{aligned}
(10 + 8i)(10 - 8i) &= 10^2 - (8i)^2 \\
&= 100 - 64i^2 \\
&= 100 - 64(-1) \\
&= 100 + 64 \\
&= 164
\end{aligned}
$$

69.
$$
\begin{aligned}
\frac{5 + 2i}{-i} &= \frac{5 + 2i}{-i} \cdot \frac{i}{i} \\
&= \frac{5i + 2i^2}{-i^2} \\
&= \frac{5i + 2(-1)}{-(-1)} \\
&= -2 + 5i
\end{aligned}
$$

73. $\dfrac{6}{-3 + 2i} = \dfrac{6}{-3 + 2i} \cdot \dfrac{(-3 - 2i)}{(-3 - 2i)}$

$\qquad = \dfrac{-18 - 12i}{(-3)^2 - (2i)^2}$

$\qquad = \dfrac{-18 - 12i}{9 - 4i^2}$

$\qquad = \dfrac{-18 - 12i}{9 - 4(-1)}$

$\qquad = \dfrac{-18 - 12i}{9 + 4}$

$\qquad = \dfrac{-18 - 12i}{13}$

$\qquad = -\dfrac{18}{13} - \dfrac{12}{13}i$

77. $\dfrac{5 + 4i}{3 + 6i} = \dfrac{5 + 4i}{3 + 6i} \cdot \dfrac{(3 - 6i)}{(3 - 6i)}$

$\qquad = \dfrac{15 - 30i + 12i - 24i^2}{(3)^2 - (6i)^2}$

$\qquad = \dfrac{15 - 18i - 24(-1)}{9 - 36i^2}$

$\qquad = \dfrac{15 - 18i + 24}{9 - 36(-1)}$

$\qquad = \dfrac{39 - 18i}{9 + 36}$

$\qquad = \dfrac{39 - 18i}{45}$

$\qquad = \dfrac{39}{45} - \dfrac{18}{45}i$

$\qquad = \dfrac{13}{15} - \dfrac{2}{5}i$

81.
$$\text{When} \quad y = 75$$
$$\text{and} \quad x = 5$$
$$\text{the equation} \quad y = Kx^2$$
$$\text{becomes} \quad 75 = K \cdot 5^2$$
$$75 = 25K$$
$$K = 3$$

The specific equation is

$\qquad y = 3x^2$

When x = 7, the last equation becomes

$\qquad y = 3 \cdot 7^2$
$\qquad y = 3 \cdot 49$
$\qquad y = 147$

85.
$$\text{When} \quad z = 40, \ x = 5 \text{ and } y = 2$$
the equation $\quad z = Kxy^2$
becomes
$$40 = K \cdot 5 \cdot 2^2$$
$$40 = K \cdot 5 \cdot 4$$
$$40 = 20K$$
$$K = 2$$

The specific equation is

$$z = 2xy^2$$

When $x = 2$ and $y = 5$, the last equation becomes

$$z = 2 \cdot 2 \cdot 5^2$$
$$z = 100$$

89. Let $x = 1 + i$ then

$$x^2 - 2x + 2 = 0 \qquad \text{becomes}$$
$$(1 + i)^2 - 2(1 + i) + 2 = 0$$
$$1 + 2i + i^2 - 2 - 2i + 2 = 0$$
$$1 + 2i - 1 - 2 - 2i + 2 = 0$$
$$0 = 0 \qquad \text{A true statement}$$

1. $\sqrt{49} = 7$

5. $16^{1/4} = \sqrt[4]{16} = 2$

9. $\sqrt[5]{32x^{15}y^{10}} = (32x^{15}y^{10})^{1/5}$ Theorem 6.1

         $= 32^{1/5}(x^{15})^{1/5}(y^{10})^{1/5}$ Property 3

         $= 2x^3y^2$

13. $x^{2/3} \cdot x^{4/3} = x^{2/3+4/3}$ Property 1

         $= x^{6/3}$ Add fractions

         $= x^2$ Reduce

17. $\dfrac{a^{3/5}}{a^{1/4}} = a^{3/5-1/4}$ Property 6

         $= a^{12/20-5/20}$ LCD is 20

         $= a^{7/20}$ Subtract fractions

21. $(3x^{1/2} + 5y^{1/2})(4x^{1/2} - 3y^{1/2})$

     $= 3x^{1/2}4x^{1/2} - 3x^{1/2}3y^{1/2} + 5y^{1/2}4x^{1/2} - 5y^{1/2}3y^{1/2}$

     $= 12x^1 - 9x^{1/2}y^{1/2} + 20x^{1/2}y^{1/2} - 15y^1$

     $= 12x + 11x^{1/2}y^{1/2} - 15y$

25. $\dfrac{28x^{5/6} + 14x^{7/6}}{7x^{1/3}} = \dfrac{28x^{5/6}}{7x^{1/3}} - \dfrac{14x^{7/6}}{7x^{1/3}}$

         $= 4x^{5/6-1/3} - 2x^{7/6-1/3}$

         $= 4x^{5/6-2/6} - 2x^{7/6-2/6}$

         $= 4x^{3/6} - 2x^{5/6}$

         $= 4x^{1/2} - 2x^{5/6}$

29. $x^{3/4} + \dfrac{5}{x^{1/4}} = \dfrac{x^{3/4}}{1} \cdot \dfrac{x^{1/4}}{x^{1/4}} + \dfrac{5}{x^{1/4}}$

         $= \dfrac{x}{x^{1/4}} + \dfrac{5}{x^{1/4}}$

         $= \dfrac{x + 5}{x^{1/4}}$

33. $\sqrt{50} = \sqrt{25 \cdot 2}$ $50 = 25 \cdot 2$

       $= \sqrt{25}\,\sqrt{2}$ Property 1

       $= 5\sqrt{2}$ $\sqrt{25} = 5$

37. $\sqrt{18x^2} = \sqrt{9x^2}\,\sqrt{2}$ Property 1

$\quad\quad\quad\quad = 3x\sqrt{2}$

41. $\sqrt[4]{32a^4b^5c^6} = \sqrt[4]{16a^4b^4c^4}\;\sqrt[4]{2bc^2}$

$\quad\quad\quad\quad\quad = 2abc^4\,\sqrt[4]{2bc^2}$

45. $\dfrac{6}{\sqrt[3]{2}} = \dfrac{6}{\sqrt[3]{2}} \cdot \dfrac{\sqrt[3]{2^2}}{\sqrt[3]{2^2}}$

$\quad\quad = \dfrac{6\,\sqrt[3]{2^2}}{\sqrt[3]{2^3}}$

$\quad\quad = \dfrac{\cancel{6}^{\,3}\,\sqrt[3]{4}}{\cancel{2}_{\,1}}$

$\quad\quad = 3\,\sqrt[3]{4}$

49. $\sqrt[3]{\dfrac{40x^2y^3}{3z}} = \dfrac{\sqrt[3]{40x^2y^3}}{\sqrt[3]{3z}}$

$\quad\quad\quad\quad = \dfrac{\sqrt[3]{8y^3}\;\sqrt[3]{5x^2}}{\sqrt[3]{3z}}$

$\quad\quad\quad\quad = \dfrac{2y\,\sqrt[3]{5x^2}}{\sqrt[3]{3z}} \cdot \dfrac{\sqrt[3]{3^2z^2}}{\sqrt[3]{3^2z^2}}$

$\quad\quad\quad\quad = \dfrac{2y\,\sqrt[3]{45x^2z^2}}{3z}$

53. $\sqrt{12} + \sqrt{3} = \sqrt{4}\;\sqrt{3} + \sqrt{3}$

$\quad\quad\quad\quad\quad = 2\sqrt{3} + 1\sqrt{3}$

$\quad\quad\quad\quad\quad = (2 + 1)\sqrt{3}$

$\quad\quad\quad\quad\quad = 3\sqrt{3}$

57. $3\sqrt{8} - 4\sqrt{72} + 5\sqrt{50}$

$\quad = 3\sqrt{4}\;\sqrt{2} - 4\sqrt{36}\;\sqrt{2} + 5\sqrt{25}\;\sqrt{2}$

$\quad = 6\sqrt{2} - 24\sqrt{2} + 25\sqrt{2}$

$\quad = (6 - 24 + 25)\sqrt{2}$

$\quad = 7\sqrt{2}$

61. $2x\,\sqrt[3]{xy^3z^2} - 6y\sqrt[3]{x^4z^2}$

$\quad = 2x\,\sqrt[3]{y^3}\;\sqrt[3]{xz^2} - 6y\,\sqrt[3]{x^3}\;\sqrt[3]{xz^2}$

$\quad = 2xy\,\sqrt[3]{xz^2} - 6xy\,\sqrt[3]{xz^2}$

$\quad = -4xy\,\sqrt[3]{xz^2}$

65. $(\sqrt{x} - 2)(\sqrt{x} - 3)$

$$\quad\quad\overset{\textbf{F}}{}\quad\quad\overset{\textbf{O}}{}\quad\quad\overset{\textbf{I}}{}\quad\overset{\textbf{L}}{}$$

$$= \sqrt{x}\ \sqrt{x} - \sqrt{x} \cdot 3 - 2\sqrt{x} + 6$$

$$= x - 3\sqrt{x} - 2\sqrt{x} + 6$$

$$= x - 5\sqrt{x} + 6$$

69. $(\sqrt{8} - \sqrt{2})(\sqrt{8} + \sqrt{2})$

$$= (\sqrt{8})^2 - (\sqrt{2})^2$$

$$= 8 - 2$$

$$= 6$$

73. $\dfrac{\sqrt{7} + \sqrt{5}}{\sqrt{7} - \sqrt{5}} = \dfrac{\sqrt{7} + \sqrt{5}}{\sqrt{7} - \sqrt{5}} \cdot \dfrac{\sqrt{7} + \sqrt{5}}{\sqrt{7} + \sqrt{5}}$

$$= \frac{7 + \sqrt{7}\ \sqrt{5} + \sqrt{5}\ \sqrt{7} + 5}{(\sqrt{7})^2 - (\sqrt{5})^2}$$

$$= \frac{\overset{6}{\cancel{12}} + \overset{1}{\cancel{2}}\sqrt{35}}{\underset{1}{\cancel{2}}}$$

$$= 6 + \sqrt{35}$$

77. $\sqrt{4a + 1} = 1$

$$(\sqrt{4a + 1})^2 = 1^2$$

$$4a + 1 = 1$$

$$4a = 0$$

$$a = 0$$

The solution is a = 0.

81. $\sqrt{3x + 1} - 3 = 1$

$$\sqrt{3x + 1} = 4$$

$$(\sqrt{3x + 1})^2 = 4^2$$

$$3x + 1 = 16$$

$$3x = 15$$

$$x = 5$$

The solution is x = 5.

85.

$$\sqrt{2y - 8} = y - 4$$

$$(\sqrt{2y - 8})^2 = (y - 4)^2 \quad\quad \text{Square both sides}$$

$$2y - 8 = y^2 - 8y + 16$$

$$0 = y^2 - 10y + 24$$

$$y^2 - 10y + 24 = 0 \quad\quad \text{Standard form}$$

$$(y - 4)(y - 6) = 0 \quad\quad \text{Factor the left side}$$

$$y - 4 = 0 \ \text{ or } \ y - 6 = 0 \quad\quad \text{Set factors equal to 0}$$

$$y = 4 \quad\quad\quad y = 6$$

Both solutions check in the original equation.

89. $y = \sqrt[3]{x} + 2$

x	y
-8	0
-1	1
0	2
1	3
8	4

See the graph on page A44 in the textbook.

93. $i^{24} = (i^2)^{12} = (-1)^{12} = 1$

97. $(3 + 5i) + (6 - 2i) = (3 + 6) + (5 - 2)i$
$$= 9 + 3i$$

101. $3i(4 + 2i) = 3i \cdot 4 + (3i)(2i)$
$$= 12i + 6i^2$$
$$= -6 + 12i$$

105. $(4 + 2i)^2 = (4)^2 + 2(4)(2i) + (2i)^2$
$$= 16 + 16i + 4i^2$$
$$= 16 + 16i - 4$$
$$= 12 + 16i$$

109. $\dfrac{3 + i}{i} = \dfrac{3 + i}{i} \cdot \dfrac{-i}{-i}$

$$= \dfrac{-3i - i^2}{-i^2}$$

$$= \dfrac{-3i - (-1)}{-(-1)}$$

$$= 1 - 3i$$

113. $\dfrac{4 - 3i}{4 + 3i} = \dfrac{4 - 3i}{4 + 3i} \cdot \mathbf{\dfrac{4 - 3i}{4 - 3i}}$

$$= \dfrac{16 - 2(4)(3i) + (3i)^2}{4^2 - (3i)^2}$$

$$= \dfrac{16 - 24i + 9i^2}{16 - 9i^2}$$

$$= \dfrac{16 - 24i + 9(-1)}{16 - 9(-1)}$$

$$= \dfrac{7 - 24i}{25}$$

$$= \dfrac{7}{25} - \dfrac{24}{25}i$$

1. $27^{-2/3} = \dfrac{1}{27^{2/3}}$

   $= \dfrac{1}{\left(27^{2/3}\right)^2}$

   $= \dfrac{1}{3^2}$

   $= \dfrac{1}{9}$

5. $\sqrt{49x^8y^{10}} = 7x^4y^5$

9. $2a^{1/2}(3a^{3/2} - 5a^{1/2})$

   $= 6a^{1/2+3/2} - 10a^{1/2+1/2}$

   $= 6a^2 - 10a$

13. $\dfrac{4}{x^{1/2}} + x^{1/2} = \dfrac{4}{x^{1/2}} + \dfrac{x^{1/2}}{1} \cdot \dfrac{x^{1/2}}{x^{1/2}}$

    $= \dfrac{4}{x^{1/2}} + \dfrac{x}{x^{1/2}}$

    $= \dfrac{x + 4}{x^{1/2}}$

17. $\sqrt{\dfrac{2}{3}} = \dfrac{\sqrt{2}}{\sqrt{3}} \cdot \dfrac{\sqrt{3}}{\sqrt{3}}$

    $= \dfrac{\sqrt{6}}{3}$

21. $(\sqrt{x} + 7)(\sqrt{x} - 4) = x - 4\sqrt{x} + 7\sqrt{x} - 28$

    $\phantom{(\sqrt{x} + 7)(\sqrt{x} - 4) = }\ $ F $\quad$ O $\quad$ I $\quad$ L

    $= x + 3\sqrt{x} - 28$

25. $\sqrt{3x + 1} = x - 3$

    $(\sqrt{3x + 1})^2 = (x - 3)^2$

    $3x + 1 = x^2 - 6x + 9$

    $0 = x^2 - 9x + 8$

    $0 = (x - 1)(x - 8)$

    $x - 1 = 0$ or $x - 8 = 0$

    $x = 1 \qquad x = 8$

    Possible solutions 1 and 8, only 8 checks; 8.

29.   Let $x = -1$

  $y = \sqrt[3]{-1} + 3$
  $y = -1 + 3$
  $y = 2$

  Let $x = 0$

  $y = \sqrt[3]{x} + 3$

  $y = \sqrt[3]{0} + 3$
  $y = 3$

  Let $x = 1$

  $y = \sqrt[3]{x} + 3$

  $y = \sqrt[3]{1} + 3$
  $y = 4$

Table	
x	y
-1	2
0	3
1	4

See the graph on page A45 in the textbook.

33. $(5 - 4i)^2 = (5 - 4i)(5 - 4i)$
  $= 25 - 40i + 16i^2$
  $= 25 - 40i + 16(-1)$
  $= 9 - 40i$

Chapter 7

Section 7.1

1. $x^2 = 25$

    $x = \pm\sqrt{25}$ Theorem 7.1
    $x = \pm 5$

    The solution set is $\{-5, 5\}$.

5. $y^2 = \dfrac{3}{4}$

    $y = \pm\sqrt{\dfrac{3}{4}}$ $\dfrac{3}{4} = \dfrac{\sqrt{3}}{\sqrt{4}} = \dfrac{\sqrt{3}}{2}$

    $y = \pm\dfrac{\sqrt{3}}{2}$

    The solution set is $\{-\dfrac{\sqrt{3}}{2}, \dfrac{\sqrt{3}}{2}\}$.

9. $4a^2 - 45 = 0$

    $4a^2 = 45$ Add 45 to both sides

    $a^2 = \dfrac{45}{4}$ Divide both sides by 4

    $a = \pm\sqrt{\dfrac{45}{4}}$ Theorem 7.1

    $a = \pm\dfrac{3\sqrt{5}}{2}$ $\sqrt{\dfrac{45}{4}} = \dfrac{\sqrt{9 \cdot 5}}{\sqrt{4}} = \dfrac{3\sqrt{5}}{2}$

    The solution set is $\{\dfrac{3\sqrt{5}}{2}, \dfrac{-3\sqrt{5}}{2}\}$.

13. $(2a + 3)^2 = -9$
    $2a + 3 = \pm\sqrt{-9}$ Theorem 7.1
    $2a + 3 = \pm 3i$
    $2a = -3 \pm 3i$ Add -3 to both sides

    $a = \dfrac{-3 \pm 3i}{2}$ Divide both sides by 2

    The solution set is $\{\dfrac{-3 + 3i}{2}, \dfrac{-3 - 3i}{2}\}$.

17. $x^2 + 8x + 16 = -27$
    $(x + 4)^2 = -27$ Write $x^2 + 8x + 16$ as $(x + 4)^2$
    $x + 4 = \pm\sqrt{-27}$ Theorem 7.1
    $x + 4 = \pm 3i\sqrt{3}$
    $x = -4 \pm 3i\sqrt{3}$

    The solution set is $\{-4 + 3i\sqrt{3}, -4 - 3i\sqrt{3}\}$.

21.
$$(x + 5)^2 + (x - 5)^2 = 52$$
$$x^2 + 10x + 25 + x^2 - 10x + 25 = 52$$
$$2x^2 + 50 = 52$$

$$2x^2 = 2 \qquad \text{Add } -50 \text{ to both sides}$$
$$x^2 = 1 \qquad \text{Divide both sides by 2}$$
$$x = \pm\sqrt{1} \qquad \text{Theorem 7.1}$$
$$x = \pm 1$$

The solution set is $\{-1, 1\}$.

25.
$$(3x + 4)(3x - 4) - (x + 2)(x - 2) = -4$$
$$9x^2 - 16 - (x^2 - 4) = -4$$
$$9x^2 - 16 - x^2 + 4 = -4$$
$$8x^2 - 12 = -4$$

$$8x^2 = 8 \qquad \text{Add } 12 \text{ to both sides}$$
$$x^2 = 1 \qquad \text{Divide both sides by 8}$$
$$x = \pm\sqrt{1} \qquad \text{Theorem 7.1}$$
$$x = \pm 1$$

29. $x^2 - 4x + \underline{\mathbf{4}} = (x - \underline{\mathbf{2}})^2$

33. $x^2 + 5x + \dfrac{25}{4} = (x + \dfrac{5}{2})^2$

37.
$$x^2 + 4x = 12$$
$$x^2 + 4x + \mathbf{4} = 12 + \mathbf{4} \qquad \text{Half of 4 is } \frac{4}{2} = 2, \text{ the square of which is 4}$$
$$(x + 2)^2 = 16$$

$$x + 2 = \pm\sqrt{16} \qquad \text{Theorem 7.1}$$
$$x + 2 = \pm 4 \qquad \text{Simplify the radical}$$
$$x = -2 \pm 4 \qquad \text{Add } -2 \text{ to both sides}$$

The last equation can be written as two separate statements:

$$x = -2 + 4 \qquad \text{or} \qquad x = -2 - 4$$
$$x = 2 \qquad\qquad\qquad x = -6$$

The solution set is $\{-6, 2\}$.

41.
$$a^2 - 2a + 5 = 0$$
$$a^2 - 2a = -5 \qquad \text{Add } -5 \text{ to each side}$$
$$a^2 - 2a + \mathbf{1} = -5 + \mathbf{1} \qquad \text{Half of 2 is } \frac{2}{2} = 1, \text{ the square of which is 1}$$
$$(a - 1)^2 = -4$$

$$a - 1 = \pm\sqrt{-4} \qquad \text{Theorem 7.1}$$
$$a - 1 = \pm 2i \qquad \sqrt{-4} = \sqrt{4}\,i = 2i$$
$$a = 1 \pm 2i$$

The solution set is $\{1 - 2i, 1 + 2i\}$.

45. $x^2 - 5x - 3 = 0$

$\qquad x^2 - 5x = 3$ Add 3 to each side

$x^2 - 5x + \dfrac{25}{4} = 3 + \dfrac{25}{4}$ Half of 5 is $\dfrac{5}{2}$, the square of which is $\dfrac{25}{4}$

$\left(x - \dfrac{5}{2}\right)^2 = \dfrac{37}{4}$ $3 + \dfrac{25}{4} = \dfrac{12}{4} + \dfrac{25}{4} = \dfrac{37}{4}$

$x - \dfrac{5}{2} = \pm\sqrt{\dfrac{37}{4}}$ Theorem 7.1

$x - \dfrac{5}{2} = \pm\dfrac{\sqrt{34}}{2}$

$x = \dfrac{5}{2} \pm \dfrac{\sqrt{34}}{2}$ Add $\dfrac{5}{2}$ to both sides

$x = \dfrac{5 \pm \sqrt{34}}{2}$ Simplify

The solution set is $\left\{\dfrac{5 - \sqrt{34}}{2}, \dfrac{5 + \sqrt{34}}{2}\right\}$.

49. $3t^2 - 8t + 1 = 0$

$\qquad 3t^2 - 8t = -1$ Add -1 to both sides

We cannot complete the square on the left side because the leading coefficient is not 1. We take an extra step and divide both sides by 3.

$\dfrac{3t^2}{3} - \dfrac{8t}{3} = -\dfrac{1}{3}$

$t^2 - \dfrac{8}{3}t = -\dfrac{1}{3}$

Half of $\dfrac{8}{3}$ is $\dfrac{4}{3}$, the square of which is $\dfrac{16}{9}$:

$t^2 - \dfrac{8}{3}t + \dfrac{16}{9} = -\dfrac{1}{3} + \dfrac{16}{9}$ Add $\dfrac{16}{9}$ to both sides

$\left(t - \dfrac{4}{3}\right)^2 = \dfrac{13}{9}$ $-\dfrac{1}{3} + \dfrac{16}{9} = -\dfrac{3}{9} + \dfrac{16}{9} = \dfrac{13}{9}$

$t - \dfrac{4}{3} = \pm\sqrt{\dfrac{13}{9}}$ Theorem 7.1

$t - \dfrac{4}{3} = \pm\dfrac{\sqrt{13}}{3}$

$t = \dfrac{4}{3} \pm \dfrac{\sqrt{13}}{3}$ Add $\dfrac{4}{3}$ to both sides

$t = \dfrac{4 \pm \sqrt{13}}{3}$ Simplify

The solution set is $\left\{\dfrac{4 - \sqrt{13}}{3}, \dfrac{4 + \sqrt{13}}{3}\right\}$.

53.   If $x = -2 + 3\sqrt{2}$
         Then $(x + 2)^2 = 18$
    Becomes $(-2 + 3\sqrt{2} + 2)^2 = 18$
                    $(3\sqrt{2})^2 = 18$
                    $9 \cdot 2 = 18$
                    $18 = 18$

57.   If $\sqrt{5} \cong 2.236$ If $\sqrt{5} \cong 2.236$

    Then $\dfrac{2 + \sqrt{5}}{2}$ Then $\dfrac{2 - \sqrt{5}}{2}$

    Becomes $\dfrac{2 + 2.236}{2}$ Becomes $\dfrac{2 - 2.236}{2}$

           $= \dfrac{4.236}{2}$ $= \dfrac{0.236}{2}$

           $= 2.118$ $= -0.118$

61. $(2x - 4)^2 = 12$

    $2x - 4 = \pm\sqrt{12}$

    $2x - 4 = \pm 2\sqrt{3}$ $\sqrt{12} = \sqrt{4}\,\sqrt{3} = 2\sqrt{3}$

    $2x = 4 \pm 2\sqrt{3}$ Add 4 to both sides

    $x = \dfrac{4 \pm 2\sqrt{3}}{2}$ Divide both sides by 2

    $x = \dfrac{2(2 \pm \sqrt{3})}{2}$

    $x = 2 \pm \sqrt{3}$

   The solution set is $\{2 - \sqrt{3}, 2 + \sqrt{3}\}$.

65. $V = \pi r^2 h$, solve for r

    $V = \pi r^2 h$

    $\dfrac{V}{\pi h} = r^2$ Divide both sides by πh

    $\pm\sqrt{\dfrac{V}{\pi h}} = r$ Theorem 7.1

69. $\sqrt{27y^5} = \sqrt{9y^4 \cdot 3y} = \sqrt{9y^4}\,\sqrt{3y} = 3y^2\sqrt{3y}$

73.   When $a = 6$, $b = 7$, and $c = -5$
   The expression $\sqrt{b^2 - 4ac}$
          Becomes $\sqrt{7^2 - 4(6)(-5)} = \sqrt{49 + 120} = \sqrt{169} = 13$

77. $\dfrac{2}{\sqrt[3]{4}} = \dfrac{2}{\sqrt[3]{4}} \quad \dfrac{\sqrt[3]{2}}{\sqrt[3]{2}}$

$= \dfrac{2 \sqrt[3]{2}}{\sqrt[3]{8}}$

$= \dfrac{2 \sqrt[3]{2}}{2}$

$= \sqrt[3]{2}$

81. $x^2 + 2ax = -a^2$

$x^2 + 2ax + \mathbf{a^2} = -a^2 + \mathbf{a^2}$ Half of 2a is $\dfrac{2a}{2} = a$, the square of which

                              is a^2

$(x + a)^2 = 0$

$\quad x + a = \pm\sqrt{0}$ Theorem 7.1

$\quad x + a = 0$

$\qquad\quad x = -a$ Add -a to both sides

The solution set is {-a}.

85. $x^2 + px + q = 0$

$\quad x^2 + px = -q$ Add -q to both sides

$x^2 + px + \dfrac{\mathbf{p^2}}{\mathbf{4}} = -q + \dfrac{\mathbf{p^2}}{\mathbf{4}}$ Half of p is $\dfrac{p}{2}$, the square of which

                           is $\dfrac{p^2}{4}$

$\left(x + \dfrac{p}{2}\right)^2 = -q + \dfrac{p^2}{4}$

$\left(x + \dfrac{p}{2}\right)^2 = \dfrac{-4q}{4} + \dfrac{p^2}{4}$ $-q = -\dfrac{q}{1} = -\dfrac{4q}{4}$

$\left(x + \dfrac{p}{2}\right)^2 = \dfrac{p^2 - 4q}{4}$

$\quad x + \dfrac{p}{2} = \pm\sqrt{\dfrac{p^2 - 4q}{4}}$ Theorem 7.1

$\qquad\quad x = -\dfrac{p}{2} \pm \dfrac{\sqrt{p^2 - 4q}}{2}$

$\qquad\quad x = \dfrac{-p \pm \sqrt{p^2 - 4q}}{2}$

Section 7.2

1. $x^2 + 5x + 6 = 0$

             $(x + 2)(x + 3) = 0$ Factor

$x + 2 = 0 \quad$ or $\quad x + 3 = 0$ Set factors to zero

$\quad x = -2 \qquad\qquad\quad = -3$

The solution set is {-2,-3}.

5.
$$\frac{1}{6}x^2 - \frac{1}{2}x + \frac{1}{3} = 0$$

$$6\left(\frac{1}{6}x^2\right) - 6\left(\frac{1}{2}x\right) + 6\left(\frac{1}{3}\right) = 0 \qquad \text{LCD is 6}$$

$$x^2 - 3x + 2 = 0$$
$$(x - 1)(x - 2) = 0 \qquad \text{Factor}$$

$x - 1 = 0 \qquad \text{or} \qquad x - 2 = 0 \qquad$ Set factors to zero
$\qquad x = 1 \qquad\qquad\qquad x = 2$

The solution set is {1,2}.

9.
$$y^2 - 5y = 0$$
$$y(y - 5) = 0 \qquad \text{Factor}$$

$y = 0 \qquad \text{or} \qquad y - 5 = 0 \qquad$ Set factors to zero
$\qquad\qquad\qquad\qquad y = 5$

The solution set is {0,5}.

13. To solve this equation, we must first put it in standard form.

$$\frac{2t^2}{3} - t = -\frac{1}{6}$$

$$6 \cdot \frac{2t^2}{3} - 6t = -\frac{1}{6} \cdot 6 \qquad \text{Multiply each by the LCD}$$

$$4t^2 - 6t = -1$$
$$4t^2 - 6t + 1 = 0 \qquad \text{Add -1 to each side}$$

This equation is not factorable; we must use the quadratic formula.
Using the coefficients a = 4, b = -6 and c = 1 in the formula

$$x = \frac{-b \pm \sqrt{b^2 - 4ac}}{2a}$$

we have $x = \dfrac{-(-6) \pm \sqrt{(-6)^2 - 4(4)(1)}}{2(4)}$

$$\text{or } x = \frac{6 \pm \sqrt{36 - 16}}{8}$$

$$= \frac{6 \pm \sqrt{20}}{8}$$

$$= \frac{6 \pm 2\sqrt{5}}{8} \qquad\qquad\qquad \sqrt{20} = \sqrt{4}\sqrt{5} = 2\sqrt{5}$$

$$= \frac{2(3 \pm \sqrt{5})}{8}$$

$$= \frac{3 \pm \sqrt{5}}{4}$$

The solution set is $\left\{\dfrac{3 + \sqrt{5}}{4}, \dfrac{3 - \sqrt{5}}{4}\right\}$.

17.
$$2x + 3 = -2x^2$$
$$2x^2 + 2x + 3 = 0 \qquad \text{Standard form}$$

This equation is not factorable; we must use the quadratic formula. Using the coefficients $a = 2$, $b = 2$ and $c = 3$ in the formula

$$x = \frac{-b \pm \sqrt{b^2 - 4ac}}{2a}$$

we have $x = \dfrac{-2 \pm \sqrt{4 - 4(2)(3)}}{2(2)}$

$$= \frac{-2 \pm \sqrt{4 - 24}}{4}$$

$$= \frac{-2 \pm \sqrt{-20}}{4}$$

$$= \frac{-2 \pm 2i\sqrt{5}}{4} \qquad\qquad \sqrt{-20} = \sqrt{4}\,\sqrt{5}\,i = 2i\sqrt{5}$$

$$= \frac{2(-1 \pm i\sqrt{5})}{2}$$

$$= -1 \pm i\sqrt{5}$$

The solution set is $\{-1 + i\sqrt{5}, -1 - i\sqrt{5}\}$.

21.
$$\frac{1}{2}r^2 = \frac{1}{6}r - \frac{2}{3}$$

$$6\left(\frac{1}{2}r^2\right) = 6\left(\frac{1}{6}r\right) - 6\left(\frac{2}{3}\right) \qquad \text{LCD is 6}$$

$$3r^2 = r - 4$$
$$3r^2 - r + 4 = 0 \qquad \text{Standard form}$$

This equation is not factorable; we must use the quadratic formula. Using the coefficients $a = 3$, $b = -1$ and $c = 4$ in the quadratic formula, we have

$$x = \frac{-(-1) \pm \sqrt{(-1)^2 - 4(3)(4)}}{2(3)}$$

$$= \frac{1 \pm \sqrt{1 - 48}}{6}$$

$$= \frac{1 \pm \sqrt{-47}}{6}$$

$$= \frac{1 \pm i\sqrt{47}}{6}$$

The solution set is $\left\{\dfrac{1 + i\sqrt{47}}{6}, \dfrac{1 - i\sqrt{47}}{6}\right\}$.

25. $(x + 3)^2 + (x - 8)(x - 1) = 16$

$x^2 + 6x + 9 + x^2 - 9x + 8 = 16$ Multiplication

$2x^2 - 3x + 17 = 16$

$2x^2 - 3x + 1 = 0$

$(2x - 1)(x - 1) = 0$

$2x - 1 = 0$ or $x - 1 = 0$

$2x = 1$ $x = 1$

$x = \dfrac{1}{2}$

The solution set is $\{\dfrac{1}{2}, 1\}$.

29.

$$\dfrac{1}{x + 1} - \dfrac{1}{x} = \dfrac{1}{2}$$

$$2x(x + 1)(\dfrac{1}{x + 1} - \dfrac{1}{x}) = \dfrac{1}{2} \cdot 2x(x + 1) \quad \text{Multiply each by the LCD}$$

$$2x(x + 1) \cdot \dfrac{1}{x + 1} - 2x(x + 1)\dfrac{1}{x} = \dfrac{1}{2} \cdot 2x(x + 1)$$

$2x - 2(x + 1) = x(x + 1)$

$2x - 2x - 2 = x^2 + x$ Multiplication

$-2 = x^2 + x$ Simplify left side

$0 = x^2 + x + 2$ Add 2 to each side

We must use the quadratic formula. Using the coefficients $a = 1$, $b = 1$ and $c = 2$ in the quadratic formula, we have

$$x = \dfrac{-1 \pm \sqrt{1 - 4(1)(2)}}{2(1)}$$

$$= \dfrac{-1 \pm \sqrt{-7}}{2}$$

$$= \dfrac{-1 \pm i\sqrt{7}}{2}$$

The solution set is $\{\dfrac{-1 + i\sqrt{7}}{2}, \dfrac{-1 - i\sqrt{7}}{2}\}$.

33.
$$\frac{1}{x+2} + \frac{1}{x+3} = 1$$

$$(x+2)(x+3)\left(\frac{1}{x+2} + \frac{1}{x+3}\right) = 1(x+2)(x+3)$$ Multiply each by the LCD

$$(x+2)(x+3)\frac{1}{x+2} + (x+2)(x+3)\frac{1}{x+3} = (x+2)(x+3)$$

$$x + 3 + x + 2 = x^2 + 5x + 6$$ Simplify and multiply

$$2x + 5 = x^2 + 5x + 6$$ Simplify

$$0 = x^2 + 3x + 1$$ Standard form

We must use the quadratic formula. Using the coefficients a = 1, b = 3 and c = 1 in the formula, we have

$$x = \frac{-3 \pm \sqrt{9 - 4(1)(1)}}{2(1)}$$

$$= \frac{-3 \pm \sqrt{5}}{2}$$

The solution set is $\{\frac{-3 + \sqrt{5}}{2}, \frac{-3 - \sqrt{5}}{2}\}$.

37.
$$x^3 - 8 = 0$$
$$(x - 2)(x^2 + 2x + 4) = 0$$ Factor as the difference of two cubes

$$x - 2 = 0 \quad \text{or} \quad x^2 + 2x + 4 = 0$$ Set each factor equal to 0

The first equation leads to a solution of x = 2. The second equation does not factor, so we must use the quadratic formula with a = 1, b = 2 and c = 4:

$$x = \frac{-2 \pm \sqrt{4 - 4(1)4}}{2(1)}$$

$$= \frac{-2 \pm \sqrt{4 - 16}}{2}$$

$$= \frac{-2 \pm \sqrt{-12}}{2}$$

$$= \frac{-2 \pm 2i\sqrt{3}}{2} \qquad \sqrt{-12} = \sqrt{4}\sqrt{3}\,i = 2i\sqrt{3}$$

$$= \frac{2(-1 \pm i\sqrt{3})}{2}$$

$$= -1 \pm i\sqrt{3}$$

The three solutions to our original equation are:

$$2, \ -1 + i\sqrt{3} \ \text{and} \ -1 - i\sqrt{3}$$

41.
$$125t^3 - 1 = 0$$
$$(5t - 1)(25t^2 + 5t + 1) = 0 \qquad \text{Factor as the difference of two cubes}$$
$$5t - 1 = 0 \quad \text{or} \quad 25t^2 + 5t + 1 = 0 \qquad \text{Set each factor equal to 0}$$

The first equation leads to a solution of $t = \frac{1}{5}$. The second equation does not factor, so we must use the quadratic formula with a = 25, b = 5 and c = 1:

$$x = \frac{-5 \pm \sqrt{25 - 4(25)(1)}}{2(25)}$$

$$= \frac{-5 \pm \sqrt{25 - 100}}{50}$$

$$= \frac{-5 \pm \sqrt{-75}}{50}$$

$$= \frac{-5 \pm 5i\sqrt{3}}{50} \qquad \sqrt{-75} = \sqrt{25}\,\sqrt{3}\,i = 5i\sqrt{3}$$

$$= \frac{5(-1 \pm i\sqrt{3})}{50}$$

$$= \frac{-1 \pm i\sqrt{3}}{10}$$

The three solutions to our original equation are:

$$\frac{1}{5}, \quad \frac{-1 + i\sqrt{3}}{10} \quad \text{and} \quad \frac{-1 - i\sqrt{3}}{10}$$

45.
$$3y^4 = 6y^3 - 6y^2$$
$$3y^4 - 6y^3 + 6y^2 = 0$$
$$3y^2(y^2 - 2y + 2) = 0$$

$$3y^2 = 0 \quad \text{or} \quad y^2 - 2y + 2 = 0 \qquad \text{Set each factor equal to 0}$$

The first equation leads to a solution of y = 0. The second equation does not factor, so we must use the quadratic formula with a = 1, b = -2 and c = 2:

$$x = \frac{-(-2) \pm \sqrt{4 - 4(1)(2)}}{2(1)}$$

$$= \frac{2 \pm \sqrt{4 - 8}}{2}$$

$$= \frac{2 \pm \sqrt{-4}}{2}$$

$$= \frac{2 \pm 2i}{2}$$

$$= \frac{2(1 \pm i)}{2}$$

$$= 1 \pm i$$

The three solutions to our original equation are:

$$0, \quad 1 + i \quad \text{and} \quad 1 - i$$

Section 7.2 continued

49. $\dfrac{-3 - 2i}{5}$

53. We let h = 4 and solve for t:

When $h = 4$
the equation $h = 20t - 16t^2$
becomes $4 = 20t - 16t^2$

$$\text{or} \quad 16t^2 - 20t + 4 = 0$$
$$4t^2 - 5t + 1 = 0 \qquad \text{Divide by 4}$$
$$(4t - 1)(t - 1) = 0$$

$$4t - 1 = 0 \quad \text{or} \quad t - 1 = 0 \qquad \text{Set factors to 0}$$
$$4t = 1 \qquad\qquad\quad t = 1$$
$$t = \frac{1}{4}$$

At $\frac{1}{4}$ and 1 seconds the object will be 4 feet off the ground.

57. To make a profit of $700 the company must take into account the revenue minus the cost.

    Profit = R - C

    If we let profit = 700, R = 10x - .002x² and c = 800 + 6.5x, we have

    $$700 = 10x - .002x^2 - (800 + 6.5x)$$
    $$700 = 10x - .002x^2 - 800 - 6.5x$$
    $$700 = 3.5x - .002x^2 - 800$$

    Putting this equation in standard form, we have

    $$.002x^2 - 3.5x + 1500 = 0$$

    Applying the quadratic formula to this equation with a = .002, b = -3.5 and c = 1500, we have

    $$x = \frac{-(-3.5) \pm \sqrt{12.25 - 4(.002)(1500)}}{2(.002)}$$

    $$= \frac{3.5 \pm \sqrt{12.25 - 12}}{2(.002)}$$

    $$= \frac{3.5 \pm \sqrt{.25}}{.004}$$

    $$= \frac{3.5 \pm .5}{.004}$$

    Writing this last expression as two expressions, we have our two solutions:

    $$x = \frac{3.5 + .5}{.004} \qquad \text{or} \qquad x = \frac{3.5 - .5}{.004}$$

    $$= \frac{4.0}{.004} \qquad\qquad\qquad = \frac{3.0}{.004}$$

    $$= 1,000 \qquad\qquad\qquad = 750$$

    The monthly profit will be $700 if they produce and sell 1,000 patterns and produce and sell 750 patterns.

61.

$$\begin{array}{r} x^2 + 7x - 12 \\ x + 2 \overline{)\ x^3 + 9x^2 + 26x + 24} \end{array}$$

Change signs

$$\begin{array}{r} -\quad - \\ +\ x^3 + 2x^2 \\ \hline 7x^2 + 26x \end{array}$$

Change signs

$$\begin{array}{r} -\quad - \\ +\ 7x^2 + 14x \\ \hline 12x + 24 \end{array}$$

Change signs

$$\begin{array}{r} +\quad + \\ -\ 12x - 24 \\ \hline 0 \end{array}$$

65. $\left(\dfrac{9}{25}\right)^{3/2} = \left[\left(\dfrac{9}{25}\right)^{1/2}\right]^3 \qquad \dfrac{3}{2} = \dfrac{1}{2} \cdot \dfrac{3}{1} = \dfrac{1}{2} \cdot 3$

$\qquad\qquad = \left(\dfrac{3}{5}\right)^3$

$\qquad\qquad = \dfrac{27}{125} \qquad\qquad 3^3 = 27,\ 5^3 = 125$

69. $\dfrac{(49x^8y^{-4})^{1/2}}{(27x^{-3}y^9)^{-1/3}} = \dfrac{(49)^{1/2}(x^8)^{1/2}(y^{-4})^{1/2}}{(27)^{-1/3}(x^{-3})^{-1/3}(y^9)^{-1/3}}$

$\qquad\qquad = \dfrac{7x^4y^{-2}}{(3)^{-1}xy^{-3}}$

$\qquad\qquad = 7 \cdot 3x^{4-1}y^{-2-(-3)}$

$\qquad\qquad = 21x^3y$

209

73. $\sqrt{2}\, x^2 + 2x - \sqrt{2} = 0$

We must use the quadratic formula. Using the coefficients $a = \sqrt{2}$, $b = 2$, and $c = -\sqrt{2}$ in the quadratic formula, we have

$$x = \frac{-2 \pm \sqrt{2^2 - 4(\sqrt{2})(-\sqrt{2})}}{2(\sqrt{2})}$$

$$= \frac{-2 \pm \sqrt{4 + 8}}{2\sqrt{2}}$$

$$= \frac{-2 \pm \sqrt{12}}{2\sqrt{2}}$$

$$= \frac{-2 \pm 2\sqrt{3}}{2\sqrt{2}}$$

$$= \frac{2(-1 \pm \sqrt{3})}{2\sqrt{2}}$$

$$= \frac{-1 \pm \sqrt{3}}{\sqrt{2}}$$

$$= \frac{-1 \pm \sqrt{3}}{\sqrt{2}} \cdot \frac{\sqrt{2}}{\sqrt{2}}$$

$$= \frac{-\sqrt{2} \pm \sqrt{6}}{2}$$

The solution set is $\{ \frac{-\sqrt{2} + \sqrt{6}}{2}, \frac{-\sqrt{2} - \sqrt{6}}{2} \}$.

77. $ix^2 + 3x + 4i = 0$

We must use the quadratic formula. Using the coefficients $a = i$, $b = 3$ and $c = 4i$ in the quadratic formula, we have

$$x = \frac{-3 \pm \sqrt{3^2 - 4(i)(4i)}}{2i}$$

$$= \frac{-3 \pm \sqrt{9 - 16i^2}}{2i}$$

$$= \frac{-3 \pm \sqrt{9 + 16}}{2i}$$

$$= \frac{-3 \pm \sqrt{25}}{2i}$$

$$= \frac{-3 \pm 5}{2i}$$

$$x = \frac{-3 + 5}{2i} \qquad \text{or} \qquad x = \frac{-3 - 5}{2i}$$

$$x = \frac{2}{2i} \qquad\qquad\qquad x = \frac{-8}{2i}$$

$$x = \frac{1}{i} \cdot \frac{i}{i} \qquad\qquad\quad x = \frac{-4}{i} \cdot \frac{i}{i}$$

$$x = \frac{i}{-1} \qquad\qquad\qquad x = \frac{-4i}{-1}$$

$$x = -i \qquad\qquad\qquad\quad x = 4i$$

The solution set is $\{-i, 4i\}$.

Section 7.3

1. $x^2 - 6x + 5 = 0$ Using $a = 1$, $b = -6$ and $c = 5$ in $b^2 - 4ac$, we have

$$(-6)^2 - 4(1)(5) = 36 - 20 = 16$$

The discriminant is a perfect square. Therefore, the equation has two rational solutions.

5. $x^2 + x - 1 = 0$ Using $a = 1$, $b = 1$ and $c = -1$ in $b^2 - 4ac$, we have

$$1^2 - 4(1)(-1) = 1 + 4 = 5$$

The discriminant is a positive number, but not a perfect square. Therefore, the equation will have two irrational solutions.

9. $$.1x^2 - .9 = 0$$
$$10(.1x^2) - 10(.9) = 10(0)$$
$$x^2 - 9 = 0$$

Using a = 1, b = 0 and c = -9 in $b^2 - 4ac$, we have

$$0^2 - 4(1)(-9) = 0 + 36 = 36$$

The discriminant is a perfect square. Therefore, the equation has two rational solutions.

13. $x^2 - kx + 25 = 0$

Using a = 1, b = -k and c = 25, we have

$$b^2 - 4ac = (-k)^2 - 4(1)(25) = k^2 - 100$$

An equation has exactly one rational solution when the discriminant is 0. We set the discriminant equal to 0 and solve:

$$k^2 - 100 = 0$$
$$k^2 = 100$$
$$k = \pm 10$$

Choosing k to be 10 or -10 will result in an equation with one rational solution.

17. $4x^2 - 12x + k = 0$

Using a = 4, b = -12 and c = k, we have

$$b^2 - 4ac = (-12)^2 - 4(4)(k) = 144 - 16k$$

An equation has exactly one rational solution when the discriminant is 0. We set the discriminant equal to 0 and solve:

$$144 - 16k = 0$$
$$-16k = -144$$
$$k = 9$$

Choosing k to be 9 will result in an equation with one rational solution.

21. $3x^2 - kx + 2 = 0$

Using $a = 3$, $b = -k$ and $c = 2$, we have

$$b^2 - 4ac = (-k)^2 - 4(3)(2) = k^2 - 24$$

An equation has exactly one rational solution when the discriminant is 0. We set the discriminant equal to 0 and solve:

$$k^2 - 24 = 0$$
$$k^2 = 24$$
$$k = \pm\sqrt{24}$$
$$k = \pm 2\sqrt{6} \qquad \sqrt{24} = \sqrt{4}\,\sqrt{6} = 2\sqrt{6}$$

Choosing k to be $2\sqrt{6}$ or $-2\sqrt{6}$ will result in an equation with one rational solution.

25. If $t = -3$ $t = 6$
Then $t + 3 = 0$ $t - 6 = 0$

Since $t + 3$ and $t - 6$ are 0, their product is also 0 by the zero-factor property. An equation with solutions -3 and 6 is

$$(t + 3)(t - 6) = 0 \qquad \text{Zero-factor property}$$
$$t^2 - 3t - 18 = 0 \qquad \text{Multiplication}$$

29. If $x = \dfrac{1}{2}$ $x = 3$
Then $2x - 1 = 0$ $x - 3 = 0$

Since $2x - 1$ and $x - 3$ are 0, their product is also 0 by the zero-factor property. An equation with solutions $\dfrac{1}{2}$ and 3 is

$$(2x - 1)(x - 3) = 0 \qquad \text{Zero-factor property}$$
$$2x^2 - 7x + 3 = 0 \qquad \text{Multiplication}$$

33. If $x = 3$ $x = -3$ $x = \dfrac{5}{6}$
Then $x - 3 = 0$ $x + 3 = 0$ $6x - 5 = 0$

Since $x - 3$, $x + 3$ and $6x - 5$ are all 0, their product is also 0 by the zero-factor property. An equation with solutions 3, -3 and $\dfrac{5}{6}$ is

$$(x - 3)(x + 3)(6x - 5) = 0 \qquad \text{Zero-factor property}$$
$$(x^2 - 9)(6x - 5) = 0 \qquad \text{Multiply first two binomials}$$
$$6x^3 - 5x^2 - 54x + 45 = 0 \qquad \text{Complete the multiplication}$$

37.   If $x = -\dfrac{2}{3}$ $x = \dfrac{2}{3}$ $x = 1$

Then $3x + 2 = 0$ $3x - 2 = 0$ $x - 1 = 0$

Since $3x + 2$, $3x - 2$ and $x - 1$ are all 0, their product is also 0 by the zero-factor property. An equation with solutions $-\dfrac{2}{3}$, $\dfrac{2}{3}$ and 1 is

$$(3x + 2)(3x - 2)(x - 1) = 0 \qquad \text{Zero-factor property}$$
$$(9x^2 - 4)(x - 1) = 0 \qquad \text{Multiply first two binomials}$$
$$9x^3 - 9x^2 - 4x + 4 = 0 \qquad \text{Complete the multiplication}$$

41.   If $x = 3$ $x = -5$ $x = -5$

Then $x - 3 = 0$ $x + 5 = 0$ $x + 5 = 0$

$$(x - 3)(x + 5)(x + 5) = 0 \qquad \text{Zero-factor property}$$
$$(x^2 - 2x - 15)(x + 5) = 0 \qquad \text{Multiply first two binomials}$$
$$x^3 + 7x^2 - 5x - 75 = 0 \qquad \text{Complete the multiplication}$$

45. If $x = -3$, then $x + 3$ is one of the factors.

$$x^3 + 6x^2 + 11x + 6 = 0$$
$$(x + 3)(x^2 + 3x + 2) = 0 \qquad \text{From long division by } x + 3$$
$$(x + 3)(x + 2)(x + 1) = 0 \qquad \text{Factoring the trinomial}$$

$x + 3 = 0$ or $x + 2 = 0$ or $x + 1 = 0$ Setting each factor to 0

$x = -3$ $x = -2$ $x = -1$ The three solutions

49. If x = 3, then x - 3 is one of the factors.

$$x^3 - 5x^2 + 8x = 6$$
$$x^3 - 5x^2 + 8x - 6 = 0 \qquad \text{Zero-factor property}$$
$$(x - 3)(x^2 - 2x + 2) = 0 \qquad \text{From long division by } x - 3$$

The trinomial is not factorable. Using the coefficients a = 1, b = -2 and c = 2 in the quadratic formula, we have

$$x = \frac{-(-2) \pm \sqrt{4 - 4(1)(2)}}{2(1)}$$

$$= \frac{2 \pm \sqrt{4 - 8}}{2}$$

$$= \frac{2 \pm \sqrt{-4}}{2}$$

$$= \frac{2 \pm 2i}{2}$$

$$= \frac{2(1 \pm i)}{2}$$

$$= 1 \pm i$$

The three solutions are 3, 1 + i and 1 - i.

53. $a^4(a^{3/2} - a^{1/2}) = a^4 a^{3/2} - a^4 a^{1/2}$

$$= a^{4+3/2} - a^{4+1/2}$$

$$= a^{11/2} - a^{9/2} \qquad\qquad 4 + \frac{3}{2} = \frac{8}{2} + \frac{3}{2} = \frac{11}{2} \text{ and}$$

$$= a^{2/2} \qquad\qquad\qquad\quad 4 + \frac{1}{2} = \frac{8}{2} + \frac{1}{2} = \frac{9}{2}$$

$$= a$$

57. $\dfrac{30x^{3/4} - 25x^{5/4}}{5x^{1/4}} = \dfrac{30x^{3/4}}{5x^{1/4}} - \dfrac{25x^{5/5}}{5x^{1/4}}$

$$= 6x^{2/4} - 5x^{4/4}$$

$$= 6x^{1/2} - 5x$$

61. $2x^{2/3} - 11x^{1/3} + 12 = (2x^{1/3} - 3)(x^{1/3} - 4)$

65. x = a is a solution of multiplicity 3.

Section 7.4

1. When
    $$y = x - 3$$
    the equation
    $$(x - 3)^2 + 3(x - 3) + 2 = 0$$
    becomes
    $$y^2 + 3y + 2 = 0$$
    $$(y + 1)(y + 2) = 0 \qquad \text{Factor}$$

    $$y + 1 = 0 \quad \text{or} \quad y + 2 = 0 \qquad \text{Set factors to 0}$$
    $$y = -1 \qquad\qquad y = -2$$

    Now we replace y with x - 3 and solve for x:

    $$x - 3 = -1 \quad \text{or} \quad x - 3 = -2$$
    $$x = 2 \qquad\qquad x = 1$$

    Another method of solving the equation is by expanding which gives us the following:

    $$(x - 3)^2 + 3(x - 3) + 2 = 0$$
    $$x^2 + 6x - 9 - 3x + 9 + 2 = 0 \qquad \text{Multiply}$$
    $$x^2 + 3x + 2 = 0 \qquad \text{Combine similar terms}$$
    $$(x + 1)(x + 2) = 0 \qquad \text{Factor}$$

    $$x + 1 = 0 \quad \text{or} \quad x + 2 = 0 \qquad \text{Set factors to 0}$$
    $$x = -1 \qquad\qquad x = -2$$

    As you can see, either method produces the same result.

5. When
    $$y = x^2$$
    the equation
    $$x - 6x^2 - 27 = 0$$
    becomes
    $$y^2 - 6y - 27 = 0$$
    $$(y + 3)(y - 9) = 0 \qquad \text{Factor}$$

    $$y + 3 = 0 \quad \text{or} \quad y - 9 = 0 \qquad \text{Set factors to 0}$$
    $$y = -3 \qquad\qquad y = 9$$

    Now we replace y with x^2 and solve for x:

    $$x^2 = -3 \quad \text{or} \quad x^2 = 9$$
    $$x = \pm \sqrt{-3} \qquad\quad x = \sqrt{9}$$
    $$x = \pm i\sqrt{3} \qquad\quad x = \pm 3$$

    The solution set is $\{3, -3, i\sqrt{3}, -i\sqrt{3}\}$.

9. Let $$x = 2a - 3$$
   the equation $$(2a - 3)^2 - 9(2a - 3) = -20$$
   becomes $$x^2 - 9x = -20$$

   $$x^2 - 9x + 20 = 0 \qquad \text{Standard form}$$
   $$(x - 4)(x - 5) = 0 \qquad \text{Factor}$$

   $$x - 4 = 0 \quad \text{or} \quad x - 5 = 0 \qquad \text{Set factors to 0}$$
   $$x = 4 \qquad\qquad x = 5$$

   Now we replace x with 2a - 3 and solve for a:

   $$2a - 3 = 4 \quad \text{or} \quad 2a - 3 = 5$$
   $$2a = 7 \qquad\qquad 2a = 8$$
   $$a = \frac{7}{2} \qquad\qquad a = 4$$

   By the expanding method, we have

   $$(2a - 3)^2 - 9(2a - 3) = -20$$
   $$(2a - 3)^2 - 9(2a - 3) + 20 = 0 \qquad \text{Standard form}$$
   $$4a^2 - 12a + 9 - 18a + 27 + 20 = 0 \qquad \text{Multiply}$$
   $$4a^2 - 30a + 56 = 0 \qquad \text{Combine terms}$$
   $$2a^2 - 15a + 28 = 0 \qquad \text{Divide 2 on both sides}$$
   $$(2a - 7)(a - 4) = 0 \qquad \text{Factor}$$

   $$2a - 7 = 0 \quad \text{or} \quad a - 4 = 0 \qquad \text{Set factors to 0}$$
   $$2a = 7 \qquad\qquad a = 4$$
   $$a = \frac{7}{2}$$

   Either method produces the solution set $\{\frac{7}{2}, 4\}$.

13. Let
$$x = t^2$$
the equation
$$6t = -t^2 + 5$$
becomes
$$6x^2 = -x + 5$$
$$6x^2 + x - 5 = 0 \qquad \text{Standard form}$$
$$(x + 1)(6x - 5) = 0 \qquad \text{Factor}$$

$$x + 1 = 0 \qquad \text{or} \qquad 6x - 5 = 0 \qquad \text{Set factors to 0}$$
$$x = -1 \qquad\qquad 6x = 5$$
$$x = \frac{5}{6}$$

Now we replace x with t^2 and solve for t:

$$t^2 = -1 \qquad \text{or} \qquad t^2 = \frac{5}{6}$$

$$t = \pm \sqrt{-1} \qquad\qquad t = \pm \sqrt{\frac{5}{6}}$$

$$t = \pm i \qquad\qquad = \pm \frac{\sqrt{5}}{\sqrt{6}} \cdot \frac{\sqrt{6}}{\sqrt{6}} \qquad \text{Rationalize the denominator}$$

$$= \pm \frac{\sqrt{30}}{6}$$

The solution set is $\{i, -i, \frac{\sqrt{30}}{6}, -\frac{\sqrt{30}}{6}\}$.

17. Let
$$y = x^{1/3}$$
the equation $\quad x^{2/3} + x^{1/3} - 6 = 0$
becomes
$$y^2 + y - 6 = 0$$
$$(y + 3)(y - 2) = 0 \qquad \text{Factor}$$

$$y + 3 = 0 \qquad \text{or} \qquad y - 2 = 0 \qquad \text{Set factors to 0}$$
$$y = -3 \qquad\qquad y = 2$$

Now we replace y with $x^{1/3}$ and solve for x:

$$x^{1/3} = -3 \qquad \text{or} \qquad x^{1/3} = 2$$
$$x = -27 \qquad\qquad x = 8$$

The solution set is $\{-27, 8\}$.

21.   Let $$y = \sqrt{x}$$

the equation $x - 7\sqrt{x} + 10 = 0$

becomes $y^2 - 7y + 10 = 0$ $(\sqrt{x})^2 = x$

                  $(y - 2)(y - 5) = 0$

$y - 2 = 0$ or $y - 5 = 0$

     $y = 2$ $y = 5$

Now we replace y with $\sqrt{x}$ and solve for x:

$\sqrt{x} = 2$ or $\sqrt{x} = 5$

    $x = 4$ $x = 25$

Since we squared both sides of each equation, we have the possibility of obtaining extraneous solutions. We have to check both solutions in our original equation:

When $x = 4$

the equation $x - 7\sqrt{x} + 10 = 0$

becomes $x - 7\sqrt{4} + 10 = 0$

                  $4 - 14 + 10 = 0$

                        $0 = 0$ A true statement.

This means 4 is a solution.

When $x = 25$

the equation $x - 7\sqrt{x} + 10 = 0$

becomes $25 - 7\sqrt{25} + 10 = 0$

               $25 - 35 + 10 = 0$

                        $0 = 0$ A true statement.

This means 5 is a solution.

The solution set is {4,5}.

25. Let $y = x^{1/5}$

the equation $2x^{2/5} - 3x^{1/5} - 2 = 0$

becomes $2y^2 - 3y - 2 = 0$

$(2y + 1)(y - 2) = 0$ Factor

$2y + 1 = 0$ or $y - 2 = 0$ Set factors to 0

$2y = -1$ $y = 2$

$y = -\dfrac{1}{2}$

Now we replace y with $x^{1/5}$ and solve for x:

$x^{1/5} = -\dfrac{1}{2}$ or $x^{1/5} = 2$

$x = -\dfrac{1}{32}$ $x = 32$

Remember, we do not have to check for extraneous solutions in this problem since we raised to an odd power.

29. Let $x = a - 2$

the equation $(a - 2) - 11\sqrt{a - 2} + 30 = 0$

becomes $x^2 - 11x + 30 = 0$ $(\sqrt{a - 2})^2 = a - 2$

$(x - 5)(x - 6) = 0$ Factor

$x - 5 = 0$ or $x - 6 = 0$ Set factors to 0

$x = 5$ $x = 6$

Now we replace x with a - 2 and solve for a - 2:

$\sqrt{a - 2} = 5$ or $\sqrt{a - 2} = 6$

$a - 2 = 25$ $a - 2 = 36$

$a = 27$ $a = 38$

Since we squared both sides of the equation, we would have to check each one in the original equation. Both solutions produce true statements. Therefore the solution set is {27,38}.

33. Let
$$y = 2x + 1$$

the equation $\quad (2x + 1) - 8\sqrt{2x + 1} + 15 = 0$
becomes $\qquad\qquad\qquad y^2 - 8y + 15 = 0 \qquad (\sqrt{2x + 1})^2$
$$= 2x - 1$$

$$(y - 3)(y - 5) = 0 \qquad \text{Factor}$$

$$
\begin{array}{lll}
y - 3 = 0 \quad\text{or}\quad & y - 5 = 0 & \text{Set factors to 0} \\
y = 3 & y = 5 &
\end{array}
$$

Now we replace y with $\sqrt{2x + 1}$ and solve for x:

$$
\begin{array}{lll}
\sqrt{2x + 1} = 3 \quad\text{or}\quad & \sqrt{2x + 1} = 5 \\
2x + 1 = 9 & 2x + 1 = 25 \\
2x = 8 & 2x = 24 \\
x = 4 & x = 12
\end{array}
$$

Since we squared both sides of the equation, we would have to check each one in the original equation. Both solutions produce true statements, therefore the solution set is {4,12}.

37. $16t^2 - vt - 20 = 0$

To solve this equation for t, we will use the quadratic formula. The coefficients are a = 16, b = -v and c = 20. Substituting these quantities into the quadratic formula, we have

$$t = \frac{-(-v) \pm \sqrt{v^2 - 4(16)(-20)}}{2(16)}$$

$$t = \frac{v \pm \sqrt{v^2 - 1280}}{32}$$

41. $kx^2 + 8x + 4 = 0$

To solve this equation, we will use the coefficients a = k, b = 8, and c = 4 in the quadratic formula. This will produce the following:

$$x = \frac{-8 \pm \sqrt{64 - 4(k)(4)}}{2(k)}$$

$$x = \frac{-8 \pm \sqrt{64 - 16k}}{2k}$$

$$x = \frac{-8 \pm \sqrt{16(4 - k)}}{2k}$$

$$x = \frac{-8 \pm 4\sqrt{4 - k}}{2k}$$

$$x = \frac{2(-4 \pm 2\sqrt{4 - k})}{2k}$$

$$x = \frac{-4 \pm 2\sqrt{4 - k}}{k}$$

45. $5\sqrt{7} - 2\sqrt{7} = (5 - 2)\sqrt{7} = 3\sqrt{7}$

49. $9x\sqrt{20x^3y^2} + 7y\sqrt{45x^5}$

$= 9x\sqrt{4x^2y^2}\,\sqrt{5x} + 7y\sqrt{9x^4}\,\sqrt{5x}$

$= 18x^2y\sqrt{5x} + 21x^2y\sqrt{5x}$

$= (18 + 21)x^2y\sqrt{5x}$

$= 39x^2y\sqrt{5x}$

53. $(\sqrt{x} + 2)^2 = (\sqrt{x})^2 + 2(2\sqrt{x}) + 2^2 \qquad (a + b)^2 = a^2 + 2ab + b^2$

$= x + 4\sqrt{x} + 4$

57. Let $y = 0$

$y = x^3 - 4x$

$0 = x^3 - 4x$

$0 = x(x^2 - 4)$

$0 = x(x + 2)(x - 2)$

$x = 0$ or $x + 2 = 0$ or $x - 2 = 0$

$\phantom{x = 0 \text{ or } x + 2} x = -2 x = 2$

The x-intercepts are 0,2,-2.

Let $x = 0$

$y = x^3 - 4x$

$y = 0^3 - 4(0)$

$y = 0$

The y-intercept is 0.

61. Let $ y = 0$

$y = 2x^3 - 7x^2 - 5x + 4$

$0 = (x - 4)(x + 1)(2x - 1)$

$x - 4 = 0$ or $x + 1 = 0$ or $2x - 1 = 0$

$ x = 4 x = -1 2x = 1$

$x = \dfrac{1}{2}$

The other x-intercepts are $-1, \dfrac{1}{2}$.

222

Section 7.5

1. $x^2 + x - 6 > 0$
$(x - 2)(x + 3) > 0$ Factor

The product is positive, so the factors must agree in sign.

```
- - - - - | - - - - - | + + + + +
←─────────────────────────────→     Sign of x - 2
- - - - - | + + + + + | + + + + +
←─────────────────────────────→     Sign of x + 3
         -3          2
```

See the graph on page A46 in your textbook.

5. $x^2 + 5x \geq -6$
$x^2 + 5x + 6 \geq 0$ Standard form
$(x + 2)(x + 3) \geq 0$

The product is positive, so the factors must agree in sign.

```
- - - - - | - - - - - | + + + + +
←─────────────────────────────→     Sign of x + 2
- - - - - | + + + + + | + + + + +
←─────────────────────────────→     Sign of x + 3
         -3          -2
```

See the graph on page A46 in your textbook.

9. $x^2 - 9 \leq 0$
$(x - 3)(x + 3) \leq 0$

The product is negative, so the factors must have opposite signs.

```
- - - - - | - - - - - | + + + + +
←─────────────────────────────→     Sign of x - 3
- - - - - | + + + + + | + + + + +
←─────────────────────────────→     Sign of x + 3
         -3          3
```

See the graph on page A46 in your textbook.

13. $200x^2 - 100x - 300 < 0$
$\quad\quad 2x^2 - x - 3 < 0$ Divide both sides by 100
$\quad (2x - 3)(x + 1) < 0$

The product is negative, so the factors must have opposite signs.

```
 - - - - - | - - - - - | + + + + +
<----------|-----------|---------->    Sign of 2x - 3
 - - - - - | + + + + + | + + + + +
<----------|-----------|---------->    Sign of x + 1
          -1          3/2
```

See the graph on page A46 in your textbook.

17. $10x^2 - 100x + 250 < 0$
$\quad\quad x^2 - 10x + 25 < 0$ Divide both sides by 10
$\quad\quad\quad (x - 5)^2 < 0$

This is a special case in which both factors are the same. Since $(x - 5)^2$ is always negative there is no real solution, $\emptyset$.

```
 - - - - - | + + + + +
<----------|---------->    Sign of x - 5
 - - - - - | + + + + +
<----------|---------->    Sign of x - 5
           5
```

This will only produce positive solutions.

21. $\quad\quad x^3 - x^2 - 9x + 9 \le 0$
$(x - 1)(x - 3)(x + 3) \le 0$

```
 - - - - - | - - - - - | - - - - - | + + + + +
<----------|-----------|-----------|---------->    Sign of x - 1
 - - - - - | - - - - - | + + + + + | + + + + +
<----------|-----------|-----------|---------->    Sign of x - 3
 - - - - - | + + + + + | + + + + + | + + + + +
<----------|-----------|-----------|---------->    Sign of x + 3
          -3          1           3
```

The original inequality indicates that the product is negative or zero. In order for this to happen, all three factors must be negative or exactly two must be positive. See the graph on page A46 in your textbook.

25. $\dfrac{3x}{x + 6} - \dfrac{8}{x + 6} < 0$

$\dfrac{3x - 8}{x + 6} < 0$

The quotient is negative, so the factors must have opposite signs.

```
- - - - - | - - - - - | + + + + +
<---------|-----------|----------->   Sign of 3x - 8
- - - - - | + + + + + | + + + + +
<---------|-----------|----------->   Sign of x + 6
         -6          8/3
```

See the graph on page A47 in your textbook.

29. $\dfrac{x - 2}{(x + 3)(x - 4)} < 0$

```
- - - - - | - - - - - | - - - - - | + + + + +
<---------|-----------|-----------|----------->   Sign of x - 2
- - - - - | - - - - - | + + + + + | + + + + +
<---------|-----------|-----------|----------->   Sign of x - 4
- - - - - | + + + + + | + + + + + | + + + + +
<---------|-----------|-----------|----------->   Sign of x + 3
         -3          2           4
```

Since the quotient is negative, the three factors must be negative or exactly two must be positive. See the graph on page A47 in your textbook.

225

33.
$$\frac{x + 7}{2x + 12} + \frac{6}{x^2 - 36} \leq 0$$

$$\frac{x + 7}{2(x + 6)} + \frac{6}{(x + 6)(x - 6)} \leq 0$$

$$\frac{x + 7}{2(x + 6)} \cdot \frac{(x - 6)}{(x - 6)} + \frac{6}{(x + 6)(x - 6)} \cdot \frac{2}{2} \leq 0 \cdot 2(x + 6)(x - 6)$$

$$\frac{x^2 + x - 42}{2(x + 6)(x - 6)} + \frac{12}{2(x + 6)(x - 6)} \leq 0$$

$$\frac{x^2 + x - 30}{2(x + 6)(x - 6)} \leq 0$$

$$\frac{(x + 6)(x - 5)}{2(x + 6)(x - 6)} \leq 0$$

$$\frac{x - 5}{2(x - 6)} \leq 0$$

Since the quotient is negative, the two factors are different. See the graph on page A47 in your textbook.

37. Since $y = x^2 - 25$, y will be positive whenever $x^2 - 25$ is positive and y will be negative whenever $x^2 - 25$ is negative. Since $x^2 - 25 = (x - 5)(x + 5)$, we have

The product $(x - 5)(x + 5)$ is   Positive   Negative   Positive

Since $y = x^2 - 25 = (x + 5)(x - 5)$
    y is positive for $x < -5$ and for $x > 5$
    y is negative for x between -5 and 5, which is the interval
    $-5 < x < 5$.

41. Let x = width, let 2x + 3 = length and area = 44,

$$A \leq L \times W$$
$$44 \leq (2x + 3)(x)$$
$$44 \leq 2x^2 + 3x$$
$$0 \leq 2x^2 + 3x - 44$$
$$0 \leq (2x + 11)(x - 4)$$

$$0 \leq 2x + 11 \qquad \text{or} \qquad 0 \leq x - 4$$
$$-11 \leq 2x \qquad\qquad\qquad 4 \leq x$$
$$-\frac{11}{2} \leq x$$

No solution

The width is at least 4 inches.

45. Let $\quad\quad y = 0 \quad\quad$ Let $\quad\quad x = 0$

$$5x - 3y = 15 \qquad\qquad 5x - 3y = 15$$
$$5x - 3(0) = 15 \qquad\qquad 5(0) - 3y = 15$$
$$5x = 15 \qquad\qquad\qquad -3y = 15$$
$$x = 3 \qquad\qquad\qquad\quad y = -5$$

The x-intercept is 3.   The y-intercept is -5.

49. $\sqrt{3t - 1} = 2$

$$3t - 1 = 4 \qquad \text{Square both sides}$$
$$3t = 5$$
$$t = \frac{5}{3}$$

53. Let $x = -7$

$y = \sqrt[3]{x - 1}$

$y = \sqrt[3]{-7 - 1}$

$y = \sqrt[3]{-8}$

$y = -2$

Let $x = 0$

$y = \sqrt[3]{x - 1}$

$y = \sqrt[3]{0 - 1}$

$y = -1$

Let $x = 2$

$y = \sqrt[3]{x - 1}$

$y = \sqrt[3]{2 - 1}$

$y = \sqrt[3]{1}$

$y = 1$

Let $x = 9$

$y = \sqrt[3]{x - 1}$

$y = \sqrt[3]{9 - 1}$

$y = \sqrt[3]{8}$

$y = 2$

Table

x	y
-7	-2
0	-1
2	1
9	2

See the graph on page A47 in the textbook.

57. $x^2 - 8x + 13 > 0$

Using the quadratic formula with $a = 1$, $b = -8$ and $c = 13$:

$$x = \frac{-(-8) \pm \sqrt{(-8)^2 - 4(1)(13)}}{2(1)}$$

$$x = \frac{8 \pm \sqrt{64 - 52}}{2}$$

$$x = \frac{8 \pm \sqrt{12}}{2}$$

$$x = \frac{8 \pm 2\sqrt{3}}{2}$$

$$x = \frac{2(4 \pm \sqrt{3})}{2}$$

$$x = 4 \pm \sqrt{3}$$

The product is positive so the factors must both be positive or negative.

```
 - - - - - |  - - - - - | + + + + +
<---------------------------------->          Sign of 4 + √3
 - - - - - | + + + + + | + + + + +
<---------------------------------->          Sign of 4 - √3
        4 - √3      4 + √3
```

See the graph on page A47 in the textbook.

Section 7.6

1. To find the x-intercepts, we let $y = 0$ and solve for x:

$$0 = x^2 + 2x - 3$$
$$0 = (x + 3)(x - 1)$$

$$x = -3 \quad \text{or} \quad x = 1$$

To find the coordinates of the vertex, we first find

$$x = \frac{-b}{2a} = \frac{-2}{2(1)} = -1$$

The x-coordinate of the vertex is -1. To find the y-coordinate we substitute -1 for x in our original equation:

$$y = (-1)^2 + 2(-1) - 3 = 1 - 2 - 3 = -4$$

The graph crosses the x-axis at -3 and 1 and has its vertex at $(-1, -4)$.

5. To find the x-intercepts, we let y = 0 and solve for x:

$$0 = x^2 - 1$$
$$0 = (x + 1)(x - 1)$$

$$x = -1 \quad \text{or} \quad x = 1$$

To find the coordinates of the vertex, we first find

$$x = \frac{-b}{2a} = \frac{-0}{2(1)} = 0$$

The x-coordinate of the vertex is 0. To find the y-coordinate we substitute 0 for x in our original equation:

$$y = (0)^2 - 1 = 0 - 1 = -1$$

The graph crosses the x-axis at -1 and 1 and has its vertex at (0,-1).

9. To find the x-intercepts

$$\begin{aligned} &\text{let} &y &= 0 \\ &\text{the equation} &y &= 2x^2 - 4x - 6 \\ &\text{becomes} &0 &= 2x^2 - 4x - 6 \\ & &0 &= x^2 - 2x - 3 \\ & &0 &= (x - 3)(x + 1) \end{aligned}$$

$$x = 3 \quad \text{or} \quad x = -1$$

To find the x-coordinate of the vertex, we find

$$x = \frac{-b}{2a} = \frac{-(-4)}{2(2)} = 1$$

To find the y-coordinate of the vertex, substitute x = 2 in the original equation.

$$y = 2(1)^2 - 4(1) - 6 = 2 - 4 - 6 = -8$$

The graph crosses the x-axis at 3 and -1 and has its vertex at (1,-8).

13. $y = x^2 - 4x - 4$

Using the formula that gives us the x-coordinate of the vertex, we have:

$$x = \frac{-b}{2a} = \frac{-(-4)}{2(1)} = 2$$

Substituting 2 for x in the equation gives us the y-coordinate of the vertex:

$$y = 2^2 - 4 \cdot 2 - 4 = -8$$

To complete the square on the right side of the equation, we add 4 to and subtract 4 from the right side of the equation.

$$y = (x^2 - 4x) - 4$$
$$y = (x^2 - 4x + \mathbf{4}) - 4 - \mathbf{4}$$
$$y = (x - 2)^2 - 8$$

In either case, the vertex is (2,-8).

Let's let x = 0 and x = 4, since each point is the same distance from x = 2, and on either side:

When x = 0
$$y = 0^2 - 4 \cdot 0 - 4$$
$$y = -4$$

When x = 4
$$y = 4^2 - 4 \cdot 4 - 4$$
$$y = -4$$

The two additional points on the graph are (0,-4) and (4,-4).

230

17. $y = x^2 + 1$

Using the formula that gives us the x-coordinate of the vertex, we have:

$$x = \frac{-b}{2a} = \frac{-0}{2(1)} = 0$$

Substituting 0 for x in the equation gives us the y-coordinate of the vertex:

$$y = 0^2 + 1 = 1$$

To complete the square on the right side of the equation, we add 0 to and subtract 0 from the right side of the equation.

$$y = (x^2 + \quad) + 1$$
$$y = (x^2 + \mathbf{0}) + 1 - \mathbf{0}$$
$$y = (x - 0)^2 + 1$$

In either case, the vertex is (0,1).

Let's let x = -2 and x = 2, since each point is the same distance from x = 0, and on either side:

When x = -2
$$y = (-2)^2 + 1$$
$$y = 4 + 1$$
$$y = 5$$

When x = 2
$$y = 2^2 + 1$$
$$y = 4 + 1$$
$$y = 5$$

The two additional points on the graph are (-2,5) and (2,5).

21. $y = 3x^2 + 4x + 1$

Using the formula that gives us the x-coordinate of the vertex, we have:

$$x = \frac{-b}{2a} = \frac{-4}{2(3)} = -\frac{4}{6} = -\frac{2}{3}$$

Substituting $-\frac{2}{3}$ for x in the equation gives us the y-coordinate of the vertex:

$$y = 3\left(-\frac{2}{3}\right)^2 + 4\left(-\frac{2}{3}\right) + 1$$

$$= \frac{4}{3} + \frac{-8}{3} - 1$$

$$= -\frac{1}{3}$$

To complete the square on the right side of the equation, we factor 3 from the first two terms, add $\frac{16}{36}$ inside the parentheses and $-\frac{16}{12}$ outside the parentheses (this amounts to adding 0 to the right side):

$$y = 3\left(x^2 + \frac{4}{3}x \qquad\right) + 1$$

$$y = 3\left(x^2 + \frac{4}{3}x + \frac{16}{36}\right) = 1 - \frac{16}{12}$$

$$y = 3\left(x + \frac{2}{3}\right)^2 - \frac{1}{3}$$

In either case, the vertex is $\left(-\frac{2}{3}, -\frac{1}{3}\right)$.

Let's let $x = -2$ and $x = 1$.

When $x = -2$
$$y = 3(-2)^2 + 4(-2) + 1$$
$$y = 12 - 8 + 1$$
$$y = 5$$

When $x = 1$
$$y = 3 \cdot 1^2 + 4 \cdot 1 + 1$$
$$y = 3 + 4 + 1$$
$$y = 8$$

The two additional points on the graph are $(-2,5)$ and $(-1,8)$.

25. $y = -x^2 + 2x + 8$

Using the formula that gives us the x-coordinate of the vertex, we have:

$$x = \frac{-b}{2a} = -\frac{2}{2(-1)} = 1$$

Substituting 1 for x in the equation gives us the y-coordinate of the vertex:

$$y = -(1)^2 + 2(1) + 8$$
$$= -1 + 2 + 8$$
$$= 9$$

The vertex is $(1,9)$.

To complete the square on the right side of the equation, we factor -1 from the first two terms, add 1 inside the parentheses and 1 on the outside of the parentheses.

$$y = -1(x^2 - 2x \qquad) + 8$$
$$y = -1(x^2 - 2x + 1) + 8 + 1$$
$$y = -1(x - 1)^2 + 9$$

The graph will be concave-down because a is negative. This means the vertex $(1,9)$ is the highest point.

29. $y = -x^2 - 8x$

Using the formula that gives us the x-coordinate of the vertex, we have:

$$x = \frac{-b}{2a} = \frac{-(-8)}{2(-1)} = -4$$

Substituting -4 for x in the equation gives us the y-coordinate of the vertex:

$$\begin{aligned} y &= -(-4)^2 - 8(-4) \\ &= -16 + 32 \\ &= 16 \end{aligned}$$

The vertex is (-4,16).

To complete the square on the right side of the equation, we factor -1 from the first two terms, add 16 inside the parentheses and 16 on the outside of the parentheses

$$\begin{aligned} y &= -(x^2 + 8x \qquad) \\ y &= -(x^2 + 8x + \mathbf{16}) + \mathbf{16} \\ y &= -(x + 4)^2 + 16 \end{aligned}$$

The graph will be concave-down because a is negative. This means the vertex (-4,16) is the highest point.

33. $P = -.002x^2 + 3.5x - 800$

We find the vertex by first finding its x-coordinate:

$$x = \frac{-b}{2a} = -\frac{3.5}{2(-.002)} = 875$$

This represents the number of patterns they need to sell each week in order to make a maximum profit. To find the maximum profit, we substitute 875 for x in the original equation.

$$\begin{aligned} P &= -.002(857)^2 + 3.5(875) - 800 \\ &= -.002(765,625) + 3,062.5 - 800 \\ &= -1,531.25 + 3,062.5 - 800 \\ &= 731.25 \end{aligned}$$

The maximum weekly profit is $731.25 and is obtained by selling 875 patterns a week.

37. $\begin{aligned}(3 - 5i) - (2 - 4i) &= 3 - 5i - 2 + 4i \\ &= (3 - 2) + (-5i + 4i) \\ &= 1 - i\end{aligned}$

41.
$$\begin{aligned} \frac{i}{3 + i} &= \frac{i}{3 + i} \cdot \frac{3 - i}{3 - i} \\\\ &= \frac{3i - i^2}{3^2 - i^2} \\\\ &= \frac{3i + 1}{9 + 1} \qquad i^2 = (\sqrt{-1})^2 = -1 \\\\ &= \frac{1 + 3i}{10} \end{aligned}$$

1. $x = y^2 + 2y - 3$

    x-intercept: When $y = 0$, $x = -3$

    y-intercept: When $x = 0$, we have $0 = y^2 + 2y - 3$, which yields:

    $$0 = y^2 + 2y - 3$$
    $$0 = (y + 3)(y - 1)$$
    $$y = -3 \quad \text{and} \quad y = 1$$

    Vertex: $y = \dfrac{-b}{2a} = \dfrac{-2}{2(1)} = -1$

    Substituting -1 for y in the equation

    $$x = (-1)^2 + 2(-1) - 3$$
    $$x = 1 - 2 - 3$$
    $$x = -4$$

    The vertex is (-4,-1).

    See the graph on page A49 in the textbook.

5. $x = y^2 - 1$

    x-intercept: When $y = 0$, $x = -1$

    y-intercept: When $x = 0$, we have $0 = y^2 - 1$, which yields:

    $$0 = y^2 - 1$$
    $$0 = (y + 1)(y - 1)$$
    $$y = -1 \quad \text{and} \quad y = 1$$

    Vertex: $y = \dfrac{-b}{2a} = -\dfrac{0}{2(1)} = 0$

    Substituting 0 for y in the equation gives us $x = -1$. The vertex is (-1,0).

    See the graph on page A49 in the textbook.

9. $x = 2y^2 - 2y - 4$

    x-intercept: When $y = 0$, $x = -4$

    y-intercept: When $x = 0$, we have

$$0 = 2y^2 - 2y - 4 \text{ which yields}$$
$$0 = 2(y^2 - y - 2)$$
$$0 = (y + 1)(y - 2)$$

$$y = -1 \quad \text{and} \quad y = 2$$

    Vertex: $y = \dfrac{-b}{2a} = \dfrac{-(-2)}{2(2)} = \dfrac{1}{8}$

        Substituting $\dfrac{1}{8}$ for y in the equation:

$$x = 2\left(\frac{1}{8}\right)^2 - 2\left(\frac{1}{8}\right) - 4$$

$$x = -4\frac{1}{8}$$

The vertex is $\left(-4\dfrac{1}{8}, \dfrac{1}{8}\right)$

See the graph on page A50 in the textbook.

1.
$$(2t - 5)^2 = 25$$
$$2t - 5 = \pm\sqrt{25} \qquad \text{Theorem 7.1}$$
$$2t - 5 = \pm 5$$
$$2t = 5 \pm 5$$
$$t = \frac{5 \pm 5}{2}$$

$$t = \frac{5 + 5}{2} \qquad \text{or} \qquad t = \frac{5 - 5}{2}$$

$$t = \frac{10}{2} \qquad\qquad t = \frac{0}{2}$$

$$t = 5 \qquad\qquad t = 0$$

The solution set is 0 and 5.

5.
$$(2x + 6)^2 = 12$$
$$2x + 6 = \pm\sqrt{12}$$
$$2x + 6 = \pm 2\sqrt{3}$$
$$2x = 6 \pm 2\sqrt{3}$$
$$x = \frac{6 \pm 2\sqrt{3}}{2}$$

$$x = \frac{6 + 2\sqrt{3}}{2} \qquad \text{or} \qquad x = \frac{6 - 2\sqrt{3}}{2}$$

$$x = \frac{2(3 + \sqrt{3})}{2} \qquad\qquad x = \frac{2(3 - \sqrt{3})}{2}$$

$$x = 3 + \sqrt{3} \qquad\qquad x = 3 - \sqrt{3}$$

The solution set is $3 + \sqrt{3}$ and $3 - \sqrt{3}$.

9.
$$a^2 + 9 = 6a$$
$$a^2 - 6a = -9 \qquad \text{Add -6a and -9 to both sides}$$
$$a^2 - 6a + \frac{36}{4} = -9 + \frac{36}{4} \qquad \text{Half of -6 is } -\frac{6}{2}, \text{ the square of which is } \frac{36}{4}$$

$$(a - 3)^2 = 0 \qquad\qquad -9 + \frac{36}{4} = -\frac{36}{4} + \frac{36}{4} = 0$$
$$a - 3 = \pm\sqrt{0} \qquad\quad \text{Theorem 7.1}$$
$$a - 3 = 0$$
$$a = 3$$

The solution set is {3}.

13.
$$\frac{1}{6}x^2 + \frac{1}{2}x - \frac{5}{3} = 0$$

$$6(\frac{1}{6}x^2) + 6(\frac{1}{2}x) - 6(\frac{5}{3}) = 6(0) \qquad \text{LCD is 6}$$

$$x^2 + 3x - 10 = 0$$
$$(x + 5)(x - 2) = 0$$

$$x + 5 = 0 \qquad \text{or} \qquad x - 2 = 0$$
$$x = -5 \qquad\qquad\qquad x = 2$$

The solution set is -5 and 2.

17. $4t^2 - 8t + 19 = 0$

Using the quadratic formula with a = 4, b = -8 and c = 19

$$x = \frac{-(-8) \pm \sqrt{(-8)^2 - 4(4)(19)}}{2(4)}$$

$$x = \frac{8 \pm \sqrt{64 - 304}}{8}$$

$$x = \frac{8 \pm \sqrt{-240}}{8}$$

$$x = \frac{8 \pm 4i\sqrt{15}}{8}$$

$$x = \frac{4(2 \pm i\sqrt{15})}{8}$$

$$x = \frac{2 \pm i\sqrt{15}}{2}$$

The solution set is $\frac{2 + i\sqrt{15}}{2}$ and $\frac{2 - i\sqrt{15}}{2}$.

21.
$$.06a^2 + .05a = .04$$
$$.06a^2 + .05a - .04 = 0$$
$$6a^2 + 5a - 4 = 0 \qquad \text{Multiply both sides by 100}$$
$$(3x + 4)(2x - 1) = 0$$

$$3x + 4 = 0 \qquad \text{or} \qquad 2x - 1 = 0$$
$$3x = -4 \qquad\qquad\qquad 2x = 1$$

$$x = -\frac{4}{3} \qquad\qquad\qquad x = \frac{1}{2}$$

The solution set is $\{-\frac{4}{3}, \frac{1}{2}\}$.

25. $(2x + 1)(x - 5) - (x + 3)(x - 2) = -17$

$$2x^2 - 9x - 5 - (x^2 + x - 6) = -17$$

$$2x^2 - 9x - 5 - x^2 - x + 6 = -17$$

$$x^2 - 10x + 1 = -17$$

$$x^2 - 10x + 18 = 0$$

Using the quadratic formula with $a = 1$, $b = -10$ and $c = 18$, we have

$$x = \frac{-(-10) \pm \sqrt{(-10)^2 - 4(1)(18)}}{2(1)}$$

$$x = \frac{10 \pm \sqrt{100 - 72}}{2}$$

$$x = \frac{10 \pm \sqrt{28}}{2}$$

$$x = \frac{10 \pm 2\sqrt{7}}{2}$$

$$x = \frac{2(5 \pm \sqrt{7})}{2}$$

$$x = 5 \pm \sqrt{5}$$

The solution set is $\{5 + \sqrt{5}, 5 - \sqrt{5}\}$.

29.
$$5x^2 = -2x + 3$$
$$5x^2 + 2x - 3 = 0 \qquad \text{Standard form}$$

Letting a = 5, b = 2 and c = -3, we have

$$x = \frac{-2 \pm \sqrt{2^2 - 4(5)(-3)}}{2(5)}$$

$$= \frac{-2 \pm \sqrt{4 + 60}}{10}$$

$$= \frac{-2 \pm \sqrt{64}}{10}$$

$$= \frac{-2 \pm \overset{4}{\cancel{8}}}{\underset{5}{\cancel{10}}}$$

$$= \frac{-1 \pm 4}{5}$$

$$x = \frac{-1 + 4}{5} \quad \text{or} \quad x = \frac{-1 - 4}{5}$$

$$x = \frac{3}{5} \qquad\qquad x = -\frac{5}{5}$$

$$\qquad\qquad\qquad\qquad x = -1$$

The solution set is $\{-1, \frac{3}{5}\}$.

33.
$$3 - \frac{2}{x} + \frac{1}{x^2} = 0$$

$$x^2(3) - x^2\left(\frac{2}{x}\right) + x^2\left(\frac{1}{x^2}\right) = x^2(0) \qquad \text{Multiply by the LCD, } x^2$$

$$3x^2 - 2x + 1 = 0$$

Letting a = 3, b = -2 and c = 1, we have

$$x = \frac{-(-2) \pm \sqrt{(-2)^2 - 4(3)(1)}}{2(3)}$$

$$= \frac{2 \pm \sqrt{4 - 12}}{6}$$

$$= \frac{2 \pm \sqrt{-8}}{6}$$

$$= \frac{2 \pm i\sqrt{4}\,\sqrt{2}}{6}$$

$$= \frac{\overset{1}{\cancel{2}} \pm \overset{1}{\cancel{2}}i\sqrt{2}}{\underset{3}{\cancel{6}}}$$

$$= \frac{1 \pm i\sqrt{2}}{3}$$

Since neither of the two solutions $\frac{1 + i\sqrt{2}}{3}$ nor $\frac{1 - i\sqrt{2}}{3}$ will make
any of the denominators in our original equation 0, they are both solutions.

37. Remember, the relationship between profit, revenue and cost is given by the formula

$$P = R - C$$

Substituting $C = 7x + 400$, $R = 34x - .1x^2$ and $P = 1300$ in the formula, we have

$$P = R - C$$
$$1300 = (34x - .1x^2) - (7x + 400)$$
$$1300 = 34x - .1x^2 - 7x - 400$$
$$1300 = -.1x^2 + 27x - 400$$
$$.1x^2 - 27x + 1700 = 0 \qquad \text{Standard form}$$

Letting $a = .1$, $b = -27$ and $c = 1700$, we have

$$x = \frac{-(-27) \pm \sqrt{(-27)^2 - 4(.1)(1700)}}{2(.1)}$$

$$= \frac{27 \pm \sqrt{729 - 680}}{.2}$$

$$= \frac{27 \pm \sqrt{49}}{.2}$$

$$= \frac{27 \pm 7}{.2}$$

$$x = \frac{27 + 7}{.2} \quad \text{or} \quad x = \frac{27 - 7}{.2}$$

$$= \frac{34}{.2} \qquad\qquad = \frac{20}{.2}$$

$$= 170 \qquad\qquad\quad = 100$$

For their profits to be $1300, they must produce and sell 100 or 170 items.

41. $2x^2 + x - 3 = 0$

Using $a = 2$, $b = 1$ and $c = -3$, we have

$$b^2 - 4ac = 1^2 - 4(2)(-3) = 1 + 24 = 25$$

The discriminant is a perfect square. Therefore, the equation has two rational solutions.

45. $3x^2 + 5x = -4$

$3x^2 + 5x + 4 = 0 \qquad \text{Standard form}$

Using $a = 3$, $b = 5$, and $c = 4$, we have

$$b^2 - 4ac = 5^2 - 4(3)(4) = 25 - 48 = -13$$

The discriminant is negative, implying the equation has two complex solutions that contain i.

49. $kx^2 + 12x + 9 = 0$

Using $a = k$, $b = 12$, and $c = 9$, we have

$$b^2 - 4ac = 12^2 - 4(k)(9) = 144 - 36k$$

An equation has exactly one rational solution when the discriminant is 0. We set the discriminant equal to 0 and solve:

$$144 - 36k = 0$$
$$144 = 36k$$
$$\frac{144}{36} = k$$
$$4 = k$$

53. If $x = 3$ $x = 5$
then $x - 3 = 0$ $x - 5 = 0$

Since $x - 3$ and $x - 5$ are all 0, their product is also 0 by the zero-factor property.

$$(x - 3)(x - 5) = 0 \qquad \text{Zero-factor property}$$
$$x^2 - 8x + 15 = 0 \qquad \text{Multiply the binomials}$$

57. If $t = 3$ $t = -3$ $t = 5$
then $t - 3 = 0$ $t + 3 = 0$ $t - 5 = 0$

Since $t - 3$, $t + 3$, and $t - 5$ are all 0, their product is also 0 by the zero-factor property.

$$(t - 3)(t + 3)(t - 5) = 0 \qquad \text{Zero-factor property}$$
$$(t^2 - 9)(t - 5) = 0 \qquad \text{Multiply first two binomials}$$
$$t^3 - 5t^2 - 9t + 45 = 0 \qquad \text{Complete the multiplication}$$

61. $(x - 2)^2 - 4(x - 2) - 60 = 0$

Replacing $x - 2$ with y, we have

$$y^2 - y - 60 = 0$$
$$(y + 6)(y - 10) = 0 \qquad \text{Factor}$$
$$y + 6 = 0 \quad \text{or} \quad y - 10 = 0 \qquad \text{Set factors to 0}$$
$$y = -6 \quad \text{or} \qquad y = 10$$

Replacing y with $x - 2$, and then solving for x, we have

$$x - 2 = -6 \quad \text{or} \quad x - 2 = 10$$
$$x = -4 \quad \text{or} \qquad x = 12$$

The solutions to our original equation are -4 and 12.

65.
$$x^4 - x^2 = 12$$
$$x^4 - x^2 - 12 = 0 \qquad \text{Standard form}$$

Replacing y with x^2, we have

$$y^2 - y - 12 = 0$$
$$(y + 3)(y - 4) = 0 \qquad \text{Factor}$$

$$y + 3 = 0 \quad \text{or} \quad y - 4 = 0 \qquad \text{Set factors to } 0$$
$$y = -3 \quad \text{or} \qquad y = 4$$

Replacing y with x^2, we have

$$x^2 = -3 \qquad \text{or} \quad x^2 = 4$$
$$x = \pm i\sqrt{3} \quad \text{or} \quad x = \pm 2$$

The solutions to our original equation are $i\sqrt{3}$, $-i\sqrt{3}$, 2 and -2.

69. $x - \sqrt{x} - 2 = 0$

Replacing $\sqrt{x}$ with y, we have

$$y^2 - y - 2 = 0$$
$$(y - 2)(y + 1) = 0 \qquad \text{Zero-factor property}$$

$$y - 2 = 0 \quad \text{or} \quad y + 1 = 0 \qquad \text{Set factors to } 0$$
$$y = 2 \quad \text{or} \qquad y = -1$$

Replacing y with $\sqrt{x}$, we have

$$\sqrt{x} = 2 \quad \text{or} \quad \sqrt{x} = -1$$
$$x = 4 \qquad \quad \text{undefined}$$

The solution to the original equation is 4.

73. $\sqrt{x + 5} = \sqrt{x} + 1$

The equation has two separate terms involving radical signs. Squaring both sides gives

$$x + 5 = x + 2\sqrt{x} + 1$$

$$4 = 2\sqrt{x} \qquad \text{Add } -x \text{ and } -1 \text{ to both sides}$$
$$16 = 4x \qquad \text{Square both sides}$$
$$4 = x \qquad \text{Divide by 4}$$

The solution set is $\{4\}$.

77. $\sqrt{y + 9} - \sqrt{y - 6} = 3$

$\qquad \sqrt{y + 9} = \sqrt{y - 6} + 3$

Squaring both sides, we have

$y + 9 = y + 3 + 6\sqrt{y - 6}$	Simplify the right side
$6 = 6\sqrt{y - 6}$	Add $-y$ and -3 to both sides
$36 = 36(y - 6)$	Square both sides
$36 = 36y - 216$	Multiply
$252 = 36y$	Add $+216$ to both sides
$7 = y$	Divide by 36

The solution set is {7}.

81. $\qquad x^2 - x - 2 < 0$

$(x + 1)(x - 2) < 0$

The product is negative, so the factors must have opposite signs.

See the graph on page A50 in the textbook.

85. $(x + 2)(x - 3)(x + 4) > 0$

The original inequality indicates that the product is positive. In order for this to happen, all three factors must be positive or exactly two must be negative. See the graph on page A50 in the textbook.

89. $\dfrac{x - 3}{(x - 2)(x - 4)} > 0$

- - - - -	+ + + + +	+ + + + +	+ + + + +

$\longleftrightarrow$ Sign of x - 2

- - - - -	- - - - -	+ + + + +	+ + + + +

$\longleftrightarrow$ Sign of x - 3

- - - - -	- - - - -	- - - - -	+ + + + +

$\longleftrightarrow$ Sign of x - 4

234

Since the quotient is positive, the three factors must be positive or exactly two must be negative. See the graph on page A50 in the textbook.

93. $y = x^2 - 4$

x-intercept: When $y = 0$,

$$0 = x^2 - 4$$
$$0 = (x + 2)(x - 2)$$

$x + 2 = 0 \qquad$ or $\qquad x - 2 = 0$

$x = -2 \qquad\qquad\qquad x = 2$

y-intercept: When $x = 0$, $y = -4$

Vertex: $x = \dfrac{-b}{2a} = \dfrac{-0}{2(1)} = 0$,

$$y = x^2 - 4$$
$$y = 0^2 - 4$$
$$y = -4$$

The vertex is $(0,-4)$.

1.
$$(2x + 4)^2 = 25$$
$$2x + 4 = \pm\sqrt{25}$$
$$2x + 4 = \pm 5$$
$$2x = -4 \pm 5$$
$$x = \frac{-4 \pm 5}{2}$$

$$x = \frac{-4 + 5}{2} \quad \text{or} \quad x = \frac{-4 - 5}{2}$$

$$x = \frac{1}{2} \qquad\qquad x = -\frac{9}{2}$$

5.
$$8t^3 - 125 = 0$$
$$(2t)^3 - 5^3 = 0$$
$$(2t - 5)(4t^2 + 10t + 25) = 0$$

$$2t - 5 = 0 \quad \text{or} \quad a = 4,\ b = 10,\ c = 25$$
$$2t = 5$$

$$t = \frac{5}{2} \qquad x = \frac{-10 \pm \sqrt{(10)^2 - 4(4)(25)}}{2(4)}$$

$$x = \frac{-10 \pm \sqrt{100 - 400}}{8}$$

$$x = \frac{-10 \pm \sqrt{-300}}{8}$$

$$x = \frac{-10 \pm 10i\sqrt{3}}{8}$$

$$x = \frac{2(-5 \pm 5i\sqrt{3})}{8}$$

$$x = \frac{-5 \pm 5i\sqrt{3}}{4}$$

The solution set is $\dfrac{5}{2}$, $\dfrac{-5 + 5i\sqrt{3}}{4}$ and $\dfrac{-5 - 5i\sqrt{3}}{4}$.

9.   Let S = 12, we have

$$S = 32t - 16t^2$$
$$12 = 32t - 16t^2$$
$$16t^2 - 32t + 12 = 0$$
$$4(4t^2 - 8t + 3) = 0$$
$$(2t - 3)(2t - 1) = 0$$

$$2t - 3 = 0 \qquad or \qquad 2t - 1 = 0$$
$$2t = 3 \qquad\qquad\qquad 2t = 1$$
$$t = \frac{3}{2} \qquad\qquad\qquad t = \frac{1}{2}$$

The time is $\frac{3}{2}$ seconds and $\frac{1}{2}$ second.

13.

$$x = 5 \qquad\qquad x = -\frac{2}{3}$$
$$x - 5 = 0 \qquad\qquad 3x = -2$$
$$3x + 2 = 0$$
$$(x - 5)(3x + 2) = 0$$
$$3x^2 - 13x - 10 = 0$$

17.   Let
$$y = x^{1/3}$$
the equation $x^{2/3} - 3x^{1/3} + 2 = 0$

becomes
$$y^2 - 3y + 2 = 0$$
$$(y - 1)(y - 2) = 0 \qquad\qquad\qquad\qquad \text{Factor}$$
$$y - 1 = 0 \quad or \quad y - 2 = 0 \qquad \text{Set factors to 0}$$
$$y = 1 \qquad\qquad y = 2$$

Now we replace y with $x^{1/3}$ and solve for x:

$$x^{1/3} = 1 \qquad or \qquad x^{1/3} = 2$$
$$x = 1 \qquad\qquad\qquad x = 8$$

The solution set is 1 and 8.

21. $x^2 - x - 6 \leq 0$

$(x + 2)(x - 3) \leq 0$

The product is negative, so the factors must have opposite signs.

```
 - - - - - | - - - - - | + + + + +
<----------|-----------|---------->      Sign of x - 3
 - - - - - | + + + + + | + + + + +
<----------|-----------|---------->      Sign of x + 2
          -2           3
```

See the graph on page A51 in the textbook.

25. $P = R - C$ when $R = 25x - .1x^2$, $C = 5x + 100$

$P = 25x - .1x^2 - (5x + 100)$
$P = 25x - .1x^2 - 5x - 100$
$P = 250x - x^2 - 50x - 1000$
$P = -x^2 + 200x - 1000$

The equation below represents the number of items they need to sell each week in order to make a maximum profit.

$$x = \frac{-b}{2a} = \frac{-200}{2(-1)} = \frac{200}{2} = 100$$

$P = -(100)^2 + 200(100) - 1000$
$P = -10,000 + 20,000 - 1,000$
$P = 900$

Maximum profit = $900 by selling 100 items per week.

1. To find the x-intercept   To find the y-intercept
    set   y = 0   let   x = 0
    the equation   3x - 2y = 6   the equation   3x - 2y = 6
    becomes   3x - 2(0) = 6   becomes   3(0) - 2y = 6
                        3x = 6   -2y = 6
                         x = 2   y = -3

    The graph with the equation 3x - 2y = 6 crosses the x-axis at (2,0)
    and the y-axis at (0,-3).

    To find the x-intercept,   To find the y-intercept
    let   y = 0   let   x = 0
    the equation x - y = 1   the equation x - y = 1
    becomes   x - 0 = 1   becomes   0 - y = 1
                      x = 1   y = -1

    The graph with the equation x - y = 1 crosses the x-axis at (1,0)
    and the y-axis at (0,-1).

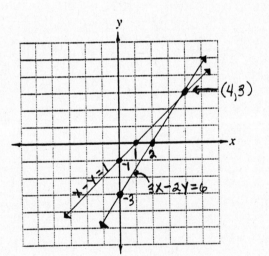

5. To find the x-intercept
   let $y = 0$

   the equation $y = \frac{1}{2}x$

   becomes $0 = \frac{1}{2}x$

                  $0 = x$

   To find another point
   let $x = 2$

   the equation $y = \frac{1}{2}x$

   becomes $y = \frac{1}{2}(2)$

                  $y = 1$

   The graph with the equation $y = \frac{1}{2}x$ crosses the x-axis and y-axis at

   (0,0) and the point (2,1).

   To find the x-intercept
   let $y = 0$

   the equation $y = -\frac{3}{4}x + 5$

   becomes $0 = -\frac{3}{4}x + 5$

                  $-5 = -\frac{3}{4}x$

                  $\frac{20}{3} = x$

   To find the y-intercept
   let $x = 0$

   the equation $y = -\frac{3}{4}x + 5$

   becomes $y = -\frac{3}{4}(0) + 5$

                  $y = 5$

   The graph with the equation $y = -\frac{3}{4}x + 5$ crosses the x-axis at $(\frac{20}{3},0)$

   and the y-axis at (0,5).

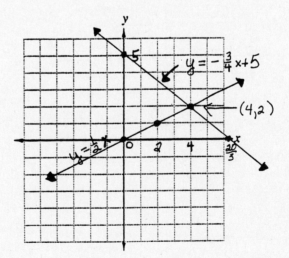

9.   To find the x-intercept
     let   y = 0
     the equation   2x - y = 5
     becomes   2x - 0 = 5
                          2x = 5

                           x = $\frac{5}{2}$

To find the y-intercept
let   x = 0
the equation   2x - y = 5
becomes   0 - y = 5
                      y = -5

The graph with the equation 2x - y = 5 crosses the x-axis at ($\frac{5}{2}$,0)

and the y-axis at (0,-5).

To find the x-intercept,
let   y = 0
the equation y = 2x - 5
becomes   0 = 2x - 5
              5 = 2x

              $\frac{5}{2}$ = x

To find the y-intercept,
let   x = 0
the equation y = 2x - 5
becomes   y = 0 - 5
              y = -5

The graph with the equation y = 2x - 5 crosses the x-axis at ($\frac{5}{2}$,0)

and the y-axis at (0,-5).

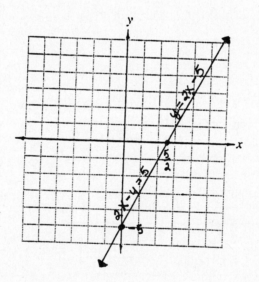

The lines coincide, any solution to one of the equations is a solution
to the other.

13.

$$3x + y = 4 \xrightarrow{\text{times } -1} -3x - y = -4$$

$$4x + y = 5 \xrightarrow{\phantom{\text{times } -1}} \underline{4x + y = 5}$$

$$\xrightarrow{\text{no change}} \quad x = 1$$

We may sutstitute x = 1 in for either equation to solve for the y-coordinate.

$$3x + y = 4$$
$$3(1) + y = 4$$
$$y = 1$$

The solution to the system is (1,1).

17.

$$x + 2y = 0 \xrightarrow{\text{multiply } -2} -2x - 4y = 0$$

$$2x - 6y = 5 \xrightarrow{\phantom{\text{multiply } -2}} \underline{2x - 6y = 5}$$

$$\xrightarrow{\text{no change}} \quad -10y = 5$$

$$y = -\frac{5}{10} = -\frac{1}{2}$$

Let $\qquad\qquad y = -\frac{1}{2}$

then $\qquad\qquad x + 2y = 0$

becomes $\quad x + 2(-\frac{1}{2}) = 0$

$$x - 1 = 0$$
$$x = 1$$

The solution to the system is $(1, -\frac{1}{2})$.

21.

$$6x + 3y = -1 \xrightarrow{\text{multiply } 3} 18x + 9y = -3$$

$$9x + 5y = 1 \xrightarrow{\phantom{\text{multiply } 3}} \underline{-18x - 10y = -2}$$

$$\xrightarrow{\text{multiply } -2} \quad -y = -5$$

$$y = 5$$

Let $\qquad\qquad y = 5$
then $\qquad\quad 6x + 3y = -1$
becomes $\quad 6x + 3(5) = -1$
$$6x = -16$$

$$x = -\frac{16}{3} = -\frac{8}{3}$$

The solution to our system is $(-\frac{8}{3}, 5)$.

25.

$$2x - 5y = 3 \xrightarrow{\text{multiply 2}} 4x - 10y = 6$$
$$-4x + 10y = 3 \xrightarrow{\text{no change}} \underline{-4x + 10y = 6}$$
$$0 = 6$$

The result is the false statement $0 = 6$, which indicates there is no solution to the system. If we were to graph the two lines, we would find that they are parallel. The solution is an empty set.

29. First clear both equations of fractions:

$$\frac{1}{2}x + \frac{1}{3}y = 13 \xrightarrow{\text{multiply 6}} 3x + 2y = 78$$

$$\frac{2}{5}x + \frac{1}{4}y = 10 \xrightarrow{\text{multiply 20}} 8x + 5y = 200$$

Now eliminate y:

$$3x + 2y = 78 \xrightarrow{\text{multiply -5}} -15x - 10y = -390$$
$$8x + 5y = 200 \xrightarrow{\text{multiply 2}} \underline{16x - 10y = 400}$$
$$x = 10$$

Let $x = 10$
then $3x + 2y = 78$
becomes $3(10) + 2y = 78$
           $30 + 2y = 78$
              $2y = 48$
               $y = 24$

The solution to our system is $(10, 24)$.

33.

Let	$x = 2y + 9$	Let	$y = -3$
then	$7x - y = 24$	then	$x = 2y + 9$
becomes	$7(2y + 9) - y = 24$	becomes	$x = 2(-3) + 9$
	$14y + 63 - y = 24$		$x = 3$
	$13y + 63 = 24$		
	$13y = -39$		
	$y = -3$		

The solution to the system is $(3, -3)$.

37.

Let	$x = y + 4$	Let	$y = 2$
then	$2x - 3y = 6$	then	$x - y = 4$
becomes	$2(y + 4) - 3y = 6$	becomes	$x - 2 = 4$
	$2y + 8 - 3y = 6$		$x = 6$
	$-y + 8 = 6$		
	$-y = -2$		
	$y = 2$		

The solution to the system is $(6, 2)$.

41. Let
$$y = 2x - 5$$
then
$$4x - 2y = 10$$
becomes $\quad 4x - 2(2x - 5) = 10$
$$4x - 4x + 10 = 10$$
$$0 = 0$$

The lines coincide, $\{(x,y)|2x - y = 5\}$.

45.
$$\begin{array}{l} 4x - 7y = 3 \\ 5x + 2y = -3 \end{array} \quad \xrightarrow[\text{multiply } -4]{\text{multiply } 5} \quad \begin{array}{r} 20x - 35y = 15 \\ -20x - 8y = 12 \\ \hline -43y = 27 \end{array}$$

$$y = -\frac{27}{43}$$

It would be a cumbersome arithmetic problem to substitute $y = -\frac{27}{43}$

so eliminate the y value instead.

$$\begin{array}{l} 4x - 7y = 3 \\ 5x + 2y = -3 \end{array} \quad \xrightarrow[\text{multiply } 7]{\text{multiply } 2} \quad \begin{array}{r} 8x - 14y = 6 \\ 35x + 14y = -21 \\ \hline 43x = -15 \end{array}$$

$$x = -\frac{15}{43}$$

The solution set is $(-\frac{15}{43}, -\frac{27}{43})$.

49.
$$\begin{array}{l} 3x - 5y = 2 \\ 7x + 2y = 1 \end{array} \quad \xrightarrow[\text{multiply } 5]{\text{multiply } 2} \quad \begin{array}{r} 6x - 10y = 4 \\ 35x + 10y = 5 \\ \hline 41x = 9 \end{array}$$

$$x = \frac{9}{41}$$

$$\begin{array}{l} 3x - 5y = 2 \\ 7x + 2y = 1 \end{array} \quad \xrightarrow[\text{multiply } -3]{\text{multiply } 7} \quad \begin{array}{r} 21x - 35y = 15 \\ -21x - 6y = -3 \\ \hline -41y = 11 \end{array}$$

$$y = -\frac{11}{43}$$

The solution set is $(\frac{9}{41}, -\frac{11}{43})$

53. For the lines to coincide the equations must be 0 = 0. First notice factoring a 3 out of the first equation gives:

$$6x - 9y = 3$$
$$3(2x - 3y) = 1(3)$$

Factoring a 2 out of the second equation gives:

$$2(2x - 3y) = \frac{c}{2}$$

The left side of both equations is the same. Since the two equations coincide, the right side must be the same. Therefore, for the second equation:

If $4x - 6y = c$
then $2(2x - 3y) = 1(2)$

The value for c = 2.

57. $(2x - 1)^2 = 25$
$$2x - 1 = \pm 5$$
$$2x = 1 \pm 5$$
$$x = \frac{1 \pm 5}{2}$$

x = 3 or x = -2

61. $x^2 - 10x + 8 = 0$
$$x^2 - 10x = -8$$
$$x^2 - 10x + 25 = -8 + 25$$
$$(x - 5)^2 = 17$$
$$x - 5 = \pm\sqrt{17}$$
$$x = 5 \pm \sqrt{17}$$

65. $ax + by = 7$ $(1,-2)$ $(3,1)$

$a(1) + b(-2) = 7$ becomes $a - 2b = 7$
$a(3) + b(1) = 7$ becomes $3a + b = 7$

To solve for a and b, we

$$\begin{array}{lll}
a - 2b = 7 & \xrightarrow{\text{multiply } -3} & -3a + 6b = -21 \\
3a + b = 7 & \xrightarrow{\text{no change}} & \underline{3a + b = 7} \\
& & 7b = -14 \\
& & b = -2
\end{array}$$

Substituting b = -2, we have

$$a - 2b = 7$$
$$a - 2(-2) = 7$$
$$a + 4 = 7$$
$$a = 3$$

254

1. $x + y + z = 4$ (1)
    $x - y + 2z = 1$ (2)
    $x - y - 3z = -4$ (3)

    First eliminate the y-variable:

    $$\begin{array}{r} x + y + z = 4 \quad (1) \\ \underline{x - y - 2z = 1} \quad (2) \\ 2x \quad\quad + 3z = 5 \quad (4) \end{array}$$

    Also:

    $$\begin{array}{r} x + y + z = 4 \quad (1) \\ \underline{x - y - 3z = -4} \quad (3) \\ 2x \quad\quad - 2z = 0 \quad (5) \end{array}$$

    Since equations (4) and (5) have only 2 variables, eliminate x from both equations.

    $$\begin{array}{lll} & \text{no change} & \\ 2x + 3z = 5 \xrightarrow{\hspace{2cm}} & 2x + 3z = 5 \\ 2x - 2z = 0 \xrightarrow{\hspace{2cm}} & \underline{-2x + 2z = 0} \\ & \text{multiply } -1 & 5z = 5 \\ & & z = 1 \end{array}$$

    Substituting $z = 1$ into equation (4), we have:

    $$\begin{aligned} 2x + 3(1) &= 5 \\ 2x + 3 &= 5 \\ 2x &= 2 \\ x &= 1 \end{aligned}$$

    Substituting $x = 1$ and $z = 1$ into equation (1) gives us:

    $$\begin{aligned} 1 + y + 1 &= 4 \\ y &= 2 \end{aligned}$$

    The solution set for the system is the ordered triple $\{(1,2,1)\}$.

5. $x + 2y + z = 3$ (1)
$2x - y + 2z = 6$ (2)
$3x + y - z = 5$ (3)

First eliminate the y-variable:

$2x - y + 2z = 6$ (2)
$3x + y - z = 5$ (3)
$\overline{5x \qquad + z = 11}$ (4)

Also multiplying equation (2) by 2, we have:

$x + 2y + z = 3$ (1)
$4x - 2y + 4z = 12$ (2) by 2
$\overline{5x \qquad + 5z = 15}$ (5)

Eliminating x using equations (4) and (5), we have:

$$
\begin{array}{llcl}
& & \text{no change} & \\
5x + z = 11 & \xrightarrow{\hspace{2cm}} & 5x + z = 11 \\
5x + 5z = 15 & \xrightarrow{\hspace{2cm}} & -5x - 5z = -15 \\
& \text{multiply } -1 & \overline{\quad -4z = -4} \\
& & z = 1
\end{array}
$$

Substituting $z = 1$ into equation (4), we have:

$5x + 1 = 11$
$5x = 10$
$x = 2$

Substituting $x = 2$ and $z = 1$ into equation (1) gives us:

$2 + 2y + 1 = 3$
$2y + 3 = 3$
$2y = 0$
$y = 0$

The solution set for the system is the ordered triple $\{(2,0,1)\}$.

9. $-x + 4y - 3z = 2$ (1)
$2x - 8y + 6z = 1$ (2)
$3x - y + z = 0$ (3)

First eliminate the x-variable by multiplying equation (1) by 2:

$-2x + 8y - 6z = 4$ (1) by 2
$2x - 8y + 6z = 1$ (2)
$\overline{\qquad\qquad 0 = 5}$

If equations (1) and (2) have no ordered triples in common, then certainly (1), (2), and (3) do not either. The equations are inconsistent and the solution set for the system is the empty set.

13. $2x - y - 3z = 1$ (1)
    $x + 2y + 4z = 3$ (2)
    $4x - 2y - 6z = 2$ (3)

First eliminate the y-variable:

$$\begin{array}{r} x + 2y + 4z = 3 \quad (2) \\ \underline{4x - 2y - 6z = 2} \quad (3) \\ 5x \qquad - 2z = 5 \quad (4) \end{array}$$

Multiply equation (1) by -2:

$$\begin{array}{r} -4x + 2y + 6z = -2 \quad (1) \text{ by } -2 \\ \underline{4x - 2y - 6z = 2} \quad (3) \\ 0 = 0 \end{array}$$

This is a dependent system.

17. $x + y = 9$ (1)
    $y + z = 7$ (2)
    $x - z = 2$ (3)

Eliminating the z-variable, we have:

$$\begin{array}{r} y + z = 7 \quad (2) \\ \underline{x \qquad - z = 2} \quad (3) \\ x + y \qquad = 9 \quad (4) \end{array}$$

Multiplying equation (4) by -1, we have:

$$\begin{array}{r} x + y = 9 \quad (1) \\ \underline{-x - y = -9} \quad (4) \text{ by } -1 \\ 0 = 0 \end{array}$$

This is a dependent system.

21. $2x - 3y = 0$ (1)
     $6y - 4z = 1$ (2)
      $x + 2z = 1$ (3)

Eliminating the x-variable by multiplying equation (3) by -2, we have:

$$
\begin{array}{ll}
2x - 3y \quad\quad = \quad 0 & (1) \\
\underline{-2x \quad\quad - 4z = -2} & (3) \\
\quad\; - 3y - 4z = -2 & (4)
\end{array}
$$

Multiplying equation (4) by -1, we have

$$
\begin{array}{ll}
6y - 4z = 1 & (2) \\
\underline{3y + 4z = 2} & (4) \text{ by } -1 \\
9y \quad\quad = 3 &
\end{array}
$$

$$y \quad\quad = \frac{3}{9}$$

$$y \quad\quad = \frac{1}{3}$$

Substituting $y = \frac{1}{3}$ into equation (1), we have:

$$2x - 3(\frac{1}{3}) = 0 \quad (1)$$

$$2x - 1 = 0$$
$$2x = 1$$

$$x = \frac{1}{2}$$

Substituting $y = \frac{1}{3}$ into equation (2), we have:

$$6(\frac{1}{3}) - 4z = 1$$

$$2 - 4z = 1$$
$$-4z = -1$$

$$z = \frac{1}{4}$$

The solution set for the system is the ordered triple $\{(\frac{1}{2}, \frac{1}{3}, \frac{1}{4})\}$.

25. First eliminate the fractions:

$\frac{1}{2}x - \frac{1}{4}y + \frac{1}{2}z = -2$ multiply 4 $2x - y + 2z = -8$ (1)

$\frac{1}{4}x - \frac{1}{12}y - \frac{1}{3}z = \frac{1}{4}$ multiply 12 $3x - y - 4z = 3$ (2)

$\frac{1}{6}x + \frac{1}{3}y - \frac{1}{2}z = \frac{3}{2}$ multiply 6 $x + 2y - 3z = 9$ (3)

Eliminating the x-variable by multiplying equation (3) by -2, we have:

$$\begin{array}{l} 2x - y + 2z = -8 \quad (1) \\ \underline{-2x - 4y + 6z = -18} \quad (3) \text{ by } -2 \\ \quad\; -5y + 8z = -26 \end{array}$$

Eliminating the x-variable by multiplying equation (3) by -3, we have:

$$\begin{array}{l} 3x - y - 4z = 3 \quad (2) \\ \underline{-3x - 6y + 9z = -27} \quad (3) \text{ by } -3 \\ \quad\; -7y + 5z = -24 \quad (5) \end{array}$$

Eliminating y-variable in equations (4) and (5), we have:

multiply -7

$$\begin{array}{ll} -5y + 8z = -26 & \quad\quad\quad 35y - 56z = 182 \\ -7y + 5z = -24 & \quad\quad\quad \underline{-35y + 25z = 120} \\ \quad\quad\quad\quad \text{multiply } 5 & \quad\quad\quad\quad\; -31z = 62 \\ & \quad\quad\quad\quad\quad\quad\; z = -2 \end{array}$$

Substituting z = -2 into equation (4), we have:

$$\begin{array}{l} -5y - 16 = -26 \\ \quad -5y = -10 \\ \quad\quad y = 2 \end{array}$$

Substituting y = 2 and z = -2 into equation (1), we have:

$$\begin{array}{l} 2x - 2 + 2(-2) = -8 \\ \quad\quad\; 2x - 6 = -8 \\ \quad\quad\quad\; 2x = -2 \\ \quad\quad\quad\;\; x = -1 \end{array}$$

The solution set for the system is the ordered triple {(-1,2,-2)}.

29. $2x - 3y = 6$

$\quad -3y = -2x + 6$ Add -2x to both sides

$\quad y = \dfrac{-2x}{-3} + \dfrac{6}{-3}$ Divide by -3

$\quad y = \dfrac{2}{3}x - 2$

33. The equation is not factorable so we must use the quadratic formula. The coefficients are a = 2, b = 4, and c = -3:

$$x = \frac{-4 \pm \sqrt{16 - 4(2)(-3)}}{2(2)}$$

$$= \frac{-4 \pm \sqrt{16 + 24}}{4}$$

$$= \frac{-4 \pm \sqrt{40}}{4}$$

$$= \frac{-4 \pm 2\sqrt{10}}{4} \qquad\qquad \sqrt{40} = \sqrt{4}\,\sqrt{10} = 2\sqrt{10}$$

$$= \frac{2(-2 \pm \sqrt{10})}{4}$$

$$= \frac{-2 \pm \sqrt{10}}{2}$$

The solutions are $\{ \frac{-2 + \sqrt{10}}{2} , \frac{-2 - 2\sqrt{10}}{2} \}$.

37.
$$t^3 - 125 = 0$$
$$(t - 5)(t^2 + 5t + 25) = 0$$

The factor t - 5 = 0 gives a solution of 5. The second factor is not factorable. Using coefficients a = 1, b = 5 and c = 25 in the quadratic formula, we have

$$t = \frac{-5 \pm \sqrt{25 - 4(1)(25)}}{2(1)}$$

$$= \frac{-5 \pm \sqrt{25 - 100}}{2}$$

$$= \frac{-5 \pm \sqrt{-75}}{2}$$

$$= \frac{-5 \pm 5i\sqrt{3}}{2} \qquad\qquad \sqrt{-75} = \sqrt{75}\,i = i\sqrt{25}\,\sqrt{3} = 5i\sqrt{3}$$

The solutions are $\{5, \frac{-5 + 5i\sqrt{3}}{2} , \frac{-5 - 5i\sqrt{3}}{2} \}$.

41. $\frac{1}{x - 3} + \frac{1}{x + 2} = 1$

Multiplying both sides by the LCD, (x - 3)(x + 2), yields

$$x + 2 + x - 3 = (x - 3)(x + 2)$$
$$2x - 1 = x^2 - x - 6$$
$$0 = x^2 - 3x - 5$$

Substituting a = 1, b = -3, and c = -5 into the quadratic formula gives us $x = \frac{3 \pm \sqrt{29}}{2}$.

45.
$$\begin{aligned}
x + y + z + w &= 10 \quad (1) \\
x + 2y - z + w &= 6 \quad (2) \\
x - y - z + 2w &= 4 \quad (3) \\
x - 2y + z - 3w &= -12 \quad (4)
\end{aligned}$$

Eliminating the x-variable and z-variable by adding equations (1) and (3), we have:

$$\begin{array}{rl}
x + y + z + w = 10 & (1) \\
\underline{x - y - z + 2w = 4} & (3) \\
2x \qquad\quad + 3w = 14 & (5)
\end{array}$$

Eliminating the x-variable and z-variable by adding equations (2) and (4), we have:

$$\begin{array}{rl}
x + 2y - z + w = 6 & (2) \\
\underline{x - 2y + z - 3w = -12} & (4) \\
2x \qquad\qquad - 4w = -6 & (6)
\end{array}$$

Eliminating the x-variable in equations (5) and (6), we have:

$$\begin{array}{lcl}
2x + 3w = 14 & \xrightarrow{\text{no change}} & 2x + 3w = 14 \\
2x - 4w = -6 & \xrightarrow{\text{multiply } -1} & \underline{-2x + 4w = 6} \\
& & 5w = 20 \\
& & w = 4
\end{array}$$

Substituting w = 4 into equation (5), we have:

$$\begin{aligned}
2x + 3(4) &= 14 \\
2x + 12 &= 14 \\
2x &= 2 \\
x &= 1
\end{aligned}$$

Substituting x = 1 and w = 4 into equations (1) and (2), we have:

$$\begin{array}{lcl}
1 + y + z + 4 = 10 & \longrightarrow & y + z = 5 \\
1 + 2y - z + 4 = 6 & & \underline{2y - z = 1} \\
& & 3y = 6 \\
& & y = 2
\end{array}$$

Substituting x = 1, y = 2 and w = 4 into equation (1), we have:

$$\begin{aligned}
1 + 2 + z + 4 &= 10 \\
z &= 3
\end{aligned}$$

The solution is (1,2,3,4).

Section 8.3

1. $\begin{vmatrix} 1 & 0 \\ 2 & 3 \end{vmatrix} = 1(3) - 2(0) = 3 - 0 = 3$

5. $\begin{vmatrix} 0 & 1 \\ 1 & 0 \end{vmatrix} = 0(0) - 1(1) = 0 - 1 = -1$

9. $\begin{vmatrix} -3 & -1 \\ 4 & -2 \end{vmatrix} = -3(-2) - 4(-1) = 6 + 4 = 10$

13. $\begin{vmatrix} 1 & 2x \\ 2 & -3x \end{vmatrix} = 21$

We expand the determinate on the left side to get:

$$1(-3x) - 2(2x) = 21$$
$$-3x - 4x = 21$$
$$-7x = 21$$
$$x = -3$$

17. $\begin{vmatrix} x^2 & 3 \\ x & 1 \end{vmatrix} = 10$

We expand the determinate on the left side to get:

$$x^2(1) - x(3) = 10$$
$$x^2 - 3x = 10$$
$$x^2 - 3x - 10 = 0$$
$$(x + 2)(x - 5) = 0$$

$$x + 2 = 0 \quad \text{or} \quad x - 5 = 0$$
$$x = -2 \qquad\qquad x = 5$$

21.

$$= 1(2)1 + 2(1)1 + 3(3)1 - 1(2)3 - 1(1)1 - 1(3)2$$
$$= 2 + 2 + 9 - 6 - 1 - 6$$
$$= 0$$

25. $\begin{vmatrix} 3 & 0 & 2 \\ 0 & -1 & -1 \\ 4 & 0 & 0 \end{vmatrix}$

The products of the three elements in row 1 with their minors are:

$$+3\begin{vmatrix} -1 & -1 \\ 0 & 0 \end{vmatrix} - 0\begin{vmatrix} 0 & -1 \\ 4 & 0 \end{vmatrix} + 2\begin{vmatrix} 0 & -1 \\ 4 & 0 \end{vmatrix}$$

We complete the problem by evaluating each of the three 2x2 determinants and then simplifying the resulting expression:

$$+3[-1(0) - 0(-1)] - 0[0(0) - 4(-1)] + 2[0(0) - 4(-1)]$$
$$= 3(0) - 0(0 - 5) + 2(0 + 4)$$
$$= 0 - 0 + 8$$
$$= 8$$

Section 8.3 continued

29. Method 1

$$\begin{vmatrix} 1 & 3 & 7 \\ -2 & 6 & 4 \\ 3 & 7 & -1 \end{vmatrix} \begin{matrix} 1 & 3 \\ -2 & 6 \\ 3 & 7 \end{matrix}$$

$$= 1(6)(-1) + 3(4)(3) + 7(-2)7 - 3(6)(7) - 7(4)1 - (-1)(-2)3$$
$$= -6 + 36 - 98 - 126 - 28 - 6$$
$$= -228$$

Method 2

$$\begin{vmatrix} 1 & 3 & 7 \\ -2 & 6 & 4 \\ 3 & 7 & -1 \end{vmatrix}$$

The products of the three elements in row 1 with their minors are:

$$+1\begin{vmatrix} 6 & 4 \\ 7 & -1 \end{vmatrix} - 3\begin{vmatrix} -2 & 4 \\ 3 & -1 \end{vmatrix} + 7\begin{vmatrix} -2 & 6 \\ 3 & 7 \end{vmatrix}$$

We complete the problem by evaluating each of the three 2x2 determinates and then simplifying the resulting expression:

$$+1[6(-1) - 7(4)] - 3[-2(-1) - 3(4)] + 7[-2(7) - 3(6)]$$
$$= 1(-6 - 28) - 3(2 - 12) + 7(-14 - 18)$$
$$= 1(-34) - 3(-10) + 7(-32)$$
$$= -34 + 30 - 224$$
$$= -228$$

33. Using a = 2, b = -5, and c = 4 in the discriminant, we have
$$b^2 - 4ac = 25 - 4(2)(4) = 25 - 32 = -7$$

Since the discriminant is -7, we will have two complex solutions.

37. If $\qquad y = \dfrac{2}{3} \qquad\quad y = 3$

then $\qquad 3y = 2 \qquad y - 3 = 0$
$$3y - 2 = 0$$

Because of the zero-factor property, we have

$$(3y - 2)(y - 3) = 0$$
$$3y^2 - 11y + 6 = 0$$

41. The products of the four elements in row 1 with their minors are:

$$+2\begin{vmatrix} 2 & 0 & 1 \\ 0 & 1 & 0 \\ 1 & 0 & 0 \end{vmatrix} - 0\begin{vmatrix} -1 & 0 & 1 \\ -3 & 1 & 0 \\ 1 & 0 & 0 \end{vmatrix} + 1\begin{vmatrix} -1 & 2 & 1 \\ -3 & 0 & 0 \\ 1 & 1 & 0 \end{vmatrix} - (-3)\begin{vmatrix} -1 & 2 & 0 \\ -3 & 0 & 1 \\ 1 & 1 & 0 \end{vmatrix}$$

(columns labeled **a**, **b**, **c**, **d**)

The products of the three elements of **a** in row 1 with their minors are:

$$+2\begin{vmatrix} 1 & 0 \\ 0 & 0 \end{vmatrix} - 0\begin{vmatrix} 0 & 0 \\ 1 & 0 \end{vmatrix} + 1\begin{vmatrix} 0 & 1 \\ 1 & 0 \end{vmatrix}$$

$$\begin{aligned}
\mathbf{a} &= +2[1(0) - 0(0)] - 0[0(0) - 1(0)] + 1[0(0) - 1(1)] \\
&= 2(0) - 0(-1) + 1(-1) \\
&= -1
\end{aligned}$$

The products of the element of **b** in row 1 with their minors are:

$$+(-1)\begin{vmatrix} 1 & 0 \\ 0 & 0 \end{vmatrix} - 0\begin{vmatrix} -3 & 0 \\ 1 & 0 \end{vmatrix} + 1\begin{vmatrix} -3 & 1 \\ 1 & 0 \end{vmatrix}$$

$$\begin{aligned}
\mathbf{b} &= -1[1(0) - 0(0)] - 0[-3(0) - 1(0)] + 1[-3(0) - 1(1)] \\
&= -1(0) - 0(0) + 1(-1) \\
&= -1
\end{aligned}$$

The products of the elements of **c** in row 1 with their minors are:

$$+(-1)\begin{vmatrix} 0 & 0 \\ 1 & 0 \end{vmatrix} - 2\begin{vmatrix} -3 & 0 \\ 1 & 0 \end{vmatrix} + 1\begin{vmatrix} -3 & 0 \\ 1 & 1 \end{vmatrix}$$

$$\begin{aligned}
\mathbf{c} &= -1[0(0) - 1(0)] - 2[-3(0) - 1(0)] + 1[-3(1) - 1(0)] \\
&= -1(0) - 2(0) + 1(-3) \\
&= -3
\end{aligned}$$

The products of the elements of **d** in row 1 with their minors are:

$$+(-1)\begin{vmatrix} 0 & 1 \\ 1 & 0 \end{vmatrix} - 2\begin{vmatrix} -3 & 1 \\ 1 & 0 \end{vmatrix} + 0\begin{vmatrix} -3 & 0 \\ 1 & 1 \end{vmatrix}$$

$$\begin{aligned}
\mathbf{d} &= -1[0(0) - 1(1)] - 2[-3(0) - 1(1)] + 0[-3(1) - 1(0)] \\
&= -1(-1) - 2(-1) + 0(-3) \\
&= 3
\end{aligned}$$

Substituting into the original product,

$$\begin{aligned}
2(1) - 0(-1) + 1(-3) &- (-3)(3) \\
&= -2 - 3 + 9 \\
&= 4
\end{aligned}$$

Section 8.4

1. $2x - 3y = 3$
    $4x - 2y = 10$

    We begin by calculating the determinates D, D_x, and D_y:

    $$D = \begin{vmatrix} 2 & -3 \\ 4 & -2 \end{vmatrix} = 2(-2) - 4(-3) = 8$$

    $$D_x = \begin{vmatrix} 3 & -3 \\ 10 & -2 \end{vmatrix} = 3(-2) - 10(-3) = 24$$

    $$D_y = \begin{vmatrix} 2 & 3 \\ 4 & 10 \end{vmatrix} = 2(10) - 4(3) = 8$$

    $$x = \frac{D_x}{D} = \frac{24}{8} = 3 \quad \text{and} \quad y = \frac{D_y}{D} = \frac{8}{8} = 1$$

    The solution set for the system is $\{(3,1)\}$.

5. $4x - 7y = 3$
    $5x + 2y = -3$

    We begin by calculating the determinates D, D_x, and D_y:

    $$D = \begin{vmatrix} 4 & -7 \\ 5 & 2 \end{vmatrix} = 4(2) - 5(-7) = 43$$

    $$D_x = \begin{vmatrix} 3 & -7 \\ -3 & 2 \end{vmatrix} = 3(2) - (-3)(-7) = -15$$

    $$D_y = \begin{vmatrix} 4 & 3 \\ 5 & -3 \end{vmatrix} = 4(-3) - 5(3) = -27$$

    $$x = \frac{D_x}{D} = -\frac{15}{43} \quad \text{and} \quad y = \frac{D_x}{D} = -\frac{27}{43}$$

    The solution set for the system is $\{(-\frac{15}{43}, -\frac{27}{43})\}$.

9.
$$x + y + z = 4$$
$$x - y - z = 2$$
$$2x + 2y - z = 2$$

We evaluate D using Method 1 from **Section 8.3**:

$$D = \begin{vmatrix} 1 & 1 & 1 \\ 1 & -1 & -1 \\ 2 & 2 & -1 \end{vmatrix} \begin{matrix} 1 & 1 \\ 1 & -1 \\ 2 & 2 \end{matrix}$$

$$= 1(-1)(-1) + 1(-1)2 + 1(1)(2) - 2(-1)1 - 2(-1)1 - (-1)1(1)$$
$$= 1 - 2 + 2 + 2 + 2 + 1$$
$$= 6$$

We evaluate D_x using Method 2 from **Section 8.3** and expanding across row 1:

$$D_x = \begin{vmatrix} 4 & 1 & 1 \\ 2 & -1 & -1 \\ 2 & 2 & -1 \end{vmatrix} = 4\begin{vmatrix} -1 & -1 \\ 2 & -1 \end{vmatrix} - 1\begin{vmatrix} 2 & -1 \\ 2 & -1 \end{vmatrix} + 1\begin{vmatrix} 2 & -1 \\ 2 & 2 \end{vmatrix}$$

$$= 4(3) - 1(0) + 1(6)$$
$$= 18$$

$$D_y = \begin{vmatrix} 1 & 4 & 1 \\ 1 & 2 & -1 \\ 2 & 2 & -1 \end{vmatrix} = -1\begin{vmatrix} 4 & 1 \\ 2 & -1 \end{vmatrix} + 2\begin{vmatrix} 1 & 1 \\ 2 & -1 \end{vmatrix} - (-1)\begin{vmatrix} 1 & 4 \\ 2 & 2 \end{vmatrix}$$

$$= -1(-6) + 2(-3) + 1(-6)$$
$$= -6$$

$$D_z = \begin{vmatrix} 1 & 1 & 4 \\ 1 & -1 & 2 \\ 2 & 2 & 2 \end{vmatrix} = 1\begin{vmatrix} -1 & 2 \\ 2 & 2 \end{vmatrix} - 1\begin{vmatrix} 1 & 4 \\ 2 & 2 \end{vmatrix} + 2\begin{vmatrix} 1 & 4 \\ -1 & 2 \end{vmatrix}$$

$$= 1(-6) - 1(-6) + 2(6)$$
$$= 12$$

$$x = \frac{D_x}{D} = \frac{18}{6} = 3 \qquad y = \frac{D_y}{D} = \frac{-6}{6} = -1 \qquad z = \frac{D_z}{D} = \frac{12}{6} = 2$$

The solution set is $\{(3,-1,2)\}$.

13. The four determinants using Method 1 used in Cramer's rule are:

$$D = \begin{vmatrix} 3 & -1 & 2 \\ 6 & -2 & 4 \\ 1 & -5 & 2 \end{vmatrix}$$

$= 3(-2)2 + -1(4((1) + 2(6)(-5) - 1(-2)(2) - (-5)4(3) - 2(6)(-1)$
$= -12 - 4 - 60 + 4 + 60 + 12$
$= 0$

$$D_x = \begin{vmatrix} 4 & -1 & 2 \\ 8 & -2 & 4 \\ 1 & -5 & 2 \end{vmatrix}$$

$= 4(-2)(2) + (-1)4(1) + 2(8)(-5) - 1(-2)(2) - (-5)(4)4 - 2(8)(-1)$
$= -16 - 4 - 80 + 4 + 80 + 16$
$= 0$

$$D_y = \begin{vmatrix} 3 & 4 & 2 \\ 6 & 8 & 4 \\ 1 & 1 & 2 \end{vmatrix}$$

$= 3(8)2 + 4(4)1 + 2(6)1 - 1(8)2 - 1(4)3 - 2(6)4$
$= 48 + 16 + 12 - 16 - 12 - 48$
$= 0$

$$D_z = \begin{vmatrix} 3 & -1 & 4 \\ 6 & -2 & 8 \\ 1 & -5 & 1 \end{vmatrix}$$

$= 3(-2)1 + (-1)8(1) + 4(6)(-5) - 1(-2)4 - (-5)8(3) - 1(6)(-1)$
$= -6 - 8 - 120 + 8 + 120 + 6$
$= 0$

The system is dependent.

17. $-x - 7y + 0z = 1$
     $x + 0y + 3z = 11$
     $0x + 2y + z = 0$

Using Method 1 to evaluate D, we have:

$$D = \begin{vmatrix} -1 & -7 & 0 \\ 1 & 0 & 3 \\ 0 & 2 & 1 \end{vmatrix}$$

$= -1(0)1 + (-7)3(0) + 0(1)2 - 0(0)0 - 2(3)(-1) - 1(1)(-7)$
$= 0 + 0 + 0 - 0 + 6 + 7$
$= 13$

Using Method 2 to evaluate D_x, D_y, and D_z, we have:

$$D_x = \begin{vmatrix} \boxed{1 & -7 & 0} \\ 11 & 0 & 3 \\ 0 & 2 & 1 \end{vmatrix} = 1\begin{vmatrix} 0 & 3 \\ 2 & 1 \end{vmatrix} - (-7)\begin{vmatrix} 11 & 3 \\ 0 & 1 \end{vmatrix} + 0\begin{vmatrix} 11 & 0 \\ 1 & 2 \end{vmatrix}$$

$= 1(-6) + 7(11) + 0(22)$
$= 71$

$$D_y = \begin{vmatrix} -1 & 1 & 0 \\ \boxed{1 & 11 & 3} \\ 0 & 0 & 1 \end{vmatrix} = 1\begin{vmatrix} 1 & 0 \\ 0 & 1 \end{vmatrix} - 11\begin{vmatrix} -1 & 0 \\ 0 & 1 \end{vmatrix} + 3\begin{vmatrix} -1 & 1 \\ 0 & 0 \end{vmatrix}$$

$= 1(1) - 11(-1) + 3(0)$
$= 12$

$$D_z = \begin{vmatrix} \boxed{-1} & -7 & 1 \\ 1 & 0 & 11 \\ 0 & 2 & 0 \end{vmatrix} = -1\begin{vmatrix} 0 & 11 \\ 2 & 0 \end{vmatrix} - 1\begin{vmatrix} -7 & 1 \\ 2 & 0 \end{vmatrix} + 0\begin{vmatrix} -7 & 1 \\ 0 & 11 \end{vmatrix}$$

$= -1(-22) - 1(-2) + -(-77)$
$= 24$

$$x = \frac{D_x}{D} = \frac{71}{13} \qquad y = \frac{D_y}{D} = \frac{12}{13} \qquad z = \frac{D_z}{D} = \frac{24}{13}$$

The solution set is $\{(\frac{71}{13}, \frac{12}{13}, \frac{24}{13})\}$.

21. Rewrite the original equations to:

$$-10x + y = 100$$
$$-12x + y = 0$$

$$D = \begin{vmatrix} -10 & 1 \\ -12 & 1 \end{vmatrix} = -10(1) - (-12)1 = 2$$

$$D_x = \begin{vmatrix} 100 & 1 \\ 0 & 1 \end{vmatrix} = 100(1) - 0(1) = 100$$

$$x = \frac{D_x}{D} = \frac{100}{2} = 50$$

The company must sell 50 items.

25. When $y = x^{1/3}$

the equation $x^{2/3} - 5x^{1/3} + 6 = 0$

becomes
$$y^2 - 5y + 6 = 0$$
$$(y - 2)(y - 3) = 0 \qquad \text{Factor}$$
$$y - 2 = 0 \text{ or } y - 3 = 0 \qquad \text{Set factors to 0}$$
$$y = 2 \qquad\qquad y = 3$$

Now we replace y with $x^{1/3}$ and solve for x:

$$x^{1/3} = 2 \text{ or } x^{1/3} = 3$$
$$x = 8 \qquad\quad x = 27 \qquad \text{Cube both sides of each equation}$$

29. $(3x + 1) - 6\sqrt{3x + 1} + 8 = 0$

$$\text{Let } y = \sqrt{3x + 1}$$
$$y^2 - 6y + 8 = 0$$
$$(y - 2)(y - 4) = 1$$
$$y = 2 \text{ or } y = 4$$

Resubstitute $\sqrt{3x + 1}$ for y.

$$\sqrt{3x + 1} = 2 \text{ or } \sqrt{3x + 1} = 4$$
$$3x + 1 = 4 \qquad\quad 3x + 1 = 16$$
$$x = 1 \qquad\qquad x = 5$$

Both solutions check in the original equation.

33.
$$D = \begin{vmatrix} a & b \\ b & a \end{vmatrix} = a(a) - b(b) = a^2 - b^2$$

$$D_x = \begin{vmatrix} -1 & b \\ 1 & a \end{vmatrix} = -1a - 1(b) = -a - b$$

$$D_y = \begin{vmatrix} a & -1 \\ b & 1 \end{vmatrix} = a(1) - b(-1) = a + b$$

$$x = \frac{D_x}{D} = \frac{-a - b}{a^2 - b^2} = \frac{-(a + b)}{(a + b)(a - b)} = -\frac{1}{a - b} = \frac{1}{b - a}$$

$$y = \frac{D_y}{D} = \frac{a + b}{a^2 - b^2} = \frac{a + b}{(a + b)(a - b)} = \frac{1}{a - b}$$

37.
$$x + 2y = 1$$
$$3x + 4y = 0$$

Section 8.5

1.
$$\begin{bmatrix} 1 & 1 & | & 5 \\ 3 & -1 & | & 3 \end{bmatrix}$$

$$\begin{bmatrix} 1 & 1 & | & 5 \\ 0 & -4 & | & -12 \end{bmatrix}$$ Multiply row 1 by -3 and add the result to row 2

$$\begin{bmatrix} 1 & 1 & | & 5 \\ 0 & 1 & | & 3 \end{bmatrix}$$ Multiply row 2 by $-\frac{1}{4}$

$$x + y = 5$$
$$y = 3$$
$$x = 2$$

The solution to our system is (2,3).

5.
$$\begin{bmatrix} 2 & -8 & | & 6 \\ 3 & -8 & | & 13 \end{bmatrix}$$

$$\begin{bmatrix} 1 & -4 & | & 3 \\ 3 & -8 & | & 13 \end{bmatrix}$$ Multiply row 1 by $\frac{1}{2}$

$$\begin{bmatrix} 1 & -4 & | & 3 \\ 0 & 4 & | & 4 \end{bmatrix}$$ Multiply row 1 by -3 and add the results to row 2

$$\begin{bmatrix} 1 & -4 & | & 3 \\ 0 & 1 & | & 1 \end{bmatrix}$$ Multiply row 2 by $\frac{1}{4}$

$$x - 4y = 3 \qquad \text{If} \qquad y = 1$$
$$y = 1 \qquad \text{then} \quad x - 4y = 3$$
$$x - 4(1) = 3$$
$$x = 7$$

The solution to our system is (7,1).

9.

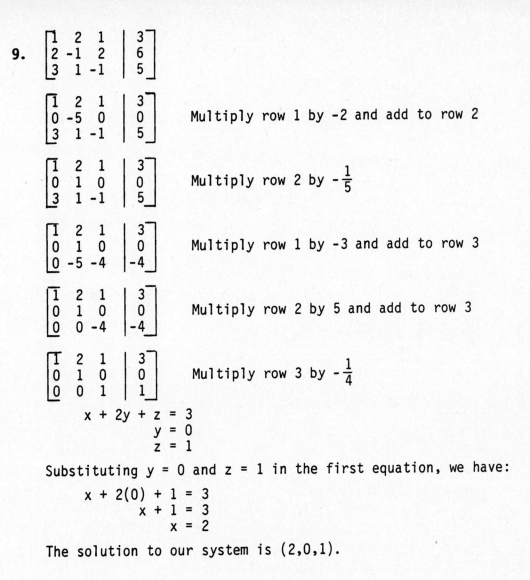

$$\begin{bmatrix} 1 & 2 & 1 & | & 3 \\ 2 & -1 & 2 & | & 6 \\ 3 & 1 & -1 & | & 5 \end{bmatrix}$$

$$\begin{bmatrix} 1 & 2 & 1 & | & 3 \\ 0 & -5 & 0 & | & 0 \\ 3 & 1 & -1 & | & 5 \end{bmatrix}$$ Multiply row 1 by -2 and add to row 2

$$\begin{bmatrix} 1 & 2 & 1 & | & 3 \\ 0 & 1 & 0 & | & 0 \\ 3 & 1 & -1 & | & 5 \end{bmatrix}$$ Multiply row 2 by $-\frac{1}{5}$

$$\begin{bmatrix} 1 & 2 & 1 & | & 3 \\ 0 & 1 & 0 & | & 0 \\ 0 & -5 & -4 & | & -4 \end{bmatrix}$$ Multiply row 1 by -3 and add to row 3

$$\begin{bmatrix} 1 & 2 & 1 & | & 3 \\ 0 & 1 & 0 & | & 0 \\ 0 & 0 & -4 & | & -4 \end{bmatrix}$$ Multiply row 2 by 5 and add to row 3

$$\begin{bmatrix} 1 & 2 & 1 & | & 3 \\ 0 & 1 & 0 & | & 0 \\ 0 & 0 & 1 & | & 1 \end{bmatrix}$$ Multiply row 3 by $-\frac{1}{4}$

$$x + 2y + z = 3$$
$$y = 0$$
$$z = 1$$

Substituting $y = 0$ and $z = 1$ in the first equation, we have:

$$x + 2(0) + 1 = 3$$
$$x + 1 = 3$$
$$x = 2$$

The solution to our system is $(2,0,1)$.

13.
$$\begin{bmatrix} 1 & 3 & 0 & | & 7 \\ 3 & 0 & -4 & | & -8 \\ 0 & 5 & -2 & | & -5 \end{bmatrix}$$

$$\begin{bmatrix} 1 & 3 & 0 & | & 7 \\ 0 & -9 & -4 & | & -29 \\ 0 & 5 & -2 & | & -5 \end{bmatrix}$$
  Multiply row 1 by -3 and add to row 2

$$\begin{bmatrix} 1 & 3 & 0 & | & 7 \\ 0 & 1 & \frac{4}{9} & | & \frac{29}{9} \\ 0 & 5 & -2 & | & -5 \end{bmatrix}$$
  Multiply row 2 by $-\frac{1}{9}$

$$\begin{bmatrix} 1 & 3 & 0 & | & 7 \\ 0 & 1 & \frac{4}{9} & | & \frac{29}{9} \\ 0 & 0 & -\frac{38}{9} & | & -\frac{190}{9} \end{bmatrix}$$
  Multiply row 2 by -5 and add to row 3

$$\begin{bmatrix} 1 & 3 & 0 & | & 7 \\ 0 & 1 & \frac{4}{9} & | & \frac{29}{9} \\ 0 & 0 & 1 & | & 5 \end{bmatrix}$$
  Multiply row 3 by $-\frac{9}{38}$

$$x + 3y = 7$$
$$y + \frac{4}{9}z = \frac{29}{9}$$
$$z = 5$$

Substituting z = 5 in the second equation, we have

$$y + \frac{4}{9}(5) = \frac{29}{9}$$
$$y = 1$$

Substituting y = 1 in the first equation, we have
$$x + 3(1) = 7$$
$$x = 4$$

The solution to our system is (4,1,5).

17. $\begin{bmatrix} 2 & -3 & | & 4 \\ 4 & -6 & | & 4 \end{bmatrix}$

$\begin{bmatrix} 1 & -\dfrac{3}{2} & | & 2 \\ 4 & -6 & | & 4 \end{bmatrix}$ Multiply row 1 by $\dfrac{1}{2}$

$\begin{bmatrix} 1 & -\dfrac{3}{2} & | & 2 \\ 0 & 0 & | & -4 \end{bmatrix}$ Multiply row 1 by -4 and add to row 2

The y value is undefined.

21. Let x = the first number. Since one of them is 1 less than twice the other, the second expression is $2x - 1$. The equation is

$$x + 2x - 1 = 11$$
$$3x - 1 = 11$$
$$3x = 12$$
$$x = 4$$

then $2x - 1 = 7$

The two numbers are 4 and 7.

25. Let one man's wages = x and the other man's wages be $x + 8$. Then the equation becomes:

$$x + x + 8 = 44$$
$$2x + 8 = 44$$
$$2x = 36$$
$$x = 18$$

then $x + 8 = 26$

One man receives \$18, and the other receives \$26.

29. $\dfrac{x - 3}{x + 2} < 0$

The inequality indicates that the quotient of $(x - 3)$ and $(x + 2)$ is negative.

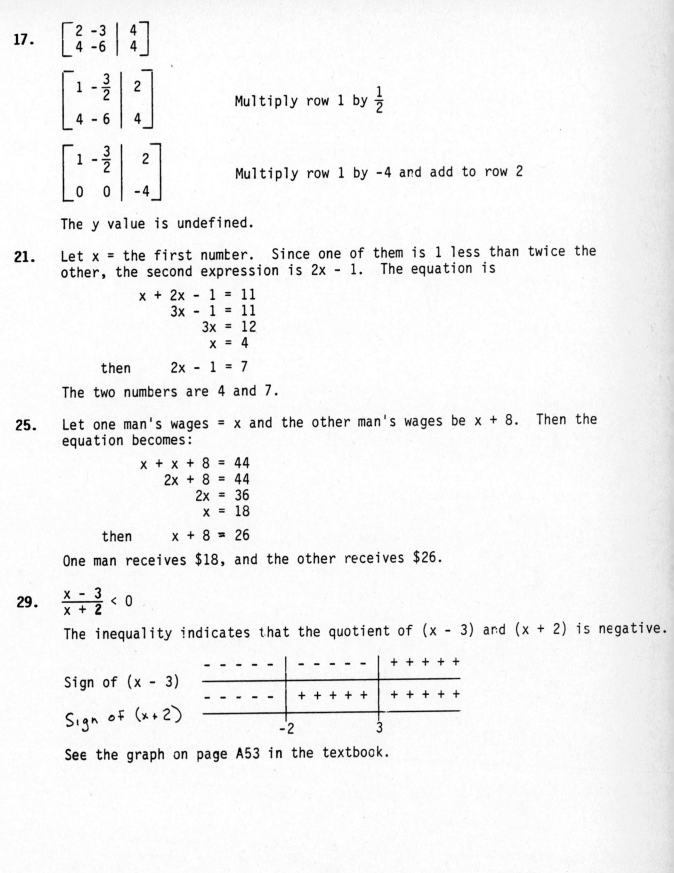

See the graph on page A53 in the textbook.

1. If we let x and y represent the two numbers, then the translation of the first sentence in the problem into an equation would be

   $$y = 2x + 3$$

   The second sentence gives us a second equation:

   $$x + y = 18$$

   The linear system that describes the situation is

   $$y = 2x + 3$$
   $$x + y = 18$$

   Substituting the expression for y from the first equation into the second yields

   $$x + (2x + 3) = 18$$
   $$3x + 3 = 18$$
   $$3x = 15$$
   $$x = 5$$

   Using x = 5 in y = 2x + 3 gives the second number:

   $$y = 2(5) + 3$$
   $$y = 13$$

   The two numbers are 5 and 13. Their sum is 18 and the second is 3 more than twice the first.

5. Let x = the smallest number, y = a number and z = the largest number. The three equations are as follows:

   $$x + y + z = 8$$
   $$2x = z - 2$$
   $$x + z = 5$$

   In linear equation form, we have:

   $$x + y + z = 8 \quad (1)$$
   $$2x + 0y - z = -2 \quad (2)$$
   $$x + 0y + z = 5 \quad (3)$$

   First eliminate to z-variable:

   $$\begin{array}{l} x + y + z = 8 \quad (1) \\ \underline{2x + 0y - z = -2 \quad (2)} \\ 3x + y = 6 \quad (4) \end{array}$$

   Also:

   $$\begin{array}{l} 2x + 0y - z = -2 \quad (2) \\ \underline{x + 0y + z = 5 \quad (3)} \\ 3x = 3 \quad (5) \\ x = 1 \end{array}$$

   Substituting x = 1 in equation (4), we have:

   $$3(1) + y = 6$$
   $$y = 3$$

   Substituting x = 1 and y = 3 in equation (1), we have:

   $$1 + 3 + z = 8$$
   $$z = 4$$

   The three numbers are 1, 3 and 4.

9. We will let x equal the amount invested at 6% and y equal the amount invested at 7%:

$$x + y = 20,000 \quad (1)$$
$$0.06x + 0.07y = 1,280 \quad (2)$$

In a table, we have:

	Dollars at 6%	Dollars at 7%	Total
Number	x	y	20,000
Interest	0.06x	0.07y	1,280

Multiplying equation (2) by 100, we have

$$x + y = 20,000 \quad (1)$$
$$6x + 7y = 128,000$$

We can eliminate y from this system by multiplying equation (1) by -7 and adding the result to equation (2):

$$
\begin{array}{rcl}
-7x - 7y &=& -140,000 \\
6x + 7y &=& 128,000 \\
\hline
-x &=& -12,000 \\
x &=& 12,000
\end{array}
$$

The amount invested at 6% is $12,000. Since the total investment is $20,000, the amount invested 7% is $8,000.

13. We will let x equal the amount invested at 6%, y equal the amount invested at 8% and z equal the amount at 9%. From the problem we can write the following equations:

$$x + y + z = 2,200 \quad (1)$$
$$3x = z \quad (2)$$
$$0.06x + 0.08y + 0.09z = 178 \quad (3)$$

Multiplying equation (3) by 100, we have:

$$6x + 8y + 9z = 17,800 \quad (4)$$

Substituting z = 3x into equations (1) and (4), we have:

$$x + y + 3x = 2,200 \quad (1)$$
$$6x + 8y + 9(3x) = 17,800 \quad (4)$$

Simplifying equations (1) and (4), we have:

$$
\begin{array}{llll}
& \text{multiply } -8 & & \\
4x + y = 2,200 & \xrightarrow{\hspace{2cm}} & -32x - 8y = -17,600 \\
33x + 8y = 17,800 & \xrightarrow{\hspace{2cm}} & \underline{33x + 8y = 17,800} \\
& \text{no change} & x = 200
\end{array}
$$

Substituting x = 200 in equation (2), we have:

$$3(200) = z$$
$$600 = z$$

Substituting x = 200 and z = 600 into equation (1), we have:

$$200 + y + 600 = 2,200$$
$$y = 800$$

The man invested $200 at 6%, $1,400 at 8% and $600 at 9%.

17. Let x = number of gallons of 20% disinfectant solution needed and y = number of gallons of 14% disinfectant solution needed. The total amount of 16% disinfectant solution is 15 gallons.

$$x + y = 15$$
$$0.02x + 0.14y = 0.16(15) = 2.4$$

In a table, we have:

	20% Solution	14% Solution	Final Solution
Total Number of Gallons	x	y	15
Gallons of Disinfectant	0.20x	0.14y	0.16(15)

Multiplying our second equation by 100 gives us an equivalent system:

$$x + y = 15$$
$$20x + 14y = 240$$

Multiplying the first equation by -14, we have:

$$-14x - 14y = -210$$
$$\underline{20x + 14y = 240}$$
$$6x = 30$$
$$x = 5$$

It takes 5 gallons of 20% disinfectant solution and 10 gallons of 14% disinfectant solution to produce 15 gallons of 16% disinfectant solution.

Section 8.6 continued

21. Let x = the speed of the plane and y = the speed of the air. Using a table, we have:

	d	r	t
With the wind	600	x + y	2
Against the wind	600	x - y	$2\frac{1}{2}$

Since d = r · t, the system we need to solve the problem is

$$600 = (x + y) \cdot 2$$
$$600 = (x - y) \cdot 2\frac{1}{2}$$

which is equivalent to

$$600 = 2x + 2y$$
$$600 = 2\frac{1}{2}x - 2\frac{1}{2}y$$

Multiplying the second equation by 2, we have:

$$
\begin{array}{llll}
600 = 2x + 2y & \xrightarrow{\text{multiply 5}} & 3000 = 10x + 10y \\
1200 = 5x - 5y & \xrightarrow{\text{multiply 2}} & \underline{2400 = 10x - 10y} \\
& & 5400 = 20x \\
& & x = 270
\end{array}
$$

Substituting x = 270 into 600 = 2x + 2y, we see that

$$600 = 2(270) + 2y$$
$$600 = 540 + 2y$$
$$60 = 2y$$
$$30 = y$$

The speed of the plane is 270 mph and the speed of the wind is 30 mph.

25. Let x = the number of nickels, y = the number of dimes, and z = the number of quarters. The three equations from the problem are:

$$
\begin{aligned}
x + y + z &= 9 \quad (1) \\
y &= z \quad (2) \\
5x + 10y + 25z &= 120 \quad (3)
\end{aligned}
$$

Substituting y = z in equations (1) and (3), we have:

$$
\begin{aligned}
x + y + z &= 9 \\
5x + 10y + 25y &= 120
\end{aligned}
$$

becomes

$$
\begin{array}{lcl}
x + 2y = 9 & \xrightarrow{\text{multiply } -5} & -5x - 10y = -45 \\
5x + 35y = 120 & \xrightarrow{\text{no change}} & 5x + 35y = 120 \\
& & \overline{\qquad 25y = 75} \\
& & \qquad\quad y = 3
\end{array}
$$

Since y = z, then z = 3. Substituting these values in equation (1), we have:

$$
\begin{aligned}
x + 3 + 3 &= 9 \\
x &= 3
\end{aligned}
$$

There are 3 of each type of coin.

29. $y = 12x - 4x^2$. We identify a = -4, b = 12, and c = 0 and use the formula $x = -\dfrac{b}{2}a$.

$$
x = \frac{-12}{2(-4)} = \frac{3}{2}
$$

To find the associated y-coordinate we substitute $\frac{3}{2}$ for x in the equation to obtain

$$
\begin{aligned}
y &= 12\left(\frac{3}{2}\right) - 4\left(\frac{3}{2}\right)^2 \\
&= 18 - 4\left(\frac{9}{4}\right) \\
&= 18 - 9 \\
&= 9
\end{aligned}
$$

The vertex is at $(\frac{3}{2}, 9)$. It is the highest point on the graph, because a is negative so the graph opens downward.

1. $$x + y = 4$$
    $$\underline{2x - y = 14}$$
    $$3x \quad\;\; = 18$$
    $$x \quad\;\;\; = 16$$

    Substituting x = 6 in the first equation:

    $$6 + y = 4$$
    $$y = -2$$

    The ordered pair (6,-2) is the solution to the system.

5. $$\qquad\qquad\qquad\;\; \text{multiply by 4}$$
    $$3x - 7y = 2 \xrightarrow{\hspace{3cm}} \quad 12x - 28y = \;\;8$$
    $$-4x + 6y = -6 \xrightarrow{\hspace{3cm}} \underline{-12x + 18y = -18}$$
    $$\qquad\qquad \text{multiply by 3} \qquad\qquad -10y = -10$$

    Substituting y = 1 in the first equation:

    $$3x - 7(1) = 2$$
    $$3x - 7 = 2$$
    $$3x = 9$$
    $$x = 3$$

    The ordered pair (3,1) is the solution to the system.

9. $$x + y = 2$$
    $$y = x - 1$$

    The second equation tells us y is x - 1. Substituting the expression
    x - 1 for y in the first equation, we have:

    $$x + x - 1 = 2$$
    $$2x - 1 = 2$$
    $$2x = 3$$
    $$x = \frac{3}{2}$$

    Putting x = $\frac{3}{2}$ in the second equation, we have

    $$y = \frac{3}{2} - 1$$

    $$y = \frac{3}{2} - \frac{2}{2} \qquad \text{Note: } 1 = \frac{1}{1} = \frac{2}{2}$$

    $$y = \frac{1}{2}$$

    The ordered pair $(\frac{3}{2}, \frac{1}{2})$ is the solution to the system.

13. $x + y = 4$
       $2x + 5y = 2$

By solving for x in the first equation, we have

   $x = 4 - y$

Substituting $x = 4 - y$ in the second equation produces

   $2(4 - y) + 5y = 2$
   $8 - 2y + 5y = 2$
   $8 + 3y = 2$
   $3y = -6$
   $y = -2$

Substituting $y = -2$ in the first equation, we have

   $x + (-2) = 4$
   $x = 6$

The ordered pair (6,-2) is the solution to the system.

17. $5x - y = 4$
       $y = 5x - 3$

The second equation tells us y is $5x - 3$. Substituting the expression $5x - 3$ for y in the first equation, we have

   $5x - (5x - 3) = 4$
   $5x - 5x + 3 = 4$
   $3 = 4$ False

The result is a false statement $3 = 4$, which indicates there is no solution to the system. We say the system is inconsistent and the lines are parallel.

21. $3x + 2y + z = 4$ (1)
$2x - 4y + z = -1$ (2)
$x + 6y + 3z = -4$ (3)

To eliminate z, we

$$
\begin{array}{ll}
-3x - 2y - z = -4 & \text{-1 times (1)} \\
\underline{2x - 4y + z = -1} & \text{(2)} \\
-x - 6y = -5 & \text{(4)}
\end{array}
$$

$$
\begin{array}{ll}
-6x + 12y - 3z = 3 & \text{-3 times (2)} \\
\underline{x + 6y + 3z = -4} & \text{(3)} \\
-5x + 18y = -1 & \text{(5)}
\end{array}
$$

Equations (4) and (5) form a linear system in two variables. We can eliminate the variable from this system as follows:

$$
\begin{array}{l}
-x - 6y = -5 \quad \xrightarrow{\text{multiply by -5}} \quad 5x + 30y = 25 \\
-5x + 18y = -1 \quad \xrightarrow{\text{no change}} \quad \underline{-5x + 18y = -1} \\
\phantom{-5x + 18y = -1 \quad \text{no change} \quad } 48y = 24 \\
\phantom{-5x + 18y = -1 \quad \text{no change} \quad } y = \frac{1}{2}
\end{array}
$$

Substituting $y = \frac{1}{2}$ into equation (4) or equation (5) and solving for x gives

$$
\begin{array}{l}
-x - 6(\frac{1}{2}) = -5 \\
-x - 3 = -5 \\
-x = -2 \\
x = 2
\end{array}
$$

Using $x = 2$ and $y = \frac{1}{2}$ in equation (1), (2) or (3) and solving for z results in

$$
\begin{array}{l}
3(2) + 2(\frac{1}{2}) + z = 4 \\
6 + 1 + z = 4 \\
7 + z = 4 \\
z = -3
\end{array}
$$

The ordered triple that satisfies all three equations is $(2, \frac{1}{2}, -3)$.

25.

$$5x - 2y + z = 4 \quad (1)$$
$$-3x + 4y - z = 2 \quad (2)$$
$$6x - 8y + 2z = -4 \quad (3)$$

To eliminate z, we

$$
\begin{array}{rl}
5x - 2y + z = 4 & (1) \\
-3x + 4y - z = 2 & (2) \\
\hline
2x + 2y \quad\quad = 6 &
\end{array}
$$

$$
\begin{array}{rl}
-6x + 8y - 2z = 4 & \text{2 times (2)} \\
6x - 8y + 2z = -8 & (3) \\
\hline
0 = -4 &
\end{array}
$$

All three variables have been eliminated, and we are left with a false statement. The two equations are inconsistent; there are no ordered triples that satisfy both equations. The solution set for the system is the empty set Ø. There is no unique solution.

29. $\begin{vmatrix} 2 & 3 \\ -5 & 4 \end{vmatrix} = 2(4) - (-5)(3) = 8 + 15 = 23$

33. Method 1

$$\begin{vmatrix} 3 & 0 & 2 \\ -1 & 4 & 0 \\ 2 & 0 & 0 \end{vmatrix} \begin{matrix} -3 & 0 \\ -1 & 4 \\ 2 & 0 \end{matrix}$$

$$= 3(4)(0) + 0(0)(2) + 2(-1)(0) - 2(4)(2) - 0(0)3 - 0(-1)(0)$$
$$= 0 + 0 + 0 - 16 - 0 - 0$$
$$= -16$$

Method 2

Given $\begin{vmatrix} 3 & 0 & 2 \\ -1 & 4 & 0 \\ 2 & 0 & 0 \end{vmatrix}$ and expanding across the first row, we have

$$3\begin{vmatrix} 4 & 0 \\ 0 & 0 \end{vmatrix} - 0\begin{vmatrix} -1 & 0 \\ 2 & 0 \end{vmatrix} + 2\begin{vmatrix} -1 & 4 \\ 2 & 0 \end{vmatrix}$$

$$= 3(0 - 0) - 0(0 - 0) + 2(0 - 8)$$
$$= 0 - 0 - 16$$
$$= -16$$

37. $\begin{vmatrix} 2 & 3x \\ -1 & 2x \end{vmatrix} = 4$

$$2(2x) - (-1)(3x) = 4$$
$$4x + 3x = 4$$
$$7x = 4$$
$$x = \frac{4}{7}$$

41. $3x - 5y = 4$
$7x - 2y = 3$

$$D = \begin{vmatrix} 3 & -5 \\ 7 & -2 \end{vmatrix} = 3(-2) - (7)(-5) = -6 + 35 = 29$$

$$D_x = \begin{vmatrix} 4 & -5 \\ 3 & -2 \end{vmatrix} = 4(-2) - (3)(-5) = -8 + 15 = 7$$

$$D_y = \begin{vmatrix} 3 & 4 \\ 7 & 3 \end{vmatrix} = 3(3) - 7(4) = 9 - 28 = -19$$

$$x = \frac{D_x}{D} = \frac{7}{29} \qquad y = \frac{D_y}{D} = -\frac{19}{29}$$

The solution set for the system is $(\frac{7}{29}, -\frac{19}{29})$.

45. $4x - 5y = -3$
$2x + 3z = 4$
$3y - z = 8$

$$D = \begin{vmatrix} 4 & -5 & 0 \\ 2 & 0 & 3 \\ 0 & 3 & -1 \end{vmatrix} = 4\begin{vmatrix} 0 & 3 \\ 3 & -1 \end{vmatrix} - (-5)\begin{vmatrix} 2 & 3 \\ 0 & -1 \end{vmatrix} + 0\begin{vmatrix} 2 & 0 \\ 0 & 3 \end{vmatrix}$$

$$= 4(0 - 9) + 5(-2 - 0) + 0(6 - 0)$$
$$= -36 - 10 + 0$$
$$= -46$$

$$D_x = \begin{vmatrix} -3 & -5 & 0 \\ 4 & 0 & 3 \\ 8 & 3 & -1 \end{vmatrix} = -3\begin{vmatrix} 0 & 3 \\ 3 & -1 \end{vmatrix} - (-5)\begin{vmatrix} 4 & 3 \\ 8 & -1 \end{vmatrix} + 0\begin{vmatrix} 4 & 0 \\ 8 & 3 \end{vmatrix}$$

$$= -3(0 - 9) + 5(-4 - 24) + 0(12 - 0)$$
$$= 27 - 140 + 0$$
$$= -113$$

$$D_y = \begin{vmatrix} 4 & -3 & 0 \\ 2 & 4 & 3 \\ 0 & 8 & -1 \end{vmatrix} = 4\begin{vmatrix} 4 & 3 \\ 8 & -1 \end{vmatrix} - (-3)\begin{vmatrix} 2 & 3 \\ 0 & -1 \end{vmatrix} + 0\begin{vmatrix} 2 & 4 \\ 0 & 8 \end{vmatrix}$$

$$= 4(-4 - 24) + 3(-2 - 0) + 0(16 - 0)$$
$$= -112 - 6 + 0$$
$$= -118$$

$$D_z = \begin{vmatrix} 4 & -5 & -3 \\ 2 & 0 & 4 \\ 0 & 3 & 8 \end{vmatrix} = 4\begin{vmatrix} 0 & 4 \\ 3 & 8 \end{vmatrix} - (-5)\begin{vmatrix} 2 & 4 \\ 0 & 8 \end{vmatrix} + (-3)\begin{vmatrix} 2 & 0 \\ 0 & 3 \end{vmatrix}$$

$$= 4(0 - 12) + 5(16 - 0) - 3(6 - 0)$$
$$= -48 + 80 - 18$$
$$= 14$$

$$x = \frac{D_x}{D} = \frac{-113}{-46} = \frac{113}{46} \qquad y = \frac{D_y}{D} = \frac{-118}{-46} = \frac{59}{23} \qquad z = \frac{D_z}{D} = \frac{14}{-46} = -\frac{7}{23}$$

The solution set for the equation is $(\frac{113}{46}, \frac{59}{23}, -\frac{7}{23})$.

49. $6x + 5y = 9$
$4x + 3y = 6$

$$\begin{bmatrix} 6 & 5 & | & 9 \\ 4 & 3 & | & 6 \end{bmatrix}$$

Divide row 1 by 6

$$\begin{bmatrix} 1 & \dfrac{5}{6} & | & \dfrac{9}{6} \\ 4 & 3 & | & 6 \end{bmatrix}$$

Divide row 2 by 4

$$\begin{bmatrix} 1 & \dfrac{5}{6} & | & \dfrac{3}{2} \\ 1 & \dfrac{3}{4} & | & \dfrac{6}{4} \end{bmatrix}$$

Subtract row 1 from row 2, and add result to row 2

$$\begin{bmatrix} 1 & \dfrac{5}{6} & | & \dfrac{3}{2} \\ 0 & -\dfrac{1}{12} & | & 0 \end{bmatrix}$$

Multiply row 2 by -12

$$\begin{bmatrix} 1 & \dfrac{5}{6} & | & \dfrac{3}{2} \\ 0 & 1 & | & 0 \end{bmatrix}$$

Multiply row 2 by $\dfrac{5}{6}$ and subtract result from row 1, and add result to row 1

$$\begin{vmatrix} 1 & 0 & | & \dfrac{3}{2} \\ 0 & 1 & | & 0 \end{vmatrix}$$

Taking this last matrix and writing the system of equations it represents, we have

$$x = \frac{3}{2}$$

$$y = 0$$

The solution to our system is $(\frac{3}{2}, 0)$.

53. Let x = one number, y = the other number

$$x = 2y + 5 \quad (1)$$
$$x + y = 11 \qquad (2)$$

$$
\begin{array}{lll}
x - 2y = 5 & \xrightarrow{\text{multiply } -1} & -x + 2y = -5 \\
x + y = 11 & \xrightarrow{\text{no change}} & \underline{x + y = 11} \\
& & 3y = 6 \\
& & y = 2
\end{array}
$$

Substituting y = 2 into the first equation, we have

$$x = 2 \cdot 2 + 5 = 9$$

The two numbers are 2 and 9.

57. We let x equal adult tickets and y equal children's tickets.

	Adult	Child	Total
Number	x	y	127
Value	2.00	1.50	214.00

The system of equations that describes this situation is given by

$$x + y = 127$$
$$2x + 1.5y = 214$$

Multiply the second equation by 10 to clear it of decimals. The system that results after doing so is

$$x + y = 127$$
$$20x + 15y = 2140$$

We can eliminate x from this system by multiplying the first equation by -20 and adding the result to the second equation:

$$
\begin{array}{rcr}
-20x - 20y & = & -2540 \\
\underline{20x + 15y} & = & \underline{2140} \\
-5y & = & -400 \\
y & = & 80 \\
x + 80 & = & 127 \\
x & = & 47
\end{array}
$$

They sold 47 adults tickets and 80 children's tickets.

61. Let x = number of ounces of 30% HCL solution needed and y = number of ounces of 70% HCL solution needed. The total amount of 50% HCL solution is 15 ounces.

	30% solution	70% solution	Final solution
Total number of ounces	x	y	15
Ounces of HCL solution	0.30x	0.70x	0.50(15)

Multiplying our second equation by 10 gives us an equivalent system:

$$x + y = 15$$
$$3x + 7y = 75$$

Multiply the first equation by -3, we have:

$$-3x - 3y = -45$$
$$\underline{3x + 7y = 75}$$
$$4y = 30$$
$$y = 7.5$$

It takes 7.5 ounces of 70% HCL solution and 7.5 ounces of 30% HCL solution to produce 15 ounces of 50% HCL solution.

Chapter 8 Test

1.
$$2x - 5y = -8 \xrightarrow{\text{no change}} 2x - 5y = -8$$
$$3x + y = 5 \xrightarrow{\text{multiply 5}} 15x + 5y = +25$$
$$\overline{17x \qquad = 17}$$
$$x \qquad = 1$$

Substituting $x = 1$ into the first equation, we have

$$2(1) - 5y = -8$$
$$-5y = -10$$
$$y = 2$$

The solution to the system is $(1,2)$.

5.
$$6x - 3y = 0$$
$$x + 2y = 5 \qquad \text{let} \qquad x = -2y + 5$$

Substituting $x = 2y + 5$ in the first equation, we have:

$$6(-2y + 5) - 3y = 0$$
$$-12y + 30 - 3y = 0$$
$$-15y + 30 = 0$$
$$-15y = -30$$
$$y = 2$$

Substituting $y = 2$ into the second equation, we have:

$$x + 2(2) = 5$$
$$x = 1$$

The solution to the system is $(1,2)$.

9.
$$5x - 4y = 2$$
$$-2x + y = 3$$

$$D = \begin{vmatrix} 5 & -4 \\ -2 & 1 \end{vmatrix} = 5(1) - (-2)(-4) = 5 - 8 = -3$$

$$D_x = \begin{vmatrix} 2 & -4 \\ 3 & 1 \end{vmatrix} = 2(1) - 3(-4) = 2 + 12 = 14$$

$$D_y = \begin{vmatrix} 5 & 2 \\ -2 & 3 \end{vmatrix} = 5(3) - (-2)(2) = 15 + 4 = 19$$

$$D_x = \frac{14}{-3} = -\frac{14}{3} \qquad D_y = \frac{19}{-3} = -\frac{19}{3}$$

The solution is $(-\frac{14}{3}, -\frac{19}{3})$.

13.
$$\begin{bmatrix} 1 & 2 & 0 & | & 4 \\ 0 & 1 & 3 & | & 18 \\ 2 & 0 & -5 & | & -29 \end{bmatrix}$$

Multiply row 1 by -2 and add to row 3, the result in row 3

$$\begin{bmatrix} 1 & 2 & 0 & | & 4 \\ 0 & 1 & 3 & | & 18 \\ 0 & -4 & -5 & | & -37 \end{bmatrix}$$

Multiply row 2 by 4 and add to row 3, the result in row 3

$$\begin{bmatrix} 1 & 2 & 0 & | & 4 \\ 0 & 1 & 3 & | & 18 \\ 0 & 0 & 7 & | & 35 \end{bmatrix}$$

Divide row 3 by 7

$$\begin{bmatrix} 1 & 2 & 0 & | & 4 \\ 0 & 1 & 3 & | & 18 \\ 0 & 0 & 1 & | & 5 \end{bmatrix}$$

$$x + 2y = 4 \quad (1)$$
$$y + 3z = 18 \quad (2)$$
$$z = 5 \quad (3)$$

Substituting (3) in (2), we have:

$$y + 3(5) = 18$$
$$y = 3 \quad (4)$$

Substituting (4) in (1), we have:

$$x + 2(3) = 4$$
$$x = -2$$

The solution to the system is (-2,3,5).

17. Let x = number of gallons of 30% alcohol solution, let y = number of gallons of 70% alcohol solution. The total amount of 60% solution is 16 gallons.

	30% solution	70% solution	Final solution
Total number of gallons	x	y	16
Gallons of alcohol solution	.30x	.70y	.60(16)

Multiplying our second equation by 10 gives us an equivalent system:

$$x + y = 16$$
$$3x + 7y = 6(16)$$

Multiplying the first equation by -3, we have:

$$-3x - 3y = -48$$
$$\underline{3x + 7y = 96}$$
$$4y = 48$$
$$y = 12$$
$$x = 4$$

It takes 4 gallons of 30% alcohol solution and 12 gallons of 70% alcohol solution.

Chapter 9

Section 9.1

1. Domain is {1,2,4}, range is {3,5,1}. This is a function.

5. Domain is {7,3}, range is {-1,4}. This is not a function.

9. Domain is {5,-3,2}, range is {-3,2}. This is a function.

13. The graph is not a function, since we can find a vertical line that crosses its graph on more than one place.

17. The graph is a function since a vertical line does not cross the graph in more than one place.

21. Given $y = \sqrt{2x - 1}$,

    then $2x - 1 \geq 0$

    $\quad\quad 2x \geq 1$

    $\quad\quad x \geq \dfrac{1}{2}$ $\quad$ The domain is $\{x \mid x \geq \dfrac{1}{2}\}$.

25. $y = \dfrac{x + 2}{x - 5}$

    The domain can be any real number that does not produce an undefined term.

    $\quad\quad x - 5 = 0$
    $\quad\quad\quad x = 5$

    The domain is all real numbers except $x \neq 5$.

25. The vertical asymptote is $x = 2$. See the graph on page A54 in the textbook.

33. The vertical asymptote is $x = 4$. See the graph on page A54 in the textbook.

37. Domain = all real numbers

    Range $= \{y \mid y \geq -4\}$.

    It is function. See the graph on page A54 in the textbook.

41. Domain $= \{x \mid x \geq 2\}$

    Range $= \{y \mid y \geq 0\}$

    It is a function. See the graph on page A55 in the textbook.

45. Domain $= \{x \mid x \geq 2\}$

    Range $=$ all real numbers

    This is not a function. See the graph on page A55 in the textbook.

49. a. yes

   b. Domain = $\{t \mid 0 \le t < 6\}$.

   Range = $\{h \mid 0 \le h \le 60\}$.

   c. 3

   d. 60

   e. 0 and 6

53.

$$4x + 5y = 5 \xrightarrow{\text{no change}} 4x + 5y = 5$$

$$\frac{6}{5}x + y = 2 \xrightarrow[\text{multiply -5}]{} \begin{array}{l} -6x - 5y = -10 \\ \hline -2x \qquad = -5 \end{array}$$

$$= -\frac{5}{2}$$

Substituting $x = \frac{5}{2}$ into the first equation, we have:

$$4\left(\frac{5}{2}\right) + 5y = 5$$

$$10 + 5y = 5$$
$$5y = -5$$
$$y = -1$$

The solution to the system is $\left(\frac{5}{2}, -1\right)$.

57. $2x - 3y = -6$ (1)
   $y = 3x - 5$ (2)

Substituting equation (2) in equation (1), we have:

$$2x - 3(3x - 5) = -6$$
$$2x - 9x + 15 = -6$$
$$-7x + 15 = -6$$
$$-7x = -21$$
$$x = 3$$

Substituting $x = 3$ into equation (2), we have:

$$y = 3(3) - 5$$
$$y = 4$$

The solution to the system is $(3,4)$.

61. $y = \dfrac{2}{x - 4}$

x	y
3.9	- 20
3.99	-200
4.01	200
4.1	20

1. $f(x) = 2x - 5$
   $f(2) = 2(2) - 5 = -1$

5. $g(x) = x^2 + 3x + 4$
   $g(-1) = (-1)^2 + 3(-1) + 4 = 1 - 3 + 4 = 2$

9. $g(x) = x^2 + 3x + 4$
   $g(4) = 4^2 + 3(4) + 4 = 32$

   $g(4) + f(4) = 32 + 3 = 35$

   $f(x) = 2x - 5$
   $f(4) = 2(4) - 5 = 3$

13. $f(1) = 4$

17. $g(-2) = 2$

21. $f(x) = 3x^2 - 4x + 1$
    $f(0) = 3(0)^2 - 4(0) + 1 = 1$

25. $f(x) = 3x^2 - 4x + 1$
    $f(a) = 3a^2 - 4a + 1$

29. $g(x) = 2x - 1$
    $g(2) = 2 \cdot 2 - 1 = 3$

    $f[g(2)] = f(3) = 16$

    $f(x) = 3x^2 - 4x + 1$
    $f(3) = 3(3)^2 - 4(3) + 1 = 16$

33. $f(x) = 3x^2 - 4x + 1$
    $f(0) = 3(0)^2 - 4(0) + 1 = 1$

    $g[f(0)] = g(1) = 1$

    $g(x) = 2x - 1$
    $g(1) = 2 \cdot 1 - 1 = 1$

37. $\qquad f(x) = 4x - 5$

$$\frac{f(x) - f(a)}{x - a} = \frac{(4x - 5) - (4a - 5)}{x - a}$$

$$= \frac{4x - 4a}{x - a}$$

$$= \frac{4(x - a)}{x - a}$$

$$= 4$$

41. $y = x^2 + 1$ means $f(x) = x^2 + 1$

$$\frac{f(x) - f(a)}{x - a} = \frac{(x^2 + 1) - (a^2 - 1)}{x - a}$$

$$= \frac{x^2 - a^2}{x - a}$$

$$= \frac{(x + a)(x - a)}{x - a}$$

$$= x + a$$

45. $y = -4x - 1$ means $f(x) = -4x - 1$

If $\qquad f(x + h) = -4(x + h) - 1$
$$= -4x - 4h - 1$$

Then $\dfrac{f(x + h) - f(x)}{h} = \dfrac{(-4x - 4h - 1) - (-4x - 1)}{h}$

$$= \frac{-4h}{h}$$

$$= -4$$

49. $g + f$ means $(g + f)(x) = g(x) + f(x)$

$g(x) + f(x) = (x - 2) + (3x - 5)$
$$= 4x - 7$$

53. $g - f$ means $(g - f)(x) = g(x) - f(x)$

$g(x) - f(x) = (x - 2) - (3x - 5)$
$$= -2x + 3$$

57. fh means $(fh)(x) = f(x)h(x)$

$f(x)h(x) = (3x - 5)(3x^2 - 11x + 10)$
$$= 9x^3 - 48x^2 + 85x - 50$$

61. $\dfrac{f}{h}$ means $\dfrac{f}{h}(x) = \dfrac{f(x)}{h(x)}$

$$\frac{f(x)}{h(x)} = \frac{3x - 5}{3x^2 - 11x + 10}$$

$$= \frac{3x - 5}{(3x - 5)(x - 2)}$$

$$= \frac{1}{x - 2}$$

65. $h + fg$ means $(h + fg)(x) = h(x) + f(x)g(x)$

$h(x) + f(x)g(x) = (3x^2 - 11x + 10) + (3x - 5)(x - 2)$
$$= (3x^2 - 11x + 10) + (3x^2 - 11x + 10)$$
$$= 6x^2 - 22x + 20$$

69.   If $x = 800 - 100p$
         then $R(p) = (800 - 100p)p$
                    $= 800p - 100p^2$

         $R(x) = 8x - .01x^2$

73. $x + y + z = 6$ (1)
         $2x - y + z = 3$ (2)
         $x + 2y - 3z = -4$ (3)

   Multiply (1) by -1 and add to (3):

$$-x - y - z = -6$$
$$\underline{x + 2y - 3z = -4}$$
$$y - 4z = -10 \quad (4)$$

   Multiply (1) by -2 and add to (2):

$$-2x - 2y - 2z = -12$$
$$\underline{2x - y + z = 3}$$
$$-3y + z = -9 \quad (5)$$

   Solving with equations (4) and (5).

77. $f(-x) = f(x)$
         $(-x)^2 - 4 = x^2 - 4$
         $x^2 - 4 = x^2 - 4$ True statement

81. $f(-x) = -f(x)$
         $3(-x) = -(3x)$
         $-3x = -3x$ True statement

Section 9.3

1. $f(x) - x^2 - 3$ is a quadratic function.

         $y = x^2 - 3$ (1)
            $= (x + 0)^2 - 3$

   The graph is a parabola with vertex $(0,-3)$.   When $x = -2$, $y = 1$ and when $x = 2$, $y = 1$ in equation (1).

5. $f(x) = 5$ is a constant function.

9. $y = x^3$ is a cubic function.

x	x^3	y
-2	-8	-8
-1	-1	-1
1	1	1
2	8	8

See the graph on page A56 in the textbook.

13. $y = (x - 2)^3$ is a cubic function.

x	$(x - 2)^3$	y
0	-8	-8
1	-1	-1
2	0	0
3	1	1
4	8	8

See the graph on page A56 in the textbook.

17. $y = \dfrac{x + 3}{x - 1}$ is a rational function.

x	$\dfrac{x + 3}{x - 1}$	y
-3	0	0
0	-3	-3
2	5	5
5	2	2

See the graph on page A56 in the textbook.

21. If $f(x) = 0$
     then $f(x) = 3x - 9$
     becomes $0 = 3x - 9$
                   $9 = 3x$
                   $x = 3$

25. If $f(x) = 0$
     then $f(x) = x^2 + 4x + 1$
     becomes $0 = x^2 + 4x + 1$

$$x = \frac{-4 \pm \sqrt{4^2 - 4(1)(1)}}{2(1)}$$

$$x = \frac{-4 \pm \sqrt{16 - 4}}{2}$$

$$x = \frac{-4 \pm 2\sqrt{3}}{2}$$

$$x = -2 \pm \sqrt{3}$$

29. Since the equation for $N(t)$ is quadratic in t, the maximum value of N will occur when $f = -\frac{b}{2a}$:

$$t = -\frac{60}{2(-10)} = -\frac{60}{-20} = 3$$

This means that the maximum number of people will be in the store 3 hours after it opens. Since the store opens at 9:00am, the maximum number of people will be in the store at 12:00 noon.

To find the number of people in the store at that time, we simply evaluate $N(3)$:

$$N(3) = 60(3) - 10(3)^2 - 180 - 90 = 90 \text{ people}$$

33. If $g(x) = \left(\frac{1}{2}\right)^x$

then $g(-1) = \left(\frac{1}{2}\right)^{-1} = (2)^1 = 2$

37. If $f(x) = 3^x$ If $g(x) = \left(\frac{1}{2}\right)^x$

then $f(2) = 3^2 = 9$ then $g(-2) = \left(\frac{1}{2}\right)^{-2}$

$$= (2)^2$$
$$= 4$$

$$f(2) + g(-2) = 9 + 4 = 13$$

41.

x	$y = 3^{-x}$	y
3	$y = 3^{-3} = \frac{1}{27}$	$\frac{1}{27}$
2	$y = 3^{-2} = \frac{1}{9}$	$\frac{1}{9}$
1	$y = 3^{-1} = \frac{1}{3}$	$\frac{1}{3}$
0	$y = 3^{0} = 1$	1
-1	$y = 3^{-(-1)} = 3$	3
-2	$y = 3^{-(-2)} = 9$	9
-3	$y = 3^{-(-3)} = 27$	27

See the graph on page A57 in your textbook.

45.

x	$y = 2^{2x}$	y
3	$y = 2^{2 \cdot 3} = 2^{6} = 64$	64
2	$y = 2^{2 \cdot 2} = 2^{4} = 16$	16
1	$y = 2^{2 \cdot 1} = 2^{2} = 4$	4
0	$y = 2^{2 \cdot 0} = 2^{0} = 1$	1
-1	$y = 2^{2(-1)} = 2^{-2} = \frac{1}{4}$	$\frac{1}{4}$
-2	$y = 2^{2(-2)} = 2^{-4} = \frac{1}{16}$	$\frac{1}{16}$
-3	$y = 2^{2(-3)} = 2^{-6} = \frac{1}{64}$	$\frac{1}{64}$

See the graph on page A57 in the textbook.

49. $\begin{vmatrix} 3 & 5 \\ -6 & 2 \end{vmatrix} = (3)(2) - (-6)(5) = 6 + 30 = 36$

53. To find the x-intercept:

    let $y = 0$

    $0 = x^3 - x$

    $0 = (x^2 - 1)x$

    $0 = (x + 1)(x - 1)x$

    $x = -1, \; x = 1, \; x = 0$

    To find the y-intercept:

    let $x = 0$

    $y = 0^3 - 0$

    $y = 0$

x	y
-2	-6
$-\dfrac{1}{2}$	$\dfrac{3}{8}$
$\dfrac{1}{2}$	$-\dfrac{3}{8}$
2	6

    See the graph on page A58 in the textbook.

57. To find the x-intercept:

    let $y = 0$

    $0 = x^3 + 3x - x - 3$

    $0 = (x + 3)(x + 1)(x - 1)$

    $x = -3, \; x = -1, \; x = 1$

    To find the y-intercept:

    let $x = 0$

    $y = 0^3 + 3(0) - 0 - 3$

    $y = -3$

x	y
-4	-15
-2	3
2	15

    See the graph on page A58 in the textbook.

Section 9.4

1. $f(x) = 3x - 1$ means $y = 3x - 1$

   The inverse of $y = 3x - 1$ is $x = 3y - 1$. We can solve the equation $x = 3y - 1$ for y in terms of x as follows:

   $$x = 3y - 1$$
   $$x + 1 = 3y$$
   $$y = \frac{x + 1}{3}$$
   $$f^{-1}(x) = \frac{x + 1}{3}$$

5. To find the inverse of $y = \frac{x - 3}{x - 1}$ we interchange x and y to get

$$x = \frac{y - 3}{y - 1}$$

$$xy - x = y - 3 \qquad \text{Multiply each side by } y - 1$$
$$xy - y = x - 3 \qquad \text{Collect terms in } y \text{ on left side}$$
$$(x - 1)y = x - 3 \qquad \text{Factor } y \text{ from left side}$$

$$y = \frac{x - 3}{x - 1} \qquad \text{Divide each side by } x - 1$$

As you can see, this function is its own inverse.

9. $f(x) = \frac{1}{2}x - 3$ means $y = \frac{1}{2}x - 3$

The inverse of $y = \frac{1}{2}x - 3$ is $x = \frac{1}{2}y - 3$. Solving the equation

$x = \frac{1}{2}y - 3$ for y in terms of x as follows:

$$x = \frac{1}{2}y - 3$$

$$x + 3 = \frac{1}{2}y$$

$$y = 2(x + 3)$$

$$f^{-1}(x) = 2(x + 3)$$

13. The inverse of $y = 2x - 1$ is $x = 2y - 1$. Solving the equation
$x = 2y - 1$ for y in terms of x as follows:

$$x = 2y - 1$$
$$x + 1 = 2y$$

$$y = \frac{x + 1}{2}$$

Substituting x = 0	Substituting x = 1
in $\quad y = 2x - 1$	in $\quad y = 2x - 1$
becomes $\quad y = 2(0) - 1$	becomes $\quad y = 2(1) - 1$
$y = -1$	$y = 1$

$f(x) = 2x - 1$ has solutions (0,-1) and (1,1).

Substituting x = 1	Substituting x = -1
in $\quad y = \frac{x + 1}{2}$	in $\quad x = \frac{-1 + 1}{2}$
becomes $\quad y = \frac{1 + 1}{2}$	$y = 0$
$y = 1$	

$f^{-1}(x) = \frac{x + 1}{2}$ has solutions (1,1) and (-1,0).

See the graph on page A59 in the textbook.

17.   The graph of

$$y = x^2 - 2x - 3$$
$$y = (x^2 - 2x + 1) - 3 - 1$$
$$y = (x - 1)^2 - 4$$

is a parabola with vertex (1,-4) and is concave-up. Other solutions are (-2,5), (-1,0), (1,3) and (4,5). Reflecting the graph about the line $y = x$ is a parabola with vertex (-4,1) and other solutions (5,-2), (0,-1), (3,1), and (5,4). The equation for the inverse is $x = y^2 - 2y - 3$. See the graph on page A59 in the textbook.

21.   The graph of $y = 4$ is a horizontal line at 4. The inverse of $y = 4$ is $x = 4$ which is a vertical line at 4. See the graph on page A59 in the textbook.

25.   The graph of $y = \frac{1}{2}x + 2$ is a linear function with y-intercept at 2 and the x-intercept at -4. Exchanging x and y in the equation, we obtain the equation of the inverse.

The inverse $x = \frac{1}{2}y + 2$

becomes $x - 2 = \frac{1}{2}y$

$$y = 2(x - 2)$$

See the graph on page A59 in the textbook.

29.   a. $f(x) = 3x - 2$
$f(2) = 3 \cdot 2 - 2 = 4$

   b. $f^{-1}(x) = \frac{x + 2}{3}$

$f^{-1}(2) = \frac{2 + 2}{3} = \frac{4}{3}$

   c. $f^{-1}(2) = \frac{4}{3}$ (from 29.b)

$f(x) = 3x - 2$

$f(\frac{4}{3}) = 3(\frac{4}{3}) - 2 = 4 - 2 = 2$

$f[f^{-1}(2)] = f(\frac{4}{3}) = 2$

   d. $f(2) = 4$ from (29.a)

$f^{-1}(x) = \frac{x + 2}{3}$

$f^{-1}(4) = \frac{4 + 2}{3} = 2$

$f^{-1}[f(2)] = f^{-1}(4) = 2$

33. Let x = geese and y = ducks. Remember 14 dimes = 1.40 and 6 dimes = .60.

$$x + y = 108 \qquad (1)$$
$$1.40x + .60y = 112.80 \quad (2)$$

Removing the decimal in equation (2) yields

$$
\begin{array}{l}
x + y = 108 \\
14x + 6x = 1,128
\end{array}
\xrightarrow[\text{no change}]{\text{multiply -6}}
\begin{array}{rcl}
-6x - 6y &=& 648 \\
15x + 6y &=& 1,128 \\ \hline
8x &=& 480 \\
x &=& 60
\end{array}
$$

The man bought 60 geese and 48 ducks.

37. $f(x) = 3x + 5$ means $y = 3x + 5$

The inverse of $y = 3x + 5$ is $x = 3y + 5$. Solving the equation $x = 3y + 5$ for y in terms of x as follows:

$$x = 3y + 5$$
$$x - 5 = 3y$$
$$\frac{x - 5}{3} = y$$
$$f^{1}(x) = \frac{x - 5}{3}$$
$$f(f^{-1}(x)) = 3\left(\frac{x - 5}{3}\right) + 5$$
$$= x - 5 + 5$$
$$= x$$

41. $f(x) = \dfrac{x - 4}{x - 2}$ means $y = \dfrac{x - 4}{x - 2}$

The inverse of $y = \dfrac{x - 4}{x - 2}$ is $x = \dfrac{y - 4}{y - 2}$. Solving the equation

$x = \dfrac{y - 4}{y - 2}$ for y in terms of x as follows:

$$x = \frac{y - 4}{y - 2}$$

$$(y - 2)x = y - 4$$
$$xy - 2x = y - 4$$
$$xy - y = 2x - 4$$
$$y(x - 1) = 2x - 4$$

$$y = \frac{2x - 4}{x - 1}$$

$$f^1(x) = \frac{2x - 4}{x - 1}$$

$$f(f^1(x)) = \frac{\left(\dfrac{2x - 4}{x - 1}\right) - 4}{\left(\dfrac{2x - 4}{x - 1}\right) - 2}$$

$$= \frac{2x - 4 - 4(x - 1)}{2x - 4 - 2(x - 1)}$$

$$= \frac{2x - 4 - 4x + 4}{2x - 4 - 2x + 2}$$

$$= \frac{-2x}{-2}$$

$$= x$$

Section 9.5

1. Let $(3,7)$ be (x_1,y_1) and $(6,3)$ be (x_2,y_2) in the distance formula, we have

$$d = \sqrt{(6 - 3)^2 + (3 - 7)^2}$$

$$= \sqrt{(3)^3 + (-4)^2}$$

$$= \sqrt{9 + 16}$$

$$= \sqrt{25}$$

$$= 5$$

Section 9.5 continued

5. Let $(3,-5)$ be (x_1,y_1) and $(-2,1)$ by (x_2,y_2) in the distance formula, we have

$$d = \sqrt{(-2-3)^2 + [1-(-5)]^2}$$
$$= \sqrt{(-5)^2 + (6)^2}$$
$$= \sqrt{25 + 36}$$
$$= \sqrt{61}$$

9. Let $(x,2)$ be (x_1,y_1) and $(1,5)$ be (x_2,y_2) and $d = \sqrt{13}$

$$\sqrt{13} = \sqrt{(1-x)^2 + (5-2)^2}$$
$$13 = (1-x)^2 + 3^2$$
$$13 = 1 - 2x + x^2 + 9$$
$$0 = x^2 - 2x - 3$$
$$0 = (x+1)(x-3)$$
$$x = -1 \quad \text{or} \quad x = 3$$

13. Let $(a,b) = (2,3)$ and $r = 4$, applying Theorem 9.1 yields
$$(x-a)^2 + (y-b)^2 = r^2$$
$$(x-2)^2 + (y-3)^2 = 4^2$$
$$(x-2)^2 + (y-3)^2 = 16$$

17. Let $(a,b) = (-5,-1)$ and $r = \sqrt{5}$, applying Theorem 9.1 yields
$$(x-a)^2 + (y-b)^2 = r^2$$
$$[x-(-5)]^2 + [y-(-1)]^2 = (\sqrt{5})^2$$
$$(x+5)^2 + (y+1)^2 = 5$$

21. Let $(a,b) = (0,0)$ and $r = 2$, applying Theorem 9.1 yields
$$(x-a)^2 + (y-b)^2 = r^2$$
$$(x-0)^2 + (y-0)^2 = 2^2$$
$$x^2 + y^2 = 4$$

25. $(x-1)^2 + (y-3)^2 = 25$
$(x-1)^2 + (y-3)^2 = 5^2$

The center is at $(1,3)$ and the radius is 5.

29. $(x+1)^2 + (y+1)^2 = 1$
$[x-(-1)]^2 + [y-(-1)]^2 = 1^2$

The center is at $(-1,-1)$ and the radius is 1.

33.
$$x^2 + y^2 + 2x = 1$$
$$x^2 + 2x + y^2 = 1$$
$$x^2 + 2x + 1 + y^2 = 1 + 1$$
$$(x+1)^2 + (y-0)^2 = 2$$
$$[x-(-1)]^2 + (y-0)^2 = (\sqrt{2})^2$$

The center is at $(-1,0)$ and the radius is $\sqrt{2}$.

37.
$$x^2 + y^2 + 2x + y = \frac{11}{4}$$

$$x^2 + 2x + y^2 + y = \frac{11}{4}$$

$$x^2 + 2x + 1 + y^2 + y + \frac{1}{4} = \frac{11}{4} + 1 + \frac{1}{4}$$

$$(x + 1)^2 + (y + \frac{1}{2})^2 = 4$$

$$[x - (-1)]^2 + [y - (-\frac{1}{2})]^2 = 2^2$$

The center is at $(-1, -\frac{1}{2})$ and the radius is 2.

41. The distance from the origin to 3 or -3 is 3. This is the radius of the circle with the center (0,0).

$$(x - a)^2 + (y - b)^2 = r^2$$
$$(x - 0)^2 + (y - 0)^2 = 3^2$$
$$x^2 + y^2 = 9$$

45. $y = \sqrt{9 - x^2}$ corresponds to the top half and $y = -\sqrt{9 - x^2}$ corresponds to the bottom half.

49. $4x - 7 = 3$
$5x + 2y = -3$

$$D = \begin{vmatrix} 4 & -7 \\ 5 & 2 \end{vmatrix} = 4(2) - 5(-7) = 8 + 35 = 43$$

$$D_x = \begin{vmatrix} 3 & -7 \\ -3 & 2 \end{vmatrix} = 3(2) - (-3)(-7) = 6 - 21 = -15$$

$$D_y = \begin{vmatrix} 4 & 3 \\ 5 & -3 \end{vmatrix} = 4(-3) - 5(3) = -12 - 15 = -27$$

$$x = \frac{D_x}{D} = -\frac{15}{43} \qquad y = \frac{D_y}{D} = \frac{-27}{-43} = \frac{27}{43}$$

The solution to the system is $(-\frac{15}{43}, \frac{27}{43})$.

53. If the circle is tangent to the y-axis, the radius is 2.
$$(x - 2)^2 + (y - 3)^2 = 2^2$$
$$(x - 2)^2 + (y - 3)^2 = 4$$

57.
$$x^2 + y^2 - 6x + 8y = 144$$
$$x^2 - 6x + y^2 + 8y = 144$$
$$x^2 - 6x + \mathbf{9} + y^2 + 8y + \mathbf{16} = 144 + \mathbf{9} + \mathbf{4}$$
$$(x - 3)^2 + (y + 4)^2 = (\sqrt{157})^2$$
$$(3,-4) \qquad r = \sqrt{157}$$

The distance from the origin is 5.

Section 9.6

1. $\dfrac{x^2}{9} + \dfrac{y^2}{16} = 1$

The original equation written in ellipse standard form:

$$\dfrac{x^2}{3^2} + \dfrac{y^2}{4^2} = 1$$

Remember a = 3, -a = -3, b = 4 and -b = -4.

The graph crosses the x-axis at (3,0), (-3,0) and the y-axis at (0,4), (0,-4).

5. $\dfrac{x^2}{3} + \dfrac{y^2}{4} = 1$

$$\dfrac{x^2}{(\sqrt{3})^2} + \dfrac{y^2}{2^2} = 1 \qquad \text{Ellipse standard form}$$

The graph crosses the x-axis at $(\sqrt{3},0)$, $(-\sqrt{3},0)$ and the y-axis at (0,2), (-,-2).

9. $x^2 + 8y^2 = 15$

$$\dfrac{x^2}{16} + \dfrac{8y^2}{16} = \dfrac{16}{16} \qquad \text{Divide each side by 16}$$

$$\dfrac{x^2}{4^2} + \dfrac{y^2}{(\sqrt{2})^2} = 1 \qquad \text{Ellipse standard form}$$

The graph crosses the x-axis at (4,0), (-4,0) and the y-axis at $(0,\sqrt{2})$, $(0,-\sqrt{2})$.

13. $\dfrac{x^2}{16} - \dfrac{y^2}{9} = 1$

$$\dfrac{x^2}{4^2} - \dfrac{x^2}{3^2} = 1 \qquad \text{Hyperbola standard form}$$

The x-intercepts are 4 and -4. The asymptotes must pass through 4 and -4 on the x-axis and 3 and -3 on the y-axis.

17. $\dfrac{y^2}{36} - \dfrac{x^2}{4} = 1$

$\dfrac{y^2}{6^2} - \dfrac{x^2}{2^2} = 1$ Hyperbola standard form

The y-intercepts are 6 and -6. The asymptotes must pass through 6 and -6 on the y-axis and 2 and -2 on the x-axis.

21. $16y^2 - 9x^2 = 144$

$\dfrac{16y^2}{144} - \dfrac{9x^2}{144} = \dfrac{144}{144}$ Divide each side by 144

$\dfrac{y^2}{9} - \dfrac{x^2}{16} = 1$

The y-intercepts are 3 and -3. The asymptotes must pass through 3 and -3 on the y-axis and 4 and -4 on the x-axis.

25. $\dfrac{x^2}{.04} - \dfrac{y^2}{.09} = 1$

The x-intercepts are .2 and -.2. There are no y-intercepts.

29. The center of the ellipse is at (4,2). The vertices are located 2 units left and right of the center (because 2 is the square root of the number below x - 4) and 3 units above and below the center (because 3 is the square root of the number below y - 2).

33. To write the equation in standard form we must complete the square on both x and y.

$$
\begin{array}{ll}
x^2 + 9y^2 + 4x - 54y + 76 = 0 & \\
x^2 + 4x + 9y^2 - 54y = -76 & \text{Similar terms together} \\
x^2 + 4x + 9(y^2 - 6y) = -76 & \text{Factor 9 from y terms} \\
x^2 + 4x + \mathbf{4} + 9(y^2 - 6y + \mathbf{9}) = -76 + \mathbf{4} + \mathbf{81} & \text{Complete the squares} \\
(x + 2)^2 + 9(y - 3)^2 = 9 & \text{Binomial squares}
\end{array}
$$

$$\dfrac{(x + 2)^2}{9} + \dfrac{(y - 3)^2}{1} = 1 \qquad \text{Divide each side by 9}$$

The last equation is in standard form. The center is at (-2,3). The vertices are 3 units left and right of the center and 1 unit above and below the center.

37.   We begin by completing the square on both x and y.

$$9y^2 - x^2 - 4x + 54y + 68 = 0$$
$$9y^2 + 54y \quad - x^2 - 4x \quad = -68 \qquad \text{Similar terms together}$$
$$9(y^2 + 6y) - 1(x^2 + 4x) = -68 \qquad \text{Factor 9 from y terms}$$
$$\text{and } -1 \text{ from x terms}$$
$$9(y^2 + 6y + \mathbf{9}) - 1(x^2 + 4x + \mathbf{4}) = -68 + \mathbf{81} - \mathbf{4} \qquad \text{Complete the squares}$$
$$9(y + 3)^2 - 1(x + 2)^2 = 9 \qquad \text{Binomial squares}$$
$$\frac{(y + 3)^2}{1} - \frac{(x + 2)^2}{9} = 1$$

This last equation is in standard form. The center of the hyperbola is at (-2,-3). The hyperbola opens up and down and has vertices 1 unit above and below the center.

41.   Let $\qquad\qquad x = 4$

the equation $\dfrac{x^2}{25} + \dfrac{y^2}{9} = 1$

becomes $\qquad \dfrac{4^2}{25} + \dfrac{y^2}{9} = 1$

$$\frac{16}{25} + \frac{y^2}{9} = 1$$

$$\frac{y^2}{9} = \frac{9}{25}$$

$$y^2 = \frac{81}{25}$$

$$y = \pm 1.8$$

45. $4x^2 + 9y^2 = 36$
$$9y^2 = 36 - 4x^2$$

$$y^2 = \frac{36 - 4x^2}{9}$$

$$y = \pm\sqrt{\frac{36 - 4x^2}{9}}$$

$$y = \pm\frac{\sqrt{36 - 4x^2}}{3}$$

$$y = \pm\frac{\sqrt{4(9 - x^2)}}{3}$$

$$y = \pm\frac{2\sqrt{9 - x^2}}{3}$$

49. $\dfrac{y^2}{9} - \dfrac{x^2}{16} = 1$

$16y^2 - 9x^2 = 144$

$\qquad 16y^2 = 9x^2 + 144$

$\qquad\quad y^2 = \dfrac{9x^2}{16} + \dfrac{144}{16}$

$\qquad\quad y^2 = \left(\dfrac{3}{4}x\right)^2 + \left(\dfrac{12}{4}\right)^2$

$\qquad\quad\, y = \pm\dfrac{3}{4}x$

53. $x + y \leqslant 5$

The boundary is the graph of $x + y = 5$. The boundary is not included since the original inequality symbol is <.

To find the x-intercept,   To find the y-intercept,
let $y = 0$ let $x = 0$
the equation $x + y = 5$ the equation $x + y = 5$
become $x + 0 = 5$ become $0 + y = 5$
                    $x = 5$ $y = 5$

Using   (0,0)
in $x + y < 5$
we have $0 + 0 < 5$
                $0 < 5$ A true statement

The region below the boundary is shaded.

57. $2x - 3y > 6$

The boundary is the broken line graph of $2x - 3y = 6$.

To find the x-intercept,   To find the y-intercept,
let $y = 0$ let $x = 0$
the equation $2x - 3y = 6$ the equation $2x - 3y = 6$
become $2x - 3(0) = 6$ become $2(0) - 3y = 6$
                     $2x = 6$ $-3y = 6$
                      $x = 3$ $y = -2$

Using   (0,0)
in $2x - 3y > 6$
we have $2(0) - 3(0) > 6$
                $0 > 6$ A false statement

The region below the boundary is shaded.

Section 9.7

1. $x^2 + y^2 \leq 49$

The boundary is $x^2 + y^2 = 49$, which is a circle with center at the origin and a radius of 7. Since the inequality sign is $\leq$, the boundary is included in the solution set and is represented by a solid line.

Testing (0,0)
in $x^2 + y^2 \leq 49$

becomes $0 + 0 \leq 49$

$0 \leq 49$ A true statement

The area inside the circle is shaded.

5. $y < x^2 - 6x + 7$

The parabola $y = x^2 - 6x + 7$ is the boundary and is not included in the solution set. The x-coordinate of the vertex is

$$x = \frac{-b}{2a} = \frac{-(-6)}{2(1)} = 3$$

Substituting 3 for x gives us the y-coordinate.

$$y = 3^2 - 6(3) + 7 = 9 - 18 + 7 = -2$$

Testing (0,0)
in $y < x^2 - 6x + 7$

becomes $0 < 0^2 - 6(0) + 7$

$0 < 7$ A true statement

See the graph on page A64 in the textbook.

9. $4x^2 + 25y^2 \leq 100$

$$\frac{4x^2}{100} + \frac{25y^2}{100} \leq \frac{100}{100}$$

$$\frac{x^2}{25} + \frac{y^2}{4} \leq 1$$

The ellipse $\frac{x^2}{25} + \frac{y^2}{4} = 1$ is the boundary and is included in the solution set. The center is at (0,0) with points at (5,0)(-5,0), (0,2) and (0,-2). Using (0,0) as the test point, we see that $0 + 0 \leq 1$ is a true statement, which means that the region containing (0,0) is in the solution set.

See the graph on page A64 in the textbook.

13. The boundary for the top equation is an ellipse with center at the origin and the solution set lies inside the boundary. The boundary for the second equation is a hyperbola centered at the origin. Since the inequality sign is >, the boundary is not included in the solution set and must therefore be represented with a broken line. The solution set lies inside the boundary. See the graph on page A64 in the textbook.

17. We have three linear inequalities, representing three sections of the coordinate plane. The graph of the solution set for this system will be the intersection of these three sections. The graph of $x + y \leq 3$

    is the section below and including the boundary $x + y = 3$. The graph of $x - 3y \leq 3$ is the section above and including the boundary line

    $x - 3y = 3$. The graph of $x \geq -2$ is all points to the right of and including the vertical line $x = -2$. The intersection of these three graphs is shown on page A65 in the textbook.

21. We have three linear inequalities, representing three sections of the coordinate plane. The graph of the solution set for this system will be the intersection of these three sections. The graph of $x + y \leq 4$

    is the section below and including the boundary $x + y = 4$. The graph of $x \geq 0$ is all points to the right of and including the vertical line

    $x = 0$. The graph of $y \geq 0$ is all points above and including the horizontal line $y = 0$. The intersection of these three graphs is shown on page A65 in the textbook.

25. $x^2 + y^2 = 16$
    $x + 2y = 8 \longrightarrow x = 8 - 2y$

    We can solve this system using the substitution method. Replacing x in the first equation with $16 - y^2$ from the second equation, we have

    $$(8 - 2y)^2 + y^2 = 16$$
    $$64 - 32y + 4y^2 + y^2 = 16$$
    $$5y^2 - 32y + 48 = 0$$
    $$(y - 4)(5y - 12) = 0$$

    $$y = 4 \quad \text{or} \quad y = \frac{12}{5}$$

    then $\quad x + 2y = 8$ $\qquad$ then $\quad x + 2y = 8$

    becomes $x + 2(4) = 8$ $\qquad$ becomes $x + 2(\frac{12}{5}) = 8$

    $$x = 0 \qquad\qquad x = -\frac{24}{5} + 8$$

    $$x = -\frac{24}{5} + \frac{40}{5}$$

    $$x = \frac{16}{5}$$

    The system has two solutions:

    $(0,4)$ and $(\frac{16}{5}, \frac{12}{5})$

312

29.
$$x^2 + y^2 = 9 \qquad\qquad y = x^2 - 3$$
then $\quad x^2 = 9 - y^2 \qquad$ then $y + 3 = x^2$

Substituting $9 - y^2$ in the second equation, we have
$$9 - y^2 = y + 3$$
$$0 = y^2 + y - 6$$
$$0 = (y + 3)(y - 2)$$

$\qquad y = -3 \qquad$ or $\qquad\qquad y = 2$

then $\quad y = x^2 - 3 \qquad$ then $\qquad y = x^2 - 3$
becomes $-3 = x^2 - 3 \qquad$ becomes $\quad 2 = x^2 - 3$
$$0 = x^2 \qquad\qquad\qquad 5 = x^2$$
$$0 = x \qquad\qquad\qquad \pm\sqrt{5} = x$$

The system has three solutions:

$\qquad (0,-3)$, $(\sqrt{5},2)$ and $(-\sqrt{5},2)$.

33.
$$3x + 2y = 10$$
$$y = x^2 - 5$$

Substituting $x^2 - 5$ for y in the first equation, we have
$$3x + 2(x^2 - 5) = 10$$
$$3x + 2x^2 - 10 = 10$$
$$2x^2 + 3x - 20 = 0$$
$$(x + 4)(2x - 5) = 0$$

$\qquad x = -4 \qquad$ or $\qquad\qquad x = \dfrac{5}{2}$

then $\quad y = x^2 - 5 \qquad$ becomes $y = \left(-\dfrac{5}{2}\right)^2 - 5$

$\qquad y = (-4)^2 - 5 \qquad\qquad y = \dfrac{25}{4} - 5$

$\qquad y = 16 - 5 \qquad\qquad\quad y = \dfrac{25}{4} - \dfrac{20}{4}$

$\qquad y = 11 \qquad\qquad\qquad\quad y = \dfrac{5}{4}$

The system has two solutions:

$\qquad (-4,11)$ and $\left(\dfrac{5}{2}, \dfrac{5}{4}\right)$.

37. $y = x^2 - 6x + 5$
     $y = x - 5$

Substituting $x - 5$ for y in the first equation, we have

$$x - 5 = x^2 - 6x + 5$$
$$0 = x - 7x + 10$$
$$0 = (x - 2)(x - 5)$$

$x = 2$ or $x = 5$

then $y = x - 5$ then $y = x - 5$
becomes $y = 2 - 5$ becomes $y = 5 - 5$
        $y = -3$ $y = 0$

The system has two solutions:

(2,-3) and (5,0).

41. $x - y = 4$
     $x^2 + y^2 = 16$

Substituting $y + 4$ for x in the second equation, we have

$$(y + 4)^2 + y^2 = 16$$
$$y^2 + 8y + 16 + y^2 = 16$$
$$2y^2 + 8y = 0$$
$$2y(y + 4) = 0$$

$y = 0$ or $y = -4$

then $x - y = 4$ then $x - y = 4$
becomes $x - 0 = 4$ becomes $x - (-4) = 4$
        $x = 4$ $x = 0$

The system has two solutions: (4,0) and (0,-4).

45. $x = y^2 - 3$
     $x + y = 9$

Substituting $9 - y$ for x in the first equation, we have

$$9 - y = y^2 - 3$$
$$0 = y^2 + y - 12$$
$$0 = (y - 3)(y + 4)$$

$y = 3$ or $y = -4$

then $x + y = 9$ then $x + y = 9$
becomes $x + 3 = 9$ becomes $x - 4 = 9$
        $x = 6$ $x = 13$

The two sets of numbers are 6,3 and 13,-4.

49. If $\qquad$ $C = 6.5x + 800$
and $\qquad$ $R = 10x - .002x^2$
and if $\qquad$ $C = R$
then $\qquad$ $6.5x + 800 = 10x - .002x^2$
$.002x^2 - 3.5x + 800 = 0$

$$x = \frac{-(-3.5) \pm \sqrt{(-3.5)^2 - 4(.002)(800)}}{2(.002)}$$

$$= \frac{3.5 \pm \sqrt{12.25 - 6.4}}{.004}$$

$$= \frac{3.5 \pm \sqrt{5.85}}{.004}$$

$$= \frac{3.5 \pm 2.42}{.004}$$

$$x = \frac{3.5 + 2.42}{.004} = 1480 \qquad x = \frac{3.5 - 2.42}{.004} - 270$$

The break-even values of x for the companies are 270 and 1480.

53.
$$\begin{bmatrix} 1 & 1 & 1 & | & 6 \\ 2 & -1 & 1 & | & 3 \\ 1 & 2 & -3 & | & -4 \end{bmatrix}$$

Multiply row 1 by -2, add to row 2 and put results in row 2.

$$\begin{bmatrix} 1 & 1 & 1 & | & 6 \\ 0 & -3 & -1 & | & -9 \\ 1 & 2 & -3 & | & -4 \end{bmatrix}$$

Multiply row 1 by -1, add to row 3 and put results in row 3.

$$\begin{bmatrix} 1 & 1 & 1 & | & 6 \\ 0 & -3 & -1 & | & -9 \\ 0 & 1 & -4 & | & -10 \end{bmatrix}$$

Multiply row 3 by 3, add to row 2 and put results in row 3.

$$\begin{bmatrix} 1 & 1 & 1 & | & 6 \\ 0 & -3 & -1 & | & -9 \\ 0 & 0 & -13 & | & -39 \end{bmatrix}$$

Divide row 2 by -3

$$\begin{bmatrix} 1 & 1 & 1 & | & 6 \\ 0 & 1 & \frac{1}{3} & | & 3 \\ 0 & 0 & -13 & | & -39 \end{bmatrix}$$

Divide row 3 by -13

$$\begin{bmatrix} 1 & 1 & 1 & | & 6 \\ 0 & 1 & \frac{1}{3} & | & 3 \\ 0 & 0 & 1 & | & 3 \end{bmatrix}$$

$$x + y + z = 6 \quad (1)$$
$$y + \frac{1}{3}z = 3 \quad (2)$$
$$z = 3 \quad (3)$$

Substitute (3) into (2), we have:

$$y + \frac{1}{3}(3) = 3$$
$$y = 2$$

Problem **53** continued on next page.

Problem **53** continued

Substitute y = 2 and z = 3 into (1), we have:

x + 2 + 3 = 6

x = 1

The solution to the system is (1,2,3).

1. Domain = {2,3,4}, Range = {4,3,2}.

5. $\{x \mid x \geq -2\}$

9. $f(0) = 3(0) - 2 = -2$

13. $f(-3) = 0$

17. $f(0) = 2(0)^2 - 4(0) + 1 = 1$

21. $f[g(0)]$
    $g(0) = 3(0) + 2 = 2$
    $f[g(0)] = f(2) = 2(2)^2 - 4(2) + 1 = 1$

25. $$f(x + h) = 2(x + h) + 1 = 2x + 2h + 1$$

$$\frac{f(x + h) - f(x)}{h} = \frac{2x + 2h + 1 - (2x + 1)}{h}$$

$$= \frac{2x + 2h + 1 - 2x - 1}{h}$$

$$= \frac{2h}{h}$$

$$= 2$$

29. $F + g = (x + 3) + (2x - 4) = 3x - 1$

33. $gg = (2x - 4)(2x - 4) = 4x^2 - 16x + 16$

37. $f(-1) \cdot g(-1) = (-1 + 3)(2(-1) - 4)$
    $= (2)(-6)$
    $= 12$

41. $f(x) = x^3 + 2x^2 - 9x - 18$
    $0 = x^3 + 2x^2 - 9x - 18$
    $0 = (x - 3)(x + 3)(x + 2)$

    $x = 3 \ , \ x = -3 \ , \ x = -2$

45. Exponential function

49.

x	y
0	1
1	2
2	4

See the graph on page A65 in the textbook.

53. $f(x) = 2^x$

$f(4) = 2^4 = 16$

57. $f(-1) + g(1) = 2^{-1} + \left(\frac{1}{3}\right)^1$

$$= \frac{1}{2} + \frac{1}{3}$$

$$= \frac{3}{6} + \frac{2}{6}$$

$$= \frac{5}{6}$$

61. $f(x) = \frac{1}{2}x + 2$ means $y = \frac{1}{2}x + 2$

The inverse of $y = \frac{1}{2}x + 2$ is $x = \frac{1}{2}y + 2$. Solving the equation

$x = \frac{1}{2}y + 2$ for y in terms of x as follows:

$$x = \frac{1}{2}y + 2$$

$$x - 2 = \frac{1}{2}y$$

$$2x - 4 = y$$

$$f^{-1}(x) = 2x - 4$$

65. $d = \sqrt{(-1 - 2)^2 + (5 - 6)^2}$

$= \sqrt{(-3)^2 + (-1)^2}$

$= \sqrt{9 + 1}$

$= \sqrt{10}$

69. $5 = \sqrt{(2 - x)^2 + [-4 - (-1)]^2}$

$5 = \sqrt{(2 - x)^2 + (-3)^2}$

73. $[x - (-5)]^2 + (y - 0)^2 = 3^2$

$(x + 5)^2 + y^2 = 9$

77. First find the distance

$$d = \sqrt{[2 - (-2)]^2 + (0 - 3)^2}$$
$$= \sqrt{4^2 + (-3)^2}$$
$$= \sqrt{25}$$
$$= 5$$

The distance is the radius of the circle.

$$[x - (-2)]^2 + (y - 3)^2 = 5^2$$
$$(x + 2)^2 + (y - 3)^2 = 25$$

81.
$$x^2 + y^2 - 6x + 4y = -4$$
$$x^2 - 6x + y^2 + 4y = -4$$
$$x^2 - 6x + \mathbf{9} + y^2 + 4y + \mathbf{4} = -4 + \mathbf{9} + \mathbf{4}$$
$$(x - 3)^2 + (y + 2)^2 = 9$$

Center (3,-2)   r = 3

See the graph on page A66 in the textbook.

85. See the graph on page A67 in the textbook.

89. To write the equation in standard form we must complete the square on both x and y.

$$9y^2 - x^2 - 4x + 54y + 68 = 0$$
$$-x^2 - 4x + 9y^2 + 54y = -68$$
$$-1(x^2 + 4x) + 9(y^2 + 6y) = -68$$
$$-1(x^2 + 4x + \mathbf{4}) + 9(y^2 + 6y + \mathbf{9}) = -68 - \mathbf{4} + 81$$
$$-1(x + 2)^2 + 9(y + 3)^2 = 9$$
$$9(y + 3)^2 - 1(x + 2)^2 = 9$$

$$\frac{(y + 3)^2}{9} - \frac{(x + 2)^2}{9} = 1$$

The last equation is in standard form. The center of the hyperbola is at (-2,-3). The hyperbola opens up and down and has vertices 1 unit above and below the center. See the graph on page A68 in the textbook.

93. $y \geq x^2 - 1$

The parabola $y = x^2 - 1$ is the boundary and is included in the solution set. The x-coordinate of the vertex is

$$x = \frac{-b}{a} = \frac{0}{1} = 0$$

Substituting 0 for x gives us the y-coordinate.

$$y = 0^2 - 1 = -1$$

Testing (0,0)
in $y \geq x^2 - 1$

becomes $0 \geq 0^2 - 1$

$$0 \geq -1 \qquad \text{True statement}$$

See the graph on page A68 in the textbook.

97. $x^2 + y^2 = 16$
$2x + y = 4 \longrightarrow y = -2x + 4$

Replacing y in the first equation with -2x + 4 from the second equation, we have:

$$x^2 + (-2x + 4)^2 = 16$$
$$x^2 + 4x^2 - 16x + 16 = 16$$
$$5x^2 - 16x + 0 = 0$$
$$x(5x - 16) = 0$$

$$x = 0 \qquad\qquad 5x - 16 = 0$$
$$5x = 16$$

$$x = 0 \quad \text{or} \quad x = \frac{16}{5}$$

Then $\quad 2x + y = 4 \qquad$ then $\quad 2x + y = 4$

$$2(0) + y = 4 \qquad\qquad 2\left(\frac{16}{5}\right) + y = 4$$

$$y = 4 \qquad\qquad\qquad \frac{32}{5} + y = 4$$

$$y = \frac{20}{5} - \frac{32}{5}$$

$$y = -\frac{12}{5}$$

The two solutions to the system are $(0,4)$ and $\left(\frac{16}{5}, -\frac{12}{5}\right)$.

1.   Domain = {-2,-3}   Range = {0,1}   Not a function

5.   f(3) + g(2) = (3 - 2) + [3(2) + 4] = 1 + 10 = 11

9.
x	y
-2	-2
-1	$-\frac{1}{4}$
0	0
1	$\frac{1}{4}$
2	2

See the graph on page A69 in the textbook.

13.   f(x) = 2x - 3   means   y = 2x - 3

The inverse of y = 2x - 3 is x = 2y - 3. Solving the equation
x = 2y - 3 for y in terms of x as follows:

$$x = 2y - 3$$
$$x + 3 = 2y$$
$$\frac{x + 3}{2} = y$$
$$f^1(x) = \frac{x + 3}{2}$$

f(x) = 2x - 3 $f^1(x) = \frac{x + 3}{2}$

x	y
0	-3
$\frac{3}{2}$	0
2	1

x	y
-3	0
0	$\frac{3}{2}$
1	2

See the graph on page A69 in the textbook.

17. $d = \sqrt{(-3 - 0)^2 + (-4 - 0)^2}$

$d = \sqrt{9 + 16}$

$d = 5$

The distance is the radius 5.

$$(x - 0)^2 + (y - 0)^2 = 5^2$$
$$x^2 + y^2 = 25$$

21.　　$(x - 2)^2 + (y + 1)^2 \leq 9$

　　　　Center $(2,-1)$　　　$r = 3$

　　　　See the graph on page A70 in the textbook.

25.　　The notation $R(x)$ indicates we are to write the revenue equation in terms of the variable x.　We need to solve the equation $x = 1200 - 100p$ for p.

$$1200 - 100p = x$$
$$-100p = x - 1200$$
$$p = \frac{x - 1200}{-100}$$
$$p = 12 - .01x$$

Now we can find $R(x)$ by substituting $12 - .01x$ for p in the formula $R(x) = xp$:

$$R(x) = xp = x(12 - .01x) = 12x - .01x^2$$

Chapter 10

Section 10.1

1. $\log_2 16 = 4$

5. $\log_{10} .01 = -2$

9. $\log_{1/2} 8 = -3$

13. $10^2 = 100$

17. $8^0 = 1$

21. $6^2 = 36$

25. $3^2 = x$
    $9 = x$

29. $2^x = 16$
    $2^4 = 16$ so $x = 4$

33. $x^2 = 4$
    $2^2 = 4$ so $x = 2$

37. $y = \log_3 x$ in exponential form is:

    $3^y = x$

when $y = 1$	when $y = 0$	when $y = -1$
$3^y = x$	$3^y = x$	$3^y = x$
$3^1 = x$	$3^0 = x$	$3^{-1} = x$
$x = 3$	$x = 1$	$x = \frac{1}{3}$

41. $y = \log_5 x$ in exponential form is:

    $5^y = x$

When $y = 1$	When $y = 0$	When $y = -1$
$5^y = x$	$5^y = x$	$5^y = x$
$5^1 = x$	$5^0 = x$	$5^{-1} = x$
$x = 5$	$x = 1$	$x = \frac{1}{5}$

45.   Substituting 2^4 for 16:

$$\log_2 16 = \log_2 2^4 = 4$$

49.   Substituting 10^3 for 1000:

$$\log_{10} 1000 = \log_{10} 10^3 = 3$$

53.   Substituting 5^0 for 1:

$$\log_5 1 = \log_5 5^0 = 0$$

57.   Substituting 2^4 for 16:

$$\log_2 16 = \log_2 2^4 = 4$$

Substituting 2^2 for 4:

$$\log_2 4 = \log_2 2^2 = 2$$

Substituting $4^{1/2}$ for 2:

$$\log_4 2 = \log_4 4^{1/2} = \frac{1}{2}$$

$$\log_4 [\log_2 (\log_2 16)] = \log_4 (\log_2 4)$$
$$= \log_4 2$$
$$= \frac{1}{2}$$

61. $pH = -\log_{10} [H^+]$ with a $pH = 6$ becomes
  $6 = -\log_{10} [H^+]$ in exponential form is
$H^+ = 10^{-6}$

65.   From the formula $m = \log_{10} T$ with $m = 8$, we have

$$8 = \log_{10} T$$

Writing this expression in exponential form, we have

$$T = 10^8 = 100,000,000$$

The earthquake that measures 8 on the Richter scale has a shockwave 100,000,000 times greater than the smallest shockwave measurable on a seismograph.

69. Domain = $\{x \mid x \geq -\frac{1}{3}\}$

$$y = \sqrt{3x + 1}$$
$$0 = \sqrt{3x + 1}$$
$$0 = 3x + 1$$
$$-1 = 3x$$
$$-\frac{1}{3} = x$$

73. $\dfrac{g(x + h) - g(x)}{h} = \dfrac{2(x + h) - 6 - (2x - 6)}{h}$

$$= \frac{2x + 2h - 6 - 2x + 6}{h}$$

$$= \frac{2h}{h}$$

$$= 2$$

Section 10.2

1. $\log_3 4x = \log_3 4 + \log_3 x$

5. $\log_2 y^5 = 5 \log_2 y$

9. $\log_6 x^2 y^3 = \log_6 x^2 + \log_6 y^3$ Property 1

                  $= 2 \log_6 x + 3 \log_6 y$ Property 3

13. $\log_b \dfrac{xy}{z} = \log_b xy - \log_b z$ Property 2

             $= \log_b x + \log_b y - \log_b z$ Property 1

17. $\log_{10} \dfrac{x^2 y}{\sqrt{z}} = \log_{10} \dfrac{x^2 y}{z^{1/2}}$ $\sqrt{z} = z^{1/2}$

             $= \log_{10} x^2 y - \log_{10} z^{1/2}$ Property 2

             $= \log_{10} x^2 + \log_{10} y - \log_{10} z^{1/2}$ Property 1

             $= 2 \log_{10} x + \log_{10} y - \frac{1}{2} \log_{10} z$ Property 3

21. $\log_b \sqrt[3]{\dfrac{x^2 y}{z}} = \log_b \left(\dfrac{x^2 y}{z}\right)^{1/3}$

    $\qquad = \dfrac{1}{3} \log_b \dfrac{x^2 y}{z^4}$ Property 3

    $\qquad = \dfrac{1}{3} (\log_b x^2 y - \log_b z^4)$ Property 2

    $\qquad = \dfrac{1}{3} (\log_b x^2 + \log_b y - \log_b z^4)$ Property 1

    $\qquad = \dfrac{1}{3} (2 \log_b x + \log_b y - 4 \log_b z)$ Property 3

    $\qquad = \dfrac{2}{3} \log_b + \dfrac{1}{3} \log_b y - \dfrac{4}{3} \log_b z$

25. $2 \log_3 x - 3 \log_3 y = \log_3 x^2 - \log_3 y^3$ Property 3

    $\qquad\qquad\qquad\quad = \log_3 \dfrac{x^2}{y^3}$ Property 2

29. $3 \log_2 x + \dfrac{1}{2} \log_2 y - \log_2 z$

    $\qquad = \log_2 x^3 + \log_2 y^{1/2} - \log_2 z$ Property 3

    $\qquad = \log_2 x^3 y^{1/2} - \log_2 z$ Property 1

    $\qquad = \log_2 \dfrac{x^3 y^{1/2}}{z}$ Property 2

    $\qquad = \log_2 \dfrac{x^3 \sqrt{y}}{z}$ $y^{1/2} = \sqrt{y}$

33. $\dfrac{3}{2} \log_{10} x - \dfrac{3}{4} \log_{10} y - \dfrac{4}{5} \log_{10} z$

    $\qquad = \log_{10} x^{3/2} - \log_{10} y^{3/4} - \log_{10} z^{4/5}$ Property 3

    $\qquad = \log_{10} x^{3/2} - (\log_{10} y^{3/4} + \log_{10} z^{4/5})$

    $\qquad = \log_{10} x^{3/2} - \log_{10} y^{3/4} z^{4/5}$ Property 1

    $\qquad = \log_{10} \dfrac{x^{3/2}}{y^{3/4} z^{4/5}}$ Property 2

37. $\log_3 x - \log_3 2 = 2$

$\log_3 \frac{x}{2} = 2$ Property 2

Writing the last line in exponential form, we have

$3^2 = \frac{x}{2}$

$9 = \frac{x}{2}$

$x = 18$

The solution set is {18}.

41. $\log_3 (x + 3) - \log_3 (x - 1) = 1$

$\log_3 \frac{x + 3}{x - 1} = 1$ Property 2

Writing the last line in exponential form, we have

$\frac{x + 3}{x - 1} = 3^1$

$\cancel{(x - 1)}\frac{x + 3}{\cancel{x - 1}} = 3(x - 1)$

$x + 3 = 3x - 3$

$3 = 2x - 3$

$6 = 2x$

$x = 3$

The solution set is {3}.

45. $\log_8 x + \log_8 (x + 3) = \frac{2}{3}$

$\log_8 [(x)(x - 3)] = \frac{2}{3}$ Property 1

Using the definition of logarithms:

$(x)(x - 3) = 8^{2/3}$

$(x)(x - 3) = 4$ $8^{2/3} = (8^{1/3})^2 = 2^2 = 4$

$x^2 - 3x - 4 = 0$

$(x - 4)(x + 1) = 0$

$x - 4 = 0$ or $x + 1 = 0$

$x = 4$ $x = -1$

Since $y = \log_8 x$ cannot be a negative number, substituting $x = -1$

onto the original equation gives a false statement.

$\log_8 -1 + \log_8 -4 = \frac{2}{3}$

The solution set is {4}

Section 10.2 continued

49. If $a = 1$ and $b = 10^{-12}$, the original equation

$$m = 0.21(\log_{10} a - \log_{10} b)$$

becomes

$$m = 0.21(\log_{10} 1 - \log_{10} 10^{-12})$$

$$m = 0.21(\log_{10} \frac{1}{10^{-12}})$$

$$m = 0.21(\log_{10} 10^{12})$$

$$\frac{m}{0.21} = \log_{10} 10^{12}$$

In exponential form, we have

$$10^{m/0.21} = 10^{12}$$

The only way the left and right sides can be equal is if the exponents are equal, that is, if

$$\frac{m}{0.21} = 12$$

$$m = 2.52$$

Substituting $a = 1$ and $b = 10^{-12}$ in the equation

$$m = .21 \log_{10} \frac{a}{b}$$

we have

$$m = .21 \log_{10} \frac{1}{10^{-12}}$$

$$m = .21 \log_{10} 10^{-12}$$

$$\frac{m}{.21} = \log_{10} 10^{12}$$

In exponential form, we have

$$10^{m/.21} = 10^{12}$$

Then

$$\frac{m}{.21} = 12$$

$$m = 2.52$$

53. $\log_{10} A = \log_{10} 100(1.06)^t$

   $\log_{10} A = \log_{10} 100 + \log_{10} (1.06)^t$ Property 1

   $\log_{10} A = \log_{10} 100 + t \log_{10} (1.06)$ Property 3

   $\log_{10} A = \log_{10} 10^2 + t \log_{10} (1.06)$ $100 = 10^2$

   $\log_{10} A = 2 + t \log_{10} (1.06)$ Identity 2

57.

x	$f(x) = 2x^2 - 4x - 6$	y
-1	$f(-1) = 2(-1)^2 - 4(-1) - 6$	0
0	$f(0) = 2 \cdot 0^2 - 4 \cdot 0 - 6$	-6
1	$f(1) = 2 \cdot 1^2 - 4 \cdot 1 - 6$	-8
3	$f(3) = 2 \cdot 3^2 - 4 \cdot 3 - 6$	0

See the graph on page A72 in the textbook.

61. $f(x) = x^2 + 2x - 1$

   $0 = x^2 + 2x - 1$

   $x = \dfrac{-2 \pm \sqrt{2^2 - 4(1)(-1)}}{2(1)}$

   $= \dfrac{-2 \pm \sqrt{8}}{2}$

   $= \dfrac{-2 \pm 2\sqrt{2}}{2}$

   $= -1 \pm \sqrt{2}$

65. $f(x) = x^2 - 4$

   $y = x^2 - 4$

   $x = y^2 - 4$ Interchange x and y

   $x + 4 = y^2$

   $\pm\sqrt{x + 4} = y$

Section 10.3

1. $\log 378 = \log (3.78) \times 10^2)$

   $= \log 3.78 + \log 10^2$

   $= .5775 + 2$

   $= 2.5775$

5. $\log 3780 = \log (3.780 \times 10^3)$

   $= \log 3.780 + \log 10^3$

   $= .5775 + 3$

   $= 3.5775$

9. $\log 37{,}800 = \log (3.78 \times 10^4)$
    $= \log 3.78 + \log 10^4$
    $= .5775 + 4$
    $= 4.5775$

13. $\log 2010 = \log (2.010 \times 10^3)$
    $= \log 2.010 + \log 10^3$
    $= .3032 + 3$
    $= 3.3032$

17. $\log .0314 = \log (3.14 \times 10^{-2})$
    $= \log 3.14 + \log 10^{-2}$
    $= .4969 - 2$
    $= -1.5031$

21. $\log x = 2.8802$
    $\log x = .8802 + 2$
    $x = 7.59 \times 10^2$
    $x = 759$

25. $\log x = 3.1553$
    $\log x = .1553 + 3$
    $x = 1.43 \times 10^3$
    $x = 1430$

29. The entries in our table are all positive numbers, so we must rewrite -7.0372 so that the decimal part (the mantissa) is positive. We add and subtract the smallest whole number that is larger than 7.0372. That number is 8:

    $\log x = -7.0372$
    $= -7.0372 + 8 - 8$
    $= .9628 - 8$
    $= 9.18 \times 10^{-8}$
    $= .00000000918$

33. $\log x = -10$
    $= 0 - 10$
    $x = 1.0 \cdot 10^{-10}$
    $x = .0000000001$

37. $\log x = -2$
    $= 0 - 2$
    $x = 1 \cdot 10^{-2}$
    $x = .01$

41. $\log x = \log 5$
    $\log x = .6990$
    $x = 5$

45. $pH = -\log [H^+]$ with $pH = 4.75$ becomes

   $4.75 = -\log (H^+]$ in exponential form is

   $H^+ = 10^{-4.75}$

   $\log H^+ = \log 10^{-4.75}$
   $= -4.75 \log 10$
   $= -4.75 + 5 - 5$
   $= .25 - 5$
   $H^+ = 1.78 \times 10^{-5}$

49. $M = \log T$ with $M = 8.3$ becomes
   $8.3 = \log T$ in exponential form is

   $T = \log 10^{8.3}$

   $\log T = \log 10^{8.3}$
   $= 8.3 \log 10$
   $= 8.3$
   $T = 2 \times 10^8$

53.   If $P = \$9000$, $t = 5$ and $w = \$4,500$ then

   $\log (1 - r) = \frac{1}{5} \log \frac{w}{p}$ becomes

   $\log (1 - r) = \frac{1}{5} \log \frac{4500}{9000}$

   $= .20 \log .5$
   $= .20 (-.3010)$
   $= .0602$ Multiplication
   $= -.0602 + 1 - 1$
   $= .9398 - 1$
   $1 - r = 8.71 \times 10^{-1}$
   $1 - r = .871$
   $-r = -.129$
   $r = .129$ or   12.9%

57. $\ln e = 1$ In exponential form, $e^1 = e$

61. $\ln e^x = x \ln e$
   $= x(1)$
   $= x$

65. $\ln Ae^{-2t} = \ln A + \ln e^{-2t}$
   $= \ln A - 2t \ln e$
   $= \ln A - 2t$ Because $\ln e = 1$

69. $\ln \frac{1}{3} = \ln 1 - \ln 3$

   $= 0 - 1.0986 \qquad \ln 1 = 0 \;$ and
   $= -1.0986 \qquad\quad \ln 3 = 1.0986$

73. $\ln 16 = \ln 2^4 \qquad 2^4 = 16$
   $= 4 \ln 2$
   $= 4(.6931)$
   $= 2.772$

77. $x^2 + 6x \qquad + y^2 - 4y \qquad = 3$
   $x^2 + 6x + \mathbf{9} + y^2 - 4y + \mathbf{4} = 3 + \mathbf{9} + \mathbf{4}$
   $(x + 3)^2 + (y - 2)^2 = 16$
   $[x - (-3)]^2 + (y - 2)^2 = 4^2$

The center is at $(-3,2)$ and the radius is 4.

81. See the graph on page A72 in the textbook.

Section 10.4

1. $3^x = 5$

   $\log 3^x = \log 5 \qquad$ Take the log of both sides
   $x \log 3 = \log 5 \qquad$ Property 3

   $x = \dfrac{\log 5}{\log 3} \qquad$ Divide by log 3

   $x = \dfrac{.6990}{.4771} \qquad$ Using the table of common logarithms

   $x = 1.4651$

5. $5^{-x} = 12$

   $\log 5^{-x} = \log 12 \qquad$ Take the log of both sides
   $-x \log 5 = \log 12 \qquad$ Property 3

   $-x = \dfrac{\log 12}{\log 5} \qquad$ Divide by log 5

   $-x = \dfrac{1.0792}{.6990} \qquad$ Using the table of common logarithms

   $-x = 1.5439$
   $x = -1.5439$

9. $\qquad 8^{x+1} = 4$

$\qquad \log 8^{x+1} = \log 4$ Take the log of both sides

$\qquad (x + 1) \log 8 = \log 4$ Property 3

$\qquad x + 1 = \dfrac{\log 4}{\log 8}$ Divide by log 8

$\qquad x = \dfrac{\log 4}{\log 8} - 1$ Add -1 to both sides

$\qquad x = \dfrac{.0621}{.9031} - 1$ Using the table of common logarithms

$\qquad x = .6667 - 1$

$\qquad x = -.3333$

13. $\qquad 3^{2x+1} = 2$

$\qquad \log 3x^{2x+1} = \log 2$ Take the log of both sides

$\qquad (2x + 1) \log 3 = \log 2$ Property 3

$\qquad 2x + 1 = \dfrac{\log 2}{\log 3}$ Divide by log 3

$\qquad 2x = \dfrac{\log 2}{\log 3} - 1$ Add -1 to both sides

$\qquad x = \dfrac{1}{2} \left(\dfrac{\log 2}{\log 3} - 1 \right)$ Multiply both sides by $\dfrac{1}{2}$

$\qquad x = \dfrac{1}{2} \left(\dfrac{.3010}{.4771} - 1 \right)$ Using the table of common logarithms

$\qquad x = \dfrac{1}{2} (.6309 - 1)$

$\qquad x = \dfrac{1}{2} (-.3691)$

$\qquad x = -.1846$

17. $$15^{3x-4} = 10$$

$$\log 15^{3x-4} = \log 10 \qquad \text{Take the log of both sides}$$
$$(3x - 4) \log 15 = \log 10 \qquad \text{Property 3}$$

$$3x - 4 = \frac{\log 10}{\log 15} \qquad \text{Divide by log 15}$$

$$3x = \frac{\log 10}{\log 15} + 4 \qquad \text{Add +4 to both sides}$$

$$x = \frac{1}{3} \left(\frac{\log 10}{\log 15} + 4 \right) \qquad \text{Mulgiply both sides by } \frac{1}{3}$$

$$x = \frac{1}{3} \left(\frac{1.0000}{1.1761} + 4 \right) \qquad \text{Using the table of common logarithms}$$

$$x = \frac{1}{3} (.8503 + 4)$$

$$x = \frac{1}{3} (4.8503)$$

$$x = 1.6168$$

21. Substituting $P = 5000$, $r = .12$, $n = 1$ and $t = 10$ into the formula:

$$A = P \left(1 + \frac{r}{n} \right)^{nt}$$

The equation becomes:

$$A = 5000 \left(\ + \frac{.12}{1} \right)^{1(10)}$$

$$A = 5000(1.12)^{10}$$
$$\log A = \log \left[5000(1.12)^{10} \right] \qquad \text{Take the log of both sides}$$
$$\log A = \log 5000 + 10 \log 1.12 \qquad \text{Properties 1 and 3}$$
$$= 3.6990 + 10(.0492)$$
$$= 3.6990 + .492$$
$$\log A = 4.1910$$
$$A = 1.552924 \times 10^4$$
$$= \$15,529.24$$

25. Substituting A = $1000 ($500 doubled), r = .06, n = 12 and p = $500 into the formula:

$$A = P\left(1 + \frac{r}{n}\right)^{nt}$$

The equation becomes

$$1000 = 500\left(1 + \frac{.06}{2}\right)^{2t}$$

$$1000 = 500(1.03)^{2t} \qquad \text{Divide by 500}$$

$$2 = (1.03)^{2t}$$

$$\log 2 = \log (1.03)^{2t} \qquad \text{Taking the log of both sides}$$
$$\log 2 = 25 \log (1.03) \qquad \text{Property 3}$$
$$.3010 = 25(.0128)$$
$$.3010 = .0256t$$
$$11.7 = t$$

It will take 11.7 years to double the investment.

29. $\log_8 16 = \dfrac{\log 16}{\log 8}$

$\log_8 16 = \dfrac{\log_{10} 16}{\log_{10} 8}$

$\log_8 16 = \dfrac{1.2041}{.9031}$

$\log_8 16 = 1.333$

33. $\log_7 15 = \dfrac{\log 15}{\log 7}$

$\log_7 15 = \dfrac{\log_{10} 15}{\log_{10} 7}$

$\log_7 15 = \dfrac{1.1761}{.8451}$

$\log_7 15 = 1.3917$

37. $\log_8 240 = \dfrac{\log 240}{\log 8}$

$\log_8 240 = \dfrac{\log_{10} 240}{\log_{10} 8}$

$\log_8 240 = \dfrac{2.3802}{.9031}$

$\log_8 240 = 2.6356$

41. $\ln 345 = 2.3026 \log 345$
$= 2.3026(2.5378)$
$= 5.8435$

45. $\ln 10 = 2.3026 \log 10$
$= 2.3026(1)$
$= 2.3026$

49. Substituting $P = 64{,}000$ in the following formula:

$$P = 32{,}000e^{.05t}$$

We have

$$64{,}000 = 32{,}000e^{.05t}$$

$$2 = e^{.05t} \qquad\qquad \frac{64{,}000}{32{,}000} = 2$$

Take the natural logarithm of each side:

$\ln 2 = \ln e^{.05t}$
$\ln 2 = .05t \ln e \qquad\qquad$ Property 3
$\ln 2 = .05t \qquad\qquad$ Because $\ln e = 1$

$$t = \frac{\ln 2}{.05}$$

$$= \frac{2.3026 \log 2}{.05}$$

$$= \frac{2.3026(.3010)}{.05}$$

$$= 13.9 \text{ years later}$$

53. $A = Pe^{rt}$

$\dfrac{A}{P} = e^{rt}$

$\ln \dfrac{A}{P} = rt \ln e \qquad$ Taking the natural log of both sides

$\ln \dfrac{A}{P} = rt \qquad$ Because $\ln e = 1$

$\dfrac{1}{r} \ln \dfrac{A}{P} = t \qquad$ Divide both sides by $\dfrac{1}{r}$

57.
$$A = P(1 - r)^t$$

$$\frac{A}{P} = (1 - r)^t$$

$$\log \frac{A}{P} = t \log (1 - r)$$

$$\log A - \log P = t \log (1 - r)$$

$$\frac{\log A - \log P}{\log (1 - r)} = t$$

61.

$$
\begin{array}{lll}
x^2 + y^2 = 16 & \xrightarrow{\text{no change}} & x^2 + y^2 = 16 \\
4x^2 + y^2 = 16 & \xrightarrow{\text{multiply by -1}} & -4x^2 - y^2 = -16 \\
& & \overline{-3x^2 \quad\quad = \quad 0} \\
& & x^2 \quad\quad = \quad 0 \\
& & x \quad\quad\; = \quad 0
\end{array}
$$

Substitute $x = 0$

$$0^2 + y^2 = 16$$

$$\sqrt{y^2} = \sqrt{16}$$

$$y = \pm 4$$

The solution to the systems is $(0,4)$ and $(0,-4)$.

1. $\log_3 81 = 4$

5. $2^3 = 8$

9. $5^2 = x$
   $x = 25$

   The solution set is $\{25\}$.

13. $x^{-2} = 0.01$
    Since $10^{-2} = 0.01$
    then $x = 10$

17. $\log_4 16 = x$
    $x^4 = 16$
    Since $2^4 = 16$
    $x = 2$

21. $\log_4(\log_3 3)$

    $\log_3 3 = x$

    $3^x = 3$
    $x = 1$
    $\log_4(\log_3 3) = \log_4 1$

    $\log_4 1 = x$

    $4^x = 1$
    $x = 0$

25. $\log_{10} \dfrac{2x}{y} = \log_{10} 2x - \log_{10} y$ Property 2

    $= \log_{10} 2 + \log_{10} x - \log_{10} y$ Property 1

29. $\log_2 x + \log_2 y = \log_2 xy$ Property 1

33. $2\log_a 5 - \dfrac{1}{2}\log_a 9 = \log_a 5^2 - \log_a 9^{1/2}$ Property 3

    $= \log_a \dfrac{5^2}{9^{1/2}}$ Property 2

    $= \log_a \dfrac{25}{3}$

37. $\log_3 x + \log_3(x - 2) = 1$

$\log_3 x(x - 2) = 1$ Property 3

The last line can be writtten in exponential form using the definition of logarithms:

$$x(x - 2) = 3^1$$
$$x^2 - 2x - 3 = 0$$
$$(x - 3)(x + 1) = 0$$
$$x = 3 \text{ or } x = -1$$

Since in the equation $y = \log_b x$, x cannot be a negative number, the solution set is {3}.

41. $\log_6(x - 1) + \log_6(x) = 1$

$\log_6(x - 1)x = 1$ Property 1

The last line can be written in exponential form using the definition of logarithms:

$$(x - 1)x = 6^1$$
$$x^2 - x - 6 = 0$$
$$(x - 3)(x + 2) = 0$$
$$x = 3 \text{ or } x = -2$$

Since the equation $y = \log_b x$, x cannot be a negative number, the solution set is {3}.

45. $\log 0.713 = \log(7.13 \times 10^{-1})$

$$= \log 7.13 + \log 10^{-1}$$
$$= 0.8531 + (-1)$$
$$= -.1469$$

49. $\log x = -1.6003$

$$= -1.6003 + 2 - 2$$
$$= -0.1469$$
$$x = 2.51 \times 10^{-2}$$
$$= 0.0251$$

53. $\ln e^2 = 2$

57. $\sqrt{936} = 936^{1/2} = n$

$$\log n = \log 936^{1/2}$$

$$= \frac{1}{2} \log 936$$

$$= \frac{1}{2}(2.9713)$$

$$\log n = 1.4857$$
$$n = 3.06 \times 10^1$$
$$n = 30.6$$

61. $pH = \log(H_3O^+)$

When $(H_3O^+) = 7.9 \times 10^{-3}$
$$pH = -\log 7.9 \times 10^{-3}$$
$$= -1[\log 7.9 + \log 10^{-3}]$$
$$= -1[0.8976 + (-3)]$$
$$= -1(-2.1024)$$
$$= 2.1024$$
$$pH = 2.1$$

65. $4^x = 8$

$$\log 4^x = \log 8$$
$$x\log 4 = \log 8$$

$$x = \frac{\log 8}{\log 4}$$

$$x = \frac{0.9031}{0.6021}$$

$$= 1.4999$$

$$= \frac{3}{2} \qquad \text{(rounded to the nearest tenth)}$$

69. $\log_{16} 8 = \dfrac{\log_{10} 8}{\log_{10} 16}$

$$= \frac{0.9031}{1.2041}$$

$$= 0.75$$

73. Substituting t = 10 years, P = $5,000, n = 1, and r = 12% into the formula:

$$A = P(1 + \frac{r}{n})^{nt}$$

The equation becomes

$$A = 5,000(1 + \frac{.12}{1})^{1 \cdot 10}$$

$$\log A = \log 5,000(1.12)^{10}$$
$$= \log 5,000 + 10 \log 1.12$$
$$= 3.6990 + 10(0.0492)$$
$$= 3.6990 + 0.4920$$
$$\log A = 4.1910$$
$$A = \$15,529.24$$

After 10 years, the account would contain $15,529.24.

1. $4^3 = x$
     $64 = x$

5. $\log_8 4 = \dfrac{\log 4}{\log 8}$

     $\log_8 4 = \dfrac{\log_{10} 4}{\log_{10} 8}$

     $\log_8 4 = \dfrac{.6021}{.9031}$

     $\log_8 4 = \dfrac{2}{3}$

9. $\ln 46.2 = 2.3026 \log 46.2$
     $= 2.3026\,(1.6646)$
     $= 3.8330$

13. $2\log_3 x - \dfrac{1}{2}\log_3 y = \log_3 x^2 - \log_3 y^{1/2}$ Property 3

     $= \log_3 \dfrac{x^2}{y^{1/2}}$ Property 2

     $= \log_3 \dfrac{x^2}{\sqrt{y}}$

17. $\log 3^{2.5} = 2.5(\log 3)$
     $\log 3^{2.5} = 2.5(.4771)$
     $\log 3^{2.5} = 1.1928$
     $3^{2.5} = 15.6$

21. $\log_5 x - \log_5 3 = 1$

     $\log_5 \dfrac{x}{3} = 1$

     $\dfrac{x}{3} = 5^1$

     $\dfrac{x}{3} = 5$

     $x = 15$

25. Substituting P = 600, r = .08, n = 4 and A = 1800 into the formula:

$$A = P\left(1 + \frac{r}{n}\right)^{nt}$$

The equation becomes:

$$1800 = 600\left(1 + \frac{.08}{4}\right)^{4t}$$

$$1800 = 600(1 + .02)^{4t}$$

$$1800 = 600(1.02)^{4t}$$

$\log 1800 = \log[600(1.02)^{4t}]$	Take the log of both sides
$\log 1800 = \log 600 + 4t \log 1.02$	Properties 1 and 3
$3.2553 = 2.7782 + 4t(.0086)$	
$.4771 = 4t(.0086)$	
$55.4767 = 4t$	
$13.869 = t$	

It will take about 13.9 years.

Chapter 11

Section 11.1

1. General term = a_n = 3n + 1
   First term = a_1 = 3(1) + 1 = 4
   Second term = a_2 = 3(2) + 1 = 7
   Third term = a_3 = 3(3) + 1 = 10
   Fourth term = a_4 = 3(4) + 1 = 13
   Fifth term = a_5 = 3(5) + 1 = 16

5. General term = a_n = n
   First term = a_1 = 1
   Second term = a_2 = 2
   Third term = a_3 = 3
   Fourth term = a_4 = 4
   Fifth term = a_5 = 5

9. General term = $a_n = \dfrac{n}{n+3}$

   First term = $a_1 = \dfrac{1}{1+3} = \dfrac{1}{4}$

   Second term = $a_2 = \dfrac{2}{2+3} = \dfrac{2}{5}$

   Third term = $a_3 = \dfrac{3}{3+3} = \dfrac{3}{6}$

   Fourth term = $a_4 = \dfrac{4}{4+3} = \dfrac{4}{7}$

   Fifth term = $a_5 = \dfrac{5}{5+3} = \dfrac{5}{8}$

13. General term = $a_n = \dfrac{1}{n^2}$

First term = $a_1 = \dfrac{1}{1^2} = 1$

Second term = $a_2 = \dfrac{1}{2^2} = \dfrac{1}{4}$

Third term = $a_3 = \dfrac{1}{3^2} = \dfrac{1}{9}$

Fourth term = $a_4 = \dfrac{1}{4^2} = \dfrac{1}{16}$

Fifth term = $a_5 = \dfrac{1}{5^2} = \dfrac{1}{25}$

17. General term = $a_n = 3^{-n}$

First term = $a_1 = 3^{-1} = \dfrac{1}{3}$

Second term = $a_2 = 3^{-2} = \dfrac{1}{9}$

Third term = $a_3 = 3^{-3} = \dfrac{1}{27}$

Fourth term = $a_4 = 3^{-4} = \dfrac{1}{81}$

Fifth term = $a_5 = 3^{-5} = \dfrac{1}{243}$

21. General term = $a_n = n - \dfrac{1}{n}$

First term = $a_1 = 1 - \dfrac{1}{1} = 0$

Second term = $a_2 = 2 - \dfrac{1}{2} = \dfrac{3}{2}$

Third term = $a_3 = 3 - \dfrac{1}{3} = \dfrac{8}{3}$

Fourth term = $a_4 = 4 - \dfrac{1}{4} = \dfrac{15}{4}$

Fifth term = $a_5 = 5 - \dfrac{1}{5} = \dfrac{24}{5}$

25. $a_1 = 3$

$a_2 = -3a_{2-1} = -3a_1 = -3(3) = -9$

$a_3 = -3a_{3-1} = -3a_2 = -3(-9) = 27$

$a_4 = -3a_{4-1} = -3a_3 = -3(27) = -81$

$a_5 = -3a_{5-1} = -3a_4 = -3(-81) = 243$

29. $a_1 = 1$

$a_2 = 2a_{2-1} + 3 = 2a_1 + 3 = 2(1) + 3 = 5$

$a_3 = 2a_{3-1} + 3 = 2a_2 + 3 = 2(5) + 3 = 13$

$a_4 = 2a_{4-1} + 3 = 2a_3 + 3 = 2(13) + 3 = 29$

$a_5 = 2a_{5-1} + 3 = 2a_4 + 3 + 2(29) + 3 = 61$

33. Look at the first four terms:

$a_1 = 2 = 1 + 1$

$a_2 = 3 = 2 + 1$

$a_3 = 4 = 3 + 1$

$a_4 = 5 = 4 + 1$

The general term is $a_n = n + 1$

37. Look at the first four terms:

$a_1 = 7 = 3(1) + 4$

$a_2 = 10 = 3(2) + 4$

$a_3 = 13 = 3(3) + 4$

$a_4 = 16 = 3(4) + 4$

The general term is $a_n = 3(n) + 4$ or recursively as

$a_1 = 7, a_n = a_{n-1} + 3$

41. Look at the first four terms:

$a_1 = 3 = 3(1)^2$

$a_2 = 12 = 3(2)^2$

$a_3 = 27 = 3(3)^2$

$a_4 = 48 = 3(4)^2$

The general term is $a_n = 3(n)^2$

Section 11.1 continued

45. Look at the first four terms:

$$a_1 = -2 = (-2)^1$$
$$a_2 = 4 = (-2)^2$$
$$a_3 = -8 = (-2)^3$$
$$a_4 = 16 = (-2)^4$$

The general term is $a_n = (-2)^n$ or recursively as

$$a_1 = -2, a_n = -2a_{n-1}$$

49. Look at the first four terms:

$$a_1 = \frac{1}{4} = \frac{1}{2^2} = \frac{1}{(1+1)^2}$$

$$a_2 = \frac{2}{9} = \frac{2}{3^2} = \frac{2}{(2+1)^2}$$

$$a_3 = \frac{3}{16} = \frac{3}{4^2} = \frac{3}{(3+1)^2}$$

$$a_4 = \frac{4}{25} = \frac{4}{5^2} = \frac{4}{(4+1)^2}$$

The general term is $a_n = \dfrac{n}{(n+1)^2}$

53. $3^4 = 81$

57. $\log_2 32 = \dfrac{\log 32}{\log 2}$

$$= \frac{\log_{10} 32}{\log_{10} 2}$$

$$= \frac{1.5051}{.3010}$$

$$= 5$$

61.

$$a_{100} = \left(1 + \frac{1}{100}\right)^{100} = (1.01)^{100} = 2.7048$$

$$a_{1000} = \left(1 + \frac{1}{1000}\right)^{1000} = (1.001)^{1000} = 2.7169$$

$$a_{10,000} = \left(1 + \frac{1}{10,000}\right)^{10,000} = (1.0001)^{10,000} = 2.7181$$

$$a_{100,000} = \left(1 + \frac{1}{100,000}\right)^{100,000} = (1.00001)^{100,000} = 2.7183$$

65.

$$1 + \frac{1}{1 + 1} = 1 + \frac{1}{2} = \frac{3}{2}$$

$$1 + \cfrac{1}{1 + \cfrac{1}{1+1}} = 1 + \frac{1}{\frac{3}{2}} = 1 + \frac{2}{3} = \frac{5}{3}$$

$$1 + \cfrac{1}{1 + \cfrac{1}{1 + \cfrac{1}{1+1}}} = 1 + \frac{1}{\frac{5}{3}} = 1 + \frac{3}{5} = \frac{8}{5}$$

$$1 + \cfrac{1}{1 + \cfrac{1}{1 + \cfrac{1}{1 + \cfrac{1}{1+1}}}} = 1 + \frac{1}{\frac{8}{5}} = 1 + \frac{5}{8} = \frac{13}{8}$$

Section 11.2

1. $\displaystyle\sum_{i=1}^{4} (2i + 4) = (2 \cdot 1 + 4) + (2 \cdot 2 + 4) + (2 \cdot 3 + 4) + (2 \cdot 4 + 4)$

$$= 6 + 8 + 10 + 12$$
$$= 36$$

5. $\displaystyle\sum_{i=2}^{4} (i^2 - 1) = (2^2 - 1) + (3^2 - 1)$

$$= 3 + 8$$
$$= 11$$

9. $\displaystyle\sum_{i=1}^{3} \frac{i^2}{2i-1} = \frac{1^2}{2 \cdot 1 - 1} + \frac{2^2}{2 \cdot 2 - 1} + \frac{3^2}{2 \cdot 3 - 1}$

$\qquad\qquad = \frac{1}{1} + \frac{4}{3} + \frac{9}{5}$

$\qquad\qquad = \frac{15}{15} + \frac{20}{15} + \frac{27}{15}$

$\qquad\qquad = \frac{62}{15}$

13. $\displaystyle\sum_{i=3}^{6} (-2)^i = (-2)^3 + (-2)^4 + (-2)^5 + (-2)^6$

$\qquad\qquad = -8 + 16 - 32 + 64$
$\qquad\qquad = 40$

17. $\displaystyle\sum_{i=2}^{7} (x+1)^i = (x+1)^2 + (x+1)^3 + (x+1)^4 + (x+1)^5 + (x+1)^6 + (x+1)^7$

21. $\displaystyle\sum_{i=3}^{8} (x+i)^i = (x+3)^3 + (x+4)^4 + (x+5)^5 + (x+6)^6 + (x+7)^7 + (x+8)^8$

25. $2 + 4 + 8 + 16$

$\qquad = 2 + 2^2 + 2^3 + 2^4$

$\qquad = \displaystyle\sum_{i=1}^{4} 2^i$

29. $5 + 10 + 17 + 26 + 37$

$\qquad = (2^2 + 1) + (3^2 + 1) + (4^2 + 1) + (5^2 + 1) + (6^2 + 1)$

$\qquad = \displaystyle\sum_{i=2}^{6} (i + 1)$

33. $\frac{1}{3} + \frac{2}{5} + \frac{3}{7} + \frac{4}{9}$

$\qquad = \frac{1}{2 \cdot 1 + 1} + \frac{2}{2 \cdot 2 + 1} + \frac{3}{2 \cdot 3 + 1} + \frac{4}{2 \cdot 4 + 1}$

$\qquad = \displaystyle\sum_{i=1}^{4} \frac{i}{2 \cdot i + 1}$

37. $\dfrac{x}{x+3} + \dfrac{x}{x+4} + \dfrac{x}{x+5}$

$= \displaystyle\sum_{i=3}^{5} \dfrac{x}{x+i}$

41.

First second $= a_1 = 16 = 1 \cdot 16 = (1 + 0)16$

Second second $= a_2 = 48 = 3 \cdot 16 = (2 + 1)16$

Third second $= a_3 = 80 = 5 \cdot 16 = (3 + 2)16$

Fourth second $= a_4 = 112 = 7 \cdot 16 = (4 + 3)16$

Fifth second $= a_5 = 114 = 9 \cdot 16 = (5 + 4)16$

Sixth second $= a_6 = 176 = 11 \cdot 16 = (6 + 5)16$

Seventh second $= a_7 = 208 = 13 \cdot 16 = (7 + 6)16$

nth second $= a_n = [n + (n - 1)]16$

The skydiver will fall 208 feet on the seventh second. The total distance he will fall in seven seconds is 784 feet.

45. $\log_{10} \dfrac{\sqrt[3]{x}}{y^2} = \log_{10} \dfrac{x^{1/3}}{y^2}$

$= \log_{10} x^{1/3} - \log_{10} y^2$ Property 2

$= \dfrac{1}{3} \log_{10} x - 2 \log_{10} y$ Property 3

49. $2 \log_3 x - 3 \log_3 y - 4 \log_3 z$

$= \log_3 x^2 - \log_3 y^3 - \log_3 z^4$ Property 3

$= \log_3 \dfrac{x^2}{y^3 z^4}$ Properties 1 and 2

53. $\log_2 x + \log_2 (x - 7) = 3$

$\log_2 x(x - 7) = 3$ Property 1

$2^3 = x(x - 7)$ Exponential form

$8 = x^2 - 7x$

$x^2 - 7x - 8 = 0$ Standard form

$(x - 8)(x + 1) = 0$

$x - 8 = 0$ or $x + 1 = 0$

$x = 8$ $x = -1$

Possible solutions are 8 and -1 but only 8 checks. The solution set is {8}.

1. The sequence 1,2,3,4,... is an example of an arithmetic progression, since each term is obtained from the preceding term by adding 1 each time. The common difference d is 1.

5. The common difference d is -5. Each term is obtained from the preceding term by adding -5 each time.

9. The common difference is $\frac{2}{3}$. Each term is obtained from the preceding

    term by adding $\frac{2}{3}$ each time.

13. Given $a_1 = 6$ and $d = -2$ and substituting them in the formula for the general term, we have

    $$a_n = a_1 + (n - 1)d$$

    $$a_n = 6 + (n - 1)(-2)$$

    To find a_{10}, we substitute $n = 10$ in the above equation:

    $$a_{10} = 6 + (10 - 1)(-2)$$
    $$= 6 + (9)(-2)$$
    $$= 6 - 18$$
    $$a_{10} = -12$$

    Substituting $a_1 = 6$, $d = -2$, $n = 10$ and $a_{10} = -12$ in the following formula,

    $$S_n = \frac{n}{2}(a_1 + a_n)$$

    we have

    $$S_{10} = \frac{10}{2}(6 - 12)$$
    $$= 5(-6)$$
    $$S_{10} = -30$$

17. The third term can be written as

$$a_3 = a_1 + (3 - 1)d$$
$$= a_1 + 2d$$

and the eighth term can be written as

$$a_8 = a_1 + (8 - 1)d$$
$$= a_1 + 7d$$

Since these terms are also equal to 16 and 26, respectively, we can write

$$a_3 = a_1 + 2d = 16$$
$$a_8 = a_1 + 7d = 26$$

To find a_1 and d, we simply solve the system:

$$a_1 + 2d = 16$$
$$a_1 + 7d = 26$$

We add the opposite of the top equation to the bottom equation. The result is

$$5d = 10$$
$$d = 2$$

To find a_1, we simply substitute 2 for d in either of the original equations and get

$$a_1 + 2(d) = 16$$
$$a_1 + 2(2) = 16$$
$$a_1 = 12$$

The general term for this progression is

$$a_n = 12 + (n - 1)2$$

which we can simplify to

$$a_n = 2n + 10$$

The first term is

$$a_1 = 2(1) + 10 = 12$$

The common difference is 2.

$$a_{20} = 2(20) + 10 = 50$$

$$S_{20} = \frac{20}{2}(12 + 50) = 10(62) = 620$$

21. The common difference is -5.

$$a_{35} = 12 + (35 - 1)(-5)$$
$$= 12 + (34)(-5)$$
$$= 12 - 170$$
$$a_{35} = -158$$

25. Her salary for each of the first five years would be:

    18,000, 18,850, 19,700, 20,550, 21,400

    The general term of this sequence is

$$a_n = 18,000 + (n - 1)(850)$$
$$= 18,000 + 850n - 850$$
$$a_n = 17,150 + 850n$$
$$a_{10} = 17,150 + 850(10)$$
$$= 17,150 + 8500$$
$$= 25,650$$

29. $\log .0576 = \log (5.76 \times 10^{-2})$
$$= \log 5.76 + \log 10^{-2}$$
$$= .7604 + (-2)$$
$$= -1.2396$$

33. $\log x = -7.3516$
$$= -7.3516 + 8 - 8$$
$$= .6484 - 8$$

    From the table, we find that the mantissa, .6484, is the logarithm of 4.45. The characteristic, -8, is the power of 10:

$$x = 4.45 \times 10^{-8}$$

Section 11.4

1. The common ratio is 5.

5. This is not a geometric progression.

9. This is not a geometric progression.

13. If $a_1 = -2$ and $r = -\frac{1}{2}$,

$$a_6 = -2\left(-\frac{1}{2}\right)^{6-1}$$

$$= -2\left(-\frac{1}{2}\right)^5$$

$$= -2\left(-\frac{1}{32}\right)$$

$$= \frac{1}{16}$$

17. If $a_1 = 10$ and $r = 2$,

$$S_{10} = \frac{10(2^{10} - 1)}{2 - 1}$$

$$= \frac{10(1024 - 1)}{1}$$

$$= 10(1023)$$
$$= 10,230$$

21. $a_1 = \frac{1}{5}$, $r = \frac{1}{2}$

$$a_8 = \frac{1}{5}\left(\frac{1}{2}\right)^{8-1}$$

$$= \frac{1}{5}\left(\frac{1}{2}\right)^7$$

$$= \frac{1}{5}\left(\frac{1}{124}\right)$$

$$= \frac{1}{640}$$

25. $a_1 = \sqrt{2}$, $r = \sqrt{2}$

$$a_{10} = \sqrt{2}\,(\sqrt{2}\,)^{10-1}$$
$$= \sqrt{2}\,(\sqrt{2}\,)^{9}$$
$$= (\sqrt{2}\,)^{10}$$

$$a_{10} = 32$$

$$s_{10} = \frac{\sqrt{2}\,(\sqrt{2}^{\,10} - 1)}{\sqrt{2} - 1}$$

$$= \frac{\sqrt{2}\,(32 - 1)}{\sqrt{2} - 1}$$

$$= \frac{31\sqrt{2}}{\sqrt{2} - 1}$$

29. $a_4 = 40$, $a_5 = ?$, $a_6 = 160$
Multiplying by ± 2 will produce the next term.

33. $4 + 2 + 1 + \ldots$, then $a_1 = 4$, $r = \frac{1}{2}$

$$S = \frac{4}{1 - \frac{1}{2}}$$

$$= \frac{4}{\frac{1}{2}}$$

$$= 8$$

37. $\frac{3}{4} + \frac{1}{4} + \frac{1}{12} + \ldots$, then $a_1 = \frac{3}{4}$, $r = \frac{1}{3}$

$$S = \frac{\frac{3}{4}}{1 - \frac{1}{3}}$$

$$= \frac{\frac{3}{4}}{\frac{2}{3}}$$

$$= \frac{9}{8}$$

41. $.272727\ldots = .27 + .0027 + .000027 \ldots$

$$= \frac{27}{100} + \frac{27}{10,000} + \frac{27}{1,000,000} + \cdots$$

$$= \frac{27}{100} + \frac{27}{100}\left(\frac{1}{100}\right) + \frac{27}{100}\left(\frac{1}{100}\right)^2 + \cdots$$

$$a_1 = \frac{27}{100}, \; r = \frac{1}{100}$$

$$S = \frac{a_1}{1 - r} = \frac{\frac{27}{100}}{1 - \frac{1}{100}} = \frac{\frac{27}{100}}{\frac{99}{100}} = \frac{27}{99} = \frac{3}{11}$$

45. $(x + y)^3 = (x + y)(x + y)(x + y)$

$$= (x^2 + 2xy + y^2)(x + y)$$

$$= x^3 + 3x^2y + 3xy^2 + y^3$$

49. $a_1 = 100, \; r = 2$

$$a_n = a_1 r^{n-1}$$

$$a_{20} = 100(2)^{20-1}$$

$$= 100(2)^{19}$$
$$= 100(524,288)$$
$$= 52,428,800$$

53. $a_1 = a, \; r = \frac{1}{2}$

$$S = \frac{a_1}{1 - r} = \frac{a}{1 - \frac{1}{2}} = \frac{a}{\frac{1}{2}} = 2a$$

Section 11.5

1. $(x + 2)^4 = \binom{4}{0}x^4(2)^0 + \binom{4}{1}x^3(2)^1 + \binom{4}{2}x^2(2)^2 + \binom{4}{3}x^1(2)^3 + \binom{4}{4}x^0(2)^4$

The coefficients $\binom{4}{0}$, $\binom{4}{1}$, $\binom{4}{2}$, $\binom{4}{3}$ and $\binom{4}{4}$ can be found in the fourth row of Pascal's triangle. They are 1, 4, 6, 4, and 1:

$$(x + 2)^4 = 1x^4(2)^0 + 4x^3(2)^1 + 6x^2(2)^2 + 4x^1(2)^3 + 1x^0(2)^4$$

$$= x^4 + 8x^3 + 24x^2 + 32x + 16$$

5. $(2x + 1)^5$

$$= \binom{5}{0}(2x)^5(1)^0 + \binom{5}{1}(2x)^4(1)^1 + \binom{5}{2}(2x)^3(1)^2 + \binom{5}{3}(2x)^2(1)^3$$

$$+ \binom{5}{4}(2x)^1(1)^4 + \binom{5}{5}(2x)^0(1)^5$$

The coefficients $\binom{5}{0}$, $\binom{5}{1}$, $\binom{5}{2}$, $\binom{5}{3}$, $\binom{5}{4}$ and $\binom{5}{5}$ can be found in the fifth row of Pascal's triangle. They are 1, 5, 10, 10, 5, and 1:

$$(2x + 1)^5 = 1(2x)^5(1)^0 + 5(2x)^4(1)^1 + 10(2x)^3(1)^2 + 10(2x)^2(1)^3$$

$$+ 5(2x)^1(1)^4 + 1(2x)^0(1)^5$$

$$= 32x^5 + 80x^4 + 80x^3 + 40x^2 + 10x + 1$$

9. $(3x - 2)^4$

$$= \binom{4}{0}(3x)^4(-2)^0 + \binom{4}{1}(3x)^3(-2)^1 + \binom{4}{2}(3x)^2(-2)^2$$

$$+ \binom{4}{3}(3x)^1(-2)^3 + \binom{4}{4}(3x)^0(-2)^4$$

The coefficients $\binom{4}{0}$, $\binom{4}{1}$, $\binom{4}{2}$, $\binom{4}{3}$ and $\binom{4}{4}$ can be found in the fourth row of Pascal's triangle. They are 1, 4, 6, 4 and 1.

$$(3x - 2)^4 = 1(3x)^4(-2)^0 + 4(3x)^3(-2)^1 + 6(3x)^2(-2)^2$$

$$+ 4(3x)^1(-2)^3 + 1(3x)^0(-2)^4$$

$$= 81x^4 - 216x^3 + 216x^2 - 96x + 16$$

13. $(x^2 + 2)^4$

$$= \binom{4}{0}(x^2)^4(2)^0 + \binom{4}{1}(x^2)^3(2)^1 + \binom{4}{2}(x^2)^2(2)^2$$

$$+ \binom{4}{1}(x^2)^1(2)^3 + \binom{4}{0}(x^2)^0(2)^4$$

The coefficients $\binom{4}{0}$, $\binom{4}{1}$, $\binom{4}{2}$, $\binom{4}{3}$ and $\binom{4}{4}$ can be found in the fourth row of Pascal's triangle. They are 1, 4, 6, 4 and 1.

$$(x^2 + 2)^4 = 1(x^2)^4(2)^0 + 4(x^2)^3(2)^1 + 6(x^2)^2(2)^2$$

$$+ 4(x^2)^1(2)^3 + 1(x^2)^0(2)^4$$

$$= x^8 + 8x^6 + 24x^4 + 32x^2 + 16$$

17. The coefficients can be found in the third row of Pascal's triangle. They are:

    1, 3, 3, 1

    Here is the expansion of $(\frac{x}{2} - 4)^3$:

    $(\frac{x}{2} - 4)^3$

    $= 1(\frac{x}{2})^3(-4)^0 + 3(\frac{x}{2})^2(-4)^1 + 3(\frac{x}{2})^1(-4)^2 + 1(\frac{x}{2})^0(-4)^3$

    $= \frac{x^3}{8} - \frac{12x^2}{4} + \frac{48x}{2} - 64$

    $= \frac{x^3}{8} - 3x^2 + 24x - 64$

21. The coefficients of the first three terms have been calculated in example 4, page 597 of your textbook. The fourth term is calculated as follows:

    $\binom{9}{3} = \frac{9!}{3!6!} = \frac{9 \cdot 8 \cdot 7 \cdot \cancel{6 \cdot 5 \cdot 4 \cdot 3 \cdot 2 \cdot 1}}{(3 \cdot 2 \cdot 1)(\cancel{6 \cdot 5 \cdot 4 \cdot 3 \cdot 2 \cdot 1})} = \frac{504}{6} = 84$

    From the binomial formula, we write the first four terms:

    $(x + 2)^9 = 1 \cdot x^9 + 9x^8(2) + 36x^7(2)^2 + 84x^6(2)^3$
    $= x^9 + 18x^8 + 144x^7 + 672x^6$

25. The coefficients of the first four terms are $\binom{10}{0}$, $\binom{10}{1}$, $\binom{10}{2}$ and and $\binom{10}{3}$, which we calculate as follows:

    $\binom{10}{0} = \frac{10!}{0!10!} = \frac{\cancel{10 \cdot 9 \cdot 8 \cdot 7 \cdot 6 \cdot 5 \cdot 4 \cdot 3 \cdot 2 \cdot 1}}{(1)(\cancel{10 \cdot 9 \cdot 8 \cdot 7 \cdot 6 \cdot 5 \cdot 4 \cdot 3 \cdot 2 \cdot 1})} = \frac{1}{1} = 1$

    $\binom{10}{1} = \frac{10!}{1!9!} = \frac{10 \cdot \cancel{9 \cdot 8 \cdot 7 \cdot 6 \cdot 5 \cdot 4 \cdot 3 \cdot 2 \cdot 1}}{(1)(\cancel{9 \cdot 8 \cdot 7 \cdot 6 \cdot 5 \cdot 4 \cdot 3 \cdot 2 \cdot 1})} = \frac{10}{1} = 10$

    $\binom{10}{2} = \frac{10!}{2!8!} = \frac{10 \cdot 9 \cdot \cancel{8 \cdot 7 \cdot 6 \cdot 5 \cdot 4 \cdot 3 \cdot 2 \cdot 1}}{(2 \cdot 1)(\cancel{8 \cdot 7 \cdot 6 \cdot 5 \cdot 4 \cdot 3 \cdot 2 \cdot 1})} = \frac{90}{2} = 45$

    $\binom{10}{3} = \frac{10!}{3!7!} = \frac{10 \cdot 9 \cdot 8 \cdot \cancel{7 \cdot 6 \cdot 5 \cdot 4 \cdot 3 \cdot 2 \cdot 1}}{(3 \cdot 2 \cdot 1)(\cancel{7 \cdot 6 \cdot 5 \cdot 4 \cdot 3 \cdot 2 \cdot 1})} = \frac{720}{6} = 120$

    From the binomial formula, we write the first four terms:

    $(x + 2)^{10} = 1 \cdot x^{10} + 10x^9(2y) + 45x^8(2y)^2 + 120x^7(2y)^3$
    $= x^{10} + 20x^9y + 180x^8y^2 + 960x^7y^3$

29. The coefficients of the first three terms are $\binom{12}{0}$, $\binom{12}{1}$ and $\binom{12}{2}$, which we calculate as follows:

$$\binom{12}{0} = \frac{12.}{0!\,12!} = \frac{\cancel{12} \cdot \cancel{11} \cdot \cancel{10} \cdot \cancel{9} \cdot \cancel{8} \cdot \cancel{7} \cdot \cancel{6} \cdot \cancel{5} \cdot \cancel{4} \cdot \cancel{3} \cdot \cancel{2} \cdot \cancel{1}}{(1)(\cancel{12} \cdot \cancel{11} \cdot \cancel{10} \cdot \cancel{9} \cdot \cancel{8} \cdot \cancel{7} \cdot \cancel{6} \cdot \cancel{5} \cdot \cancel{4} \cdot \cancel{3} \cdot \cancel{2} \cdot \cancel{1})}$$

$$= \frac{1}{1} = 1$$

$$\binom{12}{1} = \frac{12.}{1!\,11!} = \frac{12 \cdot \cancel{11} \cdot \cancel{10} \cdot \cancel{9} \cdot \cancel{8} \cdot \cancel{7} \cdot \cancel{6} \cdot \cancel{5} \cdot \cancel{4} \cdot \cancel{3} \cdot \cancel{2} \cdot \cancel{1}}{(1)(\cancel{11} \cdot \cancel{10} \cdot \cancel{9} \cdot \cancel{8} \cdot \cancel{7} \cdot \cancel{6} \cdot \cancel{5} \cdot \cancel{4} \cdot \cancel{3} \cdot \cancel{2} \cdot \cancel{1})}$$

$$= \frac{12}{1} = 12$$

$$\binom{12}{2} = \frac{12!}{2!\,10!} = \frac{12 \cdot 11 \cdot \cancel{10} \cdot \cancel{9} \cdot \cancel{8} \cdot \cancel{7} \cdot \cancel{6} \cdot \cancel{5} \cdot \cancel{4} \cdot \cancel{3} \cdot \cancel{2} \cdot \cancel{1}}{(2 \cdot 1)(\cancel{10} \cdot \cancel{9} \cdot \cancel{8} \cdot \cancel{7} \cdot \cancel{6} \cdot \cancel{5} \cdot \cancel{4} \cdot \cancel{3} \cdot \cancel{2} \cdot \cancel{1})}$$

$$= \frac{132}{3} = 66$$

From the binomial formula, we write the first three terms:

$$(x - y)^{12} = 1 \cdot x^{12} + 12x^{11}(-y)^1 + 66x^{10}(-y)^2$$
$$= x^{12} - 12x^{11}y + 66x^{10}y^2$$

33. The coefficients of the first two terms are $\binom{100}{0}$ and $\binom{100}{1}$, which we calculate as follows:

$$\binom{100}{0} = \frac{100!}{0!\,(100!)} = \frac{\cancel{100 \cdot 99 \cdot 98 \ldots 1}}{(1)(\cancel{100 \cdot 99 \cdot 98 \ldots 1})} = \frac{1}{1} = 1$$

$$\binom{100}{1} = \frac{100.}{1!\,99!} = \frac{100 \cdot \cancel{99 \cdot 98 \ldots 1}}{(1)(\cancel{99 \cdot 98 \cdot 97 \ldots 1})} = \frac{100}{1} = 100$$

From the binomial formula, we write the first three terms:

$$(x + 2)^{100} = 1 \cdot x^{100} + 100x^{99}(2)$$
$$= x^{100} + 200x^{99}$$

37. $\binom{12}{8} = \frac{12!}{4!8!} = \frac{12 \cdot 11 \cdot 10 \cdot 9 \cdot 8 \cdot 7 \cdot 6 \cdot 5 \cdot 4 \cdot 3 \cdot 2 \cdot 1}{(4 \cdot 3 \cdot 2 \cdot 1)(8 \cdot 7 \cdot 6 \cdot 5 \cdot 4 \cdot 3 \cdot 2 \cdot 1)} = \frac{11,880}{24} = 495$

$a_9 = 495(2x)^4(3y)^8$

$= 495(16x^4)(6561y^8)$

$= 51,963,120x^4y^8$

41. $\binom{9}{3} = \frac{9!}{3!6!} = \frac{9 \cdot 8 \cdot 7 \cdot 6 \cdot 5 \cdot 4 \cdot 3 \cdot 2 \cdot 1}{(3 \cdot 2 \cdot 1)(6 \cdot 5 \cdot 4 \cdot 3 \cdot 2 \cdot 1)} = \frac{504}{6} = 84$

$a_4 = 84x^6(3)^3 = 84x^6(27) = 2268x^6$

45. The third term from the binomial formula is

$\binom{7}{2}\left(\frac{1}{2}\right)^5\left(\frac{1}{2}\right)^2 = \frac{7!}{2!5!}\left(\frac{1}{2}\right)^7$

$= \frac{7 \cdot 6 \cdot 5 \cdot 4 \cdot 3 \cdot 2 \cdot 1}{(2 \cdot 1)(5 \cdot 4 \cdot 3 \cdot 2 \cdot 1)}\left(\frac{1}{2}\right)^7$

$= \frac{42}{2} \cdot \frac{1}{128}$

$= \frac{21}{128}$

49. $8^{2x+1} = 16$

$(2^3)^{2x+1} = 2^4$

$2^{6x+3} = 2^4$

$6x + 3 = 4$

$6x = 1$

$x = \frac{1}{6} \text{ or } .17$

53. $\log_4 20 = \frac{\log 20}{\log 4}$

$\log_4 20 = \frac{1.3010}{.6021}$

$= 2.16$

57. $A = 10e^{5t}$

$\ln A = \ln 10 + \ln e^{5t}$

$\ln A = \ln 10 + 5t \ln e$

$\ln A = \ln 10 + 5t$ Because $\ln e = 1$

$\dfrac{\ln A}{\ln 10} = 5t$

$t = \dfrac{1}{5} \cdot \dfrac{\ln A}{\ln 10}$

$t = \dfrac{1}{5} \ln \dfrac{A}{10}$

61. $\binom{20}{12} = \dfrac{20!}{12!8!} = \dfrac{20 \cdot 19 \cdot 18 \cdot 17 \cdot 16 \cdot 15 \cdot 14 \cdot 13 \cdot 12 \cdot 11 \cdot 10 \cdot 9 \cdot 8 \cdot 7 \cdot 6 \cdot 5 \cdot 4 \cdot 3 \cdot 2 \cdot 1}{(12 \cdot 11 \cdot 10 \cdot 9 \cdot 8 \cdot 7 \cdot 6 \cdot 5 \cdot 4 \cdot 3 \cdot 2 \cdot 1)(8 \cdot 7 \cdot 6 \cdot 5 \cdot 4 \cdot 3 \cdot 2 \cdot 1)}$

$\qquad\qquad = 125{,}970$

$\binom{20}{8} = \dfrac{20!}{8!12!} = \dfrac{20 \cdot 19 \cdot 18 \cdot 17 \cdot 16 \cdot 15 \cdot 14 \cdot 13 \cdot 12 \cdot 11 \cdot 10 \cdot 9 \cdot 8 \cdot 7 \cdot 6 \cdot 5 \cdot 4 \cdot 3 \cdot 2 \cdot 1}{(8 \cdot 7 \cdot 6 \cdot 5 \cdot 4 \cdot 3 \cdot 2 \cdot 1)(12 \cdot 11 \cdot 10 \cdot 9 \cdot 8 \cdot 7 \cdot 6 \cdot 5 \cdot 4 \cdot 3 \cdot 2 \cdot 1)}$

$\qquad\qquad = 125{,}970$

1. If the General term is $A_n = 2n + 5$

   then the First term is $A_1 = 2(1) + 5 = 7$

   the Second term is $A_2 = 2(2) + 5 = 9$

   the Third term is $A_3 = 2(3) + 5 = 11$

   the Fourth term is $A_4 = 2(4) + 5 = 13$.

5. If the General term is $A_n = 4A_{n-1}$, $n > 1$

   then the First term is $A_1 = 4$

   the Second term is $A_2 = 4(4) = 16$

   the Third term is $A_3 = 4(16) = 64$

   the Fourth term is $A_4 = 4(64) = 256$.

9. Notice $a_1 = \quad 1 = 1^4$

   $a_2 = \quad 16 = 2^4$

   $a_3 = \quad 81 = 3^4$

   $a_4 = 256 = 4^4$

   $\cdot$

   $\cdot$

   $\cdot$

   $a_n = \qquad = n^4$

   The general term is n^4.

13. $\displaystyle\sum_{i=1}^{4} (2i + 3) = [2(1) + 3] + (2(2) + 3] + [2(3) + 3] + [2(4) + 3]$

    $= 5 + 7 + 9 + 11$

    $= 32$

17. $\displaystyle\sum_{i=3}^{5} (4i + i^3) = [4(3) + 3^2] + [4(4) + 4^2] + [4(5) + 5^2]$

    $= 21 + 32 + 45$

    $= 98$

21. Given: $5 + 7 + 9 + 11 + 13$

    Writing the sum as

    $[2(1) + 3] + [2(2) + 3] + [2(3) + 3] + [2(4) + 3] + [2(5) + 3]$

    We see the formula

    $a_i = 2(i) + 3$

    Using this formula and summation notation, we have

    $\displaystyle\sum_{i=1}^{5} (2i + 3)$

25. A formula for the given terms is

$$a_i = x - 2i$$

where i assumes all integer values between 1 and 3, including 1 and 3.

The sum can be written as

$$\sum_{i=1}^{3} (x - 2i)$$

29. This sequence is an arithmetic progression because the common difference is 4.

33. This sequence is an arithmetic progression because the common difference is -3.

37. If $a_1 = -2$ and $d = 4$,

$$a_n = -2 + (n - 1)4$$

$$= -2 + 4n - 4$$
$$= 4n - 6$$

Substituting a_{10} for a_n

$$a_{10} = 4(10) - 6$$
$$= 34$$

We must first find a_1, to solve s_{10}.

$$a_1 = 4(1) - 6$$
$$= -2$$

Then $s_{10} = \dfrac{10}{2}(-2 + 34)$

$$= 5(32)$$
$$s_{10} = 160$$

41. $a_4 = a_1 + 3d = -10$

   $a_8 = a_1 + 7d = -18$

To find a_1 and d, we solve the system:

$$
\begin{array}{lll}
a_1 + 3d = -10 & \xrightarrow{\text{no change}} & a_1 + 3d = -10 \\
a_1 + 7d = -18 & \xrightarrow[\text{multiply by -1}]{} & -a_1 - 7d = 18 \\
& & \overline{\quad -4d = \;\; 8} \\
& & \quad\;\; d = -2
\end{array}
$$

Substituting d = -2 in the first equation, we have

$$a_1 + 3(-2) = -10$$

$$a_1 = -4$$

$$
\begin{aligned}
a_{20} &= a_1 + 19d \\
&= -4 + 19(-2) \\
&= -4 - 38 \\
a_{20} &= -42
\end{aligned}
$$

Now solve:

$$s_{20} = \frac{20}{2}(-4 - 42)$$

$$= 10(-46)$$

$$s_{20} = -460$$

45. $a_n = a_1 r^{n-1}$ $a_{16} = a_1 r^{16-1}$

   $= 5(-2)^{n-1}$ $= 5(-2)^{15}$

49. $a_3 = a_1 r^2 = 12$

   $a_4 = a_1 r^3 = 24$

We can solve for r by using the ratio $\dfrac{a_4}{a_3}$:

$$\frac{a_4}{a_3} = \frac{a_1 r^3}{a_1 r^2} = \frac{24}{12}$$

The common ratio is 2. To find the first term we substitute r = 2 into either of the original equations. The result is

$$a_1 r^3 = 24$$
$$a_1 (2)^3 = 24$$
$$a_1 = 3$$

Then solve

$$a_6 = a_1 r^{6-1}$$
$$= 3(2)^5$$
$$= 3(32)$$
$$= 96$$

53. $\dbinom{6}{3} = \dfrac{6!}{3!(6-3)!}$

$$= \frac{6!}{3!\,3!}$$

$$= \frac{6 \cdot 5 \cdot 4 \cdot \cancel{3 \cdot 2 \cdot 1}}{(3 \cdot 2 \cdot 1)(\cancel{3 \cdot 2 \cdot 1})}$$

$$= \frac{120}{6}$$

$$= 20$$

57. $(x - 2)^4 = \dbinom{4}{0}x^4(-2)^0 + \dbinom{4}{1}x^3(-2)^1 + \dbinom{4}{2}x^2(-2)^2 + \dbinom{4}{3}x^1(-3)^3 + \dbinom{4}{4}x^0(-2)^4$

$$= 1x^4 + 4(x^3)(-2) + 6(x^2)(4) + 4(x^1)(-8) + 1(16)$$

$$= x^4 - 8x^3 + 24x^2 - 32x + 16$$

61. $(\frac{x}{2} + 3)^4$

$$= \binom{4}{0}(\frac{x}{2})^4(3^0) + \binom{4}{1}(\frac{x}{2})^3(3^1) + \binom{4}{2}(\frac{x}{2})^2(3^2) + \binom{4}{3}(\frac{x}{2})^1(3^3) + \binom{4}{4}(\frac{x}{2})^0(3^4)$$

$$= 1(\frac{x^4}{16}) + 4(\frac{x^3}{8})(3) + 6(\frac{x^2}{4})9 + 4(\frac{x}{2})(27) + 1(81)$$

$$= \frac{x^4}{16} + \frac{3x^3}{2} + \frac{27x^2}{2} + 54x + 81$$

$$= \frac{1}{16}x^4 + \frac{3}{2}x^3 + \frac{27}{2}x^2 + 54x + 81$$

65. The coefficients of the first three terms are $\binom{11}{0}$, $\binom{11}{1}$, and $\binom{11}{2}$, which we calculate as follows:

$$\binom{11}{0} = \frac{11!}{0!11!} = \frac{11 \cdot 10 \cdot 9 \cdot 8 \cdot 7 \cdot 6 \cdot 5 \cdot 4 \cdot 3 \cdot 2 \cdot 1}{(1)(11 \cdot 10 \cdot 9 \cdot 8 \cdot 7 \cdot 6 \cdot 5 \cdot 4 \cdot 3 \cdot 2 \cdot 1)} = \frac{1}{1} = 1$$

$$\binom{11}{1} = \frac{11!}{1!10!} = \frac{11 \cdot 10 \cdot 9 \cdot 8 \cdot 7 \cdot 6 \cdot 5 \cdot 4 \cdot 3 \cdot 2 \cdot 1}{(1)(10 \cdot 9 \cdot 8 \cdot 7 \cdot 6 \cdot 5 \cdot 4 \cdot 3 \cdot 2 \cdot 1)} = \frac{11}{1} = 11$$

$$\binom{11}{2} = \frac{11!}{2!9!} = \frac{11 \cdot 10 \cdot 9 \cdot 8 \cdot 7 \cdot 6 \cdot 5 \cdot 4 \cdot 3 \cdot 2 \cdot 1}{(2 \cdot 1)(9 \cdot 8 \cdot 7 \cdot 6 \cdot 5 \cdot 4 \cdot 3 \cdot 2 \cdot 1)} = \frac{110}{2} = 55$$

From the binomial formula, we write the first three terms:

$$1 \cdot x^{11} + 11 \cdot x^{10}(y) + 55 \cdot x^9(y^2)$$

$$= x^{11} + 11x^{10}y + 55x^9y^2$$

69. $\binom{50}{0} = \frac{50!}{0!50!} = \frac{50 \cdot 49 \cdot 48 \ldots 1}{(1)50 \cdot 49 \cdot 48 \ldots 1} = \frac{1}{1} = 1$

$\binom{50}{1} = \frac{50!}{1!49!} = \frac{50 \cdot 49 \cdot 48 \ldots 1}{(1)(49 \cdot 48 \cdot 47 \ldots 1)} = \frac{50}{1} = 50$

From the binomial formula, we write the first two terms:

$$1(x^{50})(-1)^0 + 50(x^{49})(-1)^1$$

$$= x^{50} - 50x^{49}$$

1.   General term $= a_n = 3n - 5$

   First term $= a_1 = 3(1) - 5 = -2$

   Second term $= a_2 = 3(2) - 5 = 1$

   Third term $= a_3 = 3(3) - 5 = 4$

   Fourth term $= a_4 = 3(4) - 5 = 7$

   Fifth term $= a_5 = 3(5) - 5 = 10$

5.   General term $= a_n = \dfrac{n + 1}{n^2}$

   First term $= a_1 = \dfrac{1 + 1}{1^2} = \dfrac{2}{1} = 2$

   Second term $= a_2 = \dfrac{2 + 1}{2^2} = \dfrac{3}{4}$

   Third term $= a_3 = \dfrac{3 + 1}{3^2} = \dfrac{4}{9}$

   Fourth term $= a_4 = \dfrac{4 + 1}{4^2} = \dfrac{5}{16}$

   Fifth term $= a_5 = \dfrac{5 + 1}{5^2} = \dfrac{6}{25}$

9.   Look at the first four terms:

   $$a_1 = \frac{1}{2} = \left(\frac{1}{2}\right)^1$$

   $$a_2 = \frac{1}{4} = \left(\frac{1}{2}\right)^2$$

   $$a_3 = \frac{1}{8} = \left(\frac{1}{2}\right)^3$$

   $$a_4 = \frac{1}{16} = \left(\frac{1}{2}\right)^4$$

   The general term is $a_n = \left(\frac{1}{2}\right)^n = \dfrac{1}{2^n}$

13. $a_3 = a_1 r^2 = 18$

$a_5 = a_1 r^4 = 162$

$$\frac{a_5}{a_3} = \frac{a_1 r^4}{a_1 r^2} = \frac{162}{18}$$

$$r^2 = 9$$
$$r = 3$$

$$a_3 = a_1 3^2 = 18$$
$$a_1 = 2$$

So,

$$a_2 = a_1 r^1 = 2 \cdot 3 = 6$$

17. $s = \dfrac{a_1}{1-r}, \ a_1 = \dfrac{1}{2}, \ r = \dfrac{1}{3}$

$$s = \frac{\frac{1}{2}}{1 - \frac{1}{3}} = \frac{\frac{1}{2}}{\frac{2}{3}} = \frac{3}{4}$$

21. $a_6 = \binom{8}{6}(2x)^8(-3y)^4$

$$= \frac{8!}{5!3!}(2x)^3(-3y)^5$$
$$= 56(8x^3)(-243y^5)$$
$$= 108,867x^3y^5$$

Appendix A

1.
$$-2 \,|\quad \begin{array}{rrr} 1 & -5 & 6 \\ & -2 & 14 \\ \hline 1 & -7 & \underline{|\,20} \end{array}$$

From the synthetic division, we have

$$\frac{x^2 - 5x + 6}{x + 2} = x - 7 + \frac{20}{x + 2}$$

5.
$$+2 \,|\quad \begin{array}{rrrr} 1 & 2 & 3 & 4 \\ & 2 & 8 & 22 \\ \hline 1 & 4 & 11 & \underline{|\,26} \end{array}$$

From the synthetic division, we have

$$\frac{x^3 + 2x^2 + 3x + 4}{x - 2} = x^2 + 4x + 11 + \frac{26}{x - 2}$$

9.
$$1 \,|\quad \begin{array}{rrrr} 2 & 0 & 1 & -3 \\ & 2 & 2 & 3 \\ \hline 2 & 2 & 3 & \underline{|\,0} \end{array}$$

From the synthetic division, we have

$$\frac{2x^3 - x^2 + 2x + 5}{x - 3} = 2x^2 + 2x + 3$$

13.
$$2 \,|\quad \begin{array}{rrrrrr} 1 & -2 & 1 & -3 & -1 & 1 \\ & 2 & 0 & 2 & -2 & -6 \\ \hline 1 & 0 & 1 & -1 & -3 & \underline{|\,-5} \end{array}$$

From the synthetic division, we have

$$\frac{x^5 - 2x^4 + x^3 - 3x^2 - x + 1}{x - 3} = x^4 + 0x^3 + x^2 - x - 3 - \frac{5}{x - 3}$$

$$= x^4 + x^2 - x - 3 - \frac{5}{x - 3}$$

17.
$$-1 \,|\quad \begin{array}{rrrrr} 1 & 0 & 0 & 0 & -1 \\ & -1 & 1 & -1 & 1 \\ \hline 1 & -1 & 1 & -1 & \underline{|\,0} \end{array}$$

From the synthetic division, we have

$$\frac{x^4 - 1}{x + 1} = x^3 - x^2 + x - 1$$

370

Appendix B

1. $A \cup B$ indicate all the elements in A or B are shaded. The union of $A \cup B$ with C indicates all the elements in A, B or C are shaded. See the Venn diagram on page A76 in the textbook.

5. See example 1 on page A6 in the textbook.

9. $(A \cup B) \cap (A \cup C) = (A \cap A) \cup (A \cap C) \cup (B \cap A) \cup (B \cap C)$

    See the Venn diagram on page A77 in the textbook.

13. If $\{x | x \in A$ and $x \in B\}$ then $\{x | x \in A \cap B\}$. See the Venn diagram on page A77 in the textbook.

17. If $\{x | x \notin A$ or $x \in B\}$ then $\{x | x$ as of the universal set except $x \notin A \cap B\}$. See the Venn diagram on page A77 in the textbook.

C 1
D 2
E 3
F 4
G 5
H 6
I 7
J 8